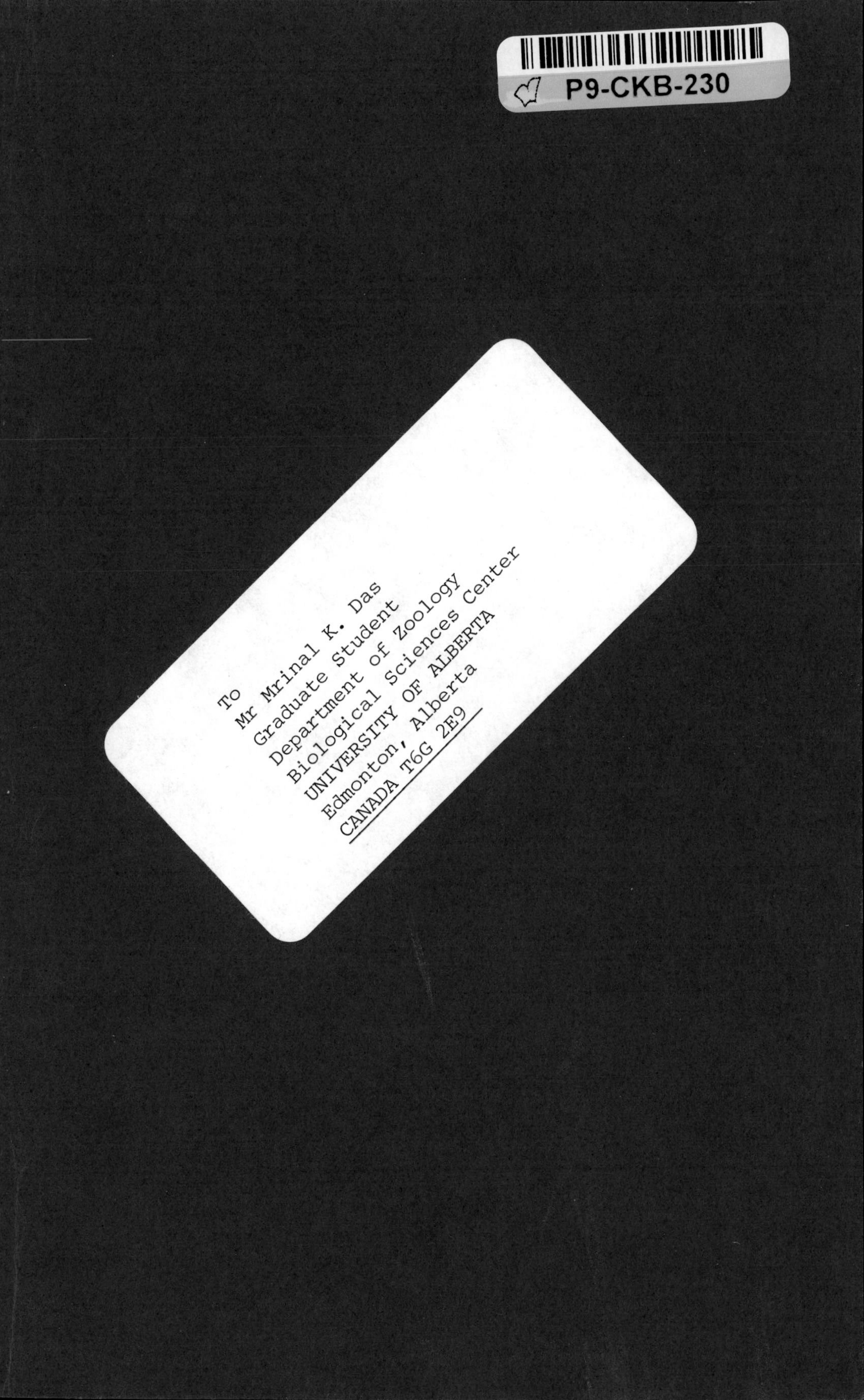
P9-CKB-230
To
Mr Mrinal K. Das
Graduate Student
Department of Zoology
Biological Sciences Center
UNIVERSITY OF ALBERTA
Edmonton, Alberta
CANADA T6G 2E9

INTRODUCTION TO

BIOSTATISTICS

A series of books in BIOLOGY

INTRODUCTION TO
BIOSTATISTICS

Robert R. Sokal and F. James Rohlf
STATE UNIVERSITY OF NEW YORK AT STONY BROOK

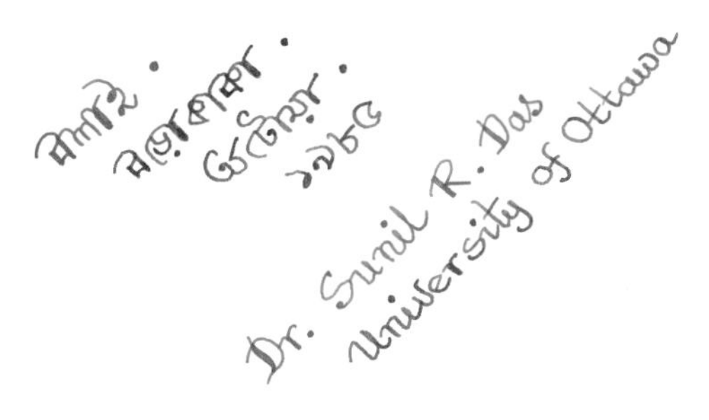

W. H. Freeman and Company
San Francisco

Library of Congress Cataloging in Publication Data

Sokal, Robert R
Introduction to biostatistics.

(A Series of books in biology)
Bibliography: p.
1. Biometry. I. Rohlf, F. James, 1936– joint author. II. Title. [DNLM: 1. Biometry. QH 405 S683i 1973]
QH405.S62 574'.01'82 71–178257
ISBN 0-7167-0693-8

Printed in the United States of America

International Standard Book Number: 0-7167-0693-8

6 7 8 9

to Julie and Pat

Contents

Preface

The favorable reception that our comprehensive *Biometry* has received from teachers and students and the suggestions of numerous colleagues have encouraged us to write this briefer *Introduction to Biostatistics.* This book is aimed at the undergraduate student of biological statistics for whom it should present a thorough foundation in the subject, requiring no more than an elementary preparation in mathematics. We hope that the book will also be useful in short courses in biostatistics such as are often found in medical and other professional schools. In spite of the need for brevity, the informal style of our more comprehensive volume has been retained here, and we hope that the various pedagogical features that have proven themselves in our biometry book will be of value in this volume also.

Many of the approaches detailed in the preface of our earlier volume are followed again here; however, some have been modified in keeping with the different audience at which we are aiming. Although we furnish detailed outlines for computation of all the methods discussed in the book, we have placed less emphasis on the computational aspects of the material. This was done for two reasons: in many undergraduate courses the students have relatively little opportunity and motivation for carrying out lengthy computations with biological research material. Also, the development of electronic desk calculators has so revolutionized the methodologies of statistical computations that a comprehensive treatment of the various strategies for each type of calculator available would be beyond the scope of this book. It would also be outdated soon after publication. Thus, we shall rely on the instructor in the course to advise his students on the best computational procedures to be followed using the facilities of any given institution.

Subject matter is arranged in chapters and sections, numbered by the conventional decimal numbering system; the number before the decimal point refers to the chapter, the numbering after the point to the section. Topics can be referred to in the Contents as well as found in the Index. The Tables are numbered with the same decimal system, the first number denoting the chapter, the second the number of the table within the chapter. Certain special tables are entitled "Boxes" and are numbered as such, again by means of the decimal system. These boxes serve a double function. They illustrate computational methods for solving various types of biostatistical problems and can therefore be used as convenient patterns for computation by persons using the book. They usually contain all the steps necessary—from the initial setup of the problem to the final result; hence to students familiar with the book the boxes can serve as quick summary reminders of the technique. A second important use of the boxes relates to their origin as mimeographed sheets handed out in the authors' biometry courses. In the lecture time at the disposal of the instructor it would have been impossible to convey even as much as half the subject matter covered in the course if the material contained in the sheets had to be put on the blackboard. By asking the students to refer in class to what are now the boxes and some of the tables in this book, much time was saved and, incidentally, students were able to devote their attention to understanding the contents of these sheets rather than to copying them. Instructors who employ this book as a text may wish to avail themselves of the boxes in a similar manner.

Figures are numbered by the same decimal system as tables and boxes. A similar numbering system is also used for (sampling) Experiments, (mathematical) Expressions, and (homework) Exercises. Numbers for appendixes carry the prefix letter A.

Since the emphasis in this book is on practical applications of statistics to biology, discussions of statistical theory are deliberately kept to a minimum. Derivations are given for a number of formulas but these are consigned to Appendix A1, where they should be studied and reworked by the student.

The statistical tables necessary for the methods discussed in this book are found in Appendix A2. We assume that students will be able to lay their hands on ordinary mathematical tables containing common logarithms, square roots, and trigonometric functions. Our tables have been extracted from a more extensive volume of tables published separately (which includes the mathematical tables already referred to) entitled *Statistical Tables* by Rohlf and Sokal, W. H. Freeman and Company, 1969. We are indebted to the Literary Executor of the late Sir Ronald A. Fisher, F.R.S., to Dr. Frank Yates, F.R.S., and to Oliver & Boyd, Edinburgh, for permission to reprint

Table III (our Table **III**) from their book *Statistical Tables for Biological, Agricultural and Medical Research.*

Homework exercises for students reading this book in connection with a biostatistics course (or studying on their own) are given at the end of each chapter. In keeping with our own convictions, these are largely real research problems. Some of them, therefore, require a fair amount of computation for their solution.

Most of what is presented in this text is extracted from our more comprehensive textbook of biometry. A new section on probability has been added at the suggestion of several colleagues. Our sequence of presentation progresses in conventional fashion from descriptive statistics to the fundamental distributions and the testing of elementary statistical hypotheses, but we then proceed immediately to the analysis of variance. The familiar and time-honored t-test is treated merely as a special case of the analysis of variance and is relegated to several sections in various appropriate chapters of the book. We have taken this step deliberately for two reasons. (1) It is urgent that students become acquainted with analysis of variance at the earliest possible moment, since a thorough foundation in analysis of variance is essential nowadays to every biologist. (2) If analysis of variance is presented and understood early, the necessity for an employment of the t-distribution is materially reduced, except for the setting of confidence limits and a few other special situations. All t-tests can be carried out as analyses of variance and many are more informative when carried out as such. The amount of computation is generally equivalent.

In other topics we have also been concerned with introducing new and improved techniques, stressing these over earlier methods that we feel are less suitable. Notable instances of such innovations are the adoption of the simultaneous test procedure for a posteriori multiple comparisons tests and the employment of the G-statistic for the analysis of frequencies in lieu of the traditional chi-square tests, which are therefore less extensively discussed than might otherwise have been the case.

We are grateful to Professors K. R. Gabriel, R. C. Lewontin, F. J. Sonleitner, Theodore J. Crovello, and Albert J. Rowell for their extensive comments on the earlier version of the manuscript. Professors Arnold B. Larson and Gunther Schlager furnished valuable comments on a draft copy of this book. We are indebted to Edwin Bryant, David Fisher, Koichi Fujii, John Kishpaugh, and David Wool for carefully checking the numerical accuracy of tables and boxes. Our wives, Julie and Pat, assisted us with much of the editorial work, and our secretary, Mrs. Ethel Savarese, has been invaluable in getting the manuscript to press.

Stony Brook, New York
October, 1972

Robert R. Sokal
F. James Rohlf

INTRODUCTION TO BIOSTATISTICS

1 Introduction

This chapter sets the stage for your study of biostatistics. We shall first of all define the field itself (Section 1.1). We shall then cast a necessarily brief glance at its historical development (Section 1.2). Section 1.3 concludes the chapter with a discussion of the attitudes that the person trained in statistics brings to biological research.

1.1 Some definitions

We shall define *biostatistics* as *the application of statistical methods to the solution of biological problems*. Biostatistics is also called *biological statistics* or *biometry*.

The definition of biostatistics leaves us somewhat up in the air—"statistics" has not been defined. *Statistics* is a science well known by name even to the layman. The number of definitions you can find for it is limited only by the number of books you wish to consult. In its modern sense we might define statistics as *the scientific study of numerical data based on natural phenomena.* All parts of this definition are important and deserve emphasis.

Scientific study: We are concerned with the commonly accepted criteria of validity of scientific evidence. Objectivity in presentation and in evaluation of data and the general ethical code of scientific methodology must constantly be in evidence if the old canard that "figures never lie, only statisticians do" is not to be revived.

Data: Statistics generally deals with populations or groups of individuals; hence it deals with *quantities* of information, not with a single *datum.* Thus the measurement of a single animal or the response from a single biochemical test will generally not be of interest.

Numerical: Unless the data of a study can be quantified in one way or the other, they would not be amenable to statistical analysis. Numerical data can be measurements—the length or width of a structure or the amount of a chemical in a body fluid—or counts—the number of bristles or teeth. The different kinds of variables will be discussed in greater detail in Chapter 2.

Natural phenomena: We use this term in a wide sense, including all those events that happen in animate and inanimate nature not under the control of man, plus those evoked by the scientist and partly under his control, as in an experiment. Different biologists will concern themselves with different levels of natural phenomena; other kinds of scientists, with yet different ones. But all would agree that the chirping of crickets, the number of peas in a pod, and the age at maturity of a chicken are natural phenomena. The heartbeat of rats in response to adrenalin or the mutation rate in maize after irradiation may still be considered natural, even though man has interfered with the phenomenon through his experiment. However, the average biologist would not consider the number of stereo hi-fi sets bought by persons in different states in a given year to be a natural phenomenon, although sociologists or human ecologists might so consider it and deem it worthy of study. The qualification "natural phenomena" is included in the definition of statistics largely to make certain that the phenomena studied are not arbitrary ones that are entirely under the will and control of the researcher, such as the number of animals employed in an experiment.

The word statistics is also used in another, though related, meaning. It is used as the plural of the noun *statistic,* which refers to any one of many computed or estimated statistical quantities, such as the mean, the standard deviation, or the correlation coefficient. Each one of these is a statistic.

1.2 The development of biostatistics

The origin of modern statistics can be traced back into the 17th century, in which it derived from two sources. The first of these related to political science and developed as a quantitative description of the various aspects of the affairs of a government or state (hence the term statistics). This subject also became known as political arithmetic. Taxes and insurance caused people to become interested in problems of censuses, longevity, and mortality. Such considerations assumed increasing importance, especially in England, as the country prospered during the development of its empire. John Graunt (1620–1674) and William Petty (1623–1687) were early students of vital statistics and others followed in their footsteps.

At about the same time there developed the second root of modern statistics: the mathematical theory of probability engendered by the interest in games of chance among the leisure classes of the time. Important contributions to this theory were made by Blaise Pascal (1623–1662) and Pierre de Fermat (1601–1665), both Frenchmen. Jacques Bernoulli (1654–1705), a Swiss, laid the foundation of modern probability theory in *Ars Conjectandi*, published posthumously. Abraham de Moivre (1667–1754), a Frenchman living in England, was the first to combine the statistics of his day with probability theory in working out annuity values. De Moivre also was the first to approximate the important normal distribution through the expansion of the binomial.

A later stimulus for the development of statistics came from the science of astronomy, in which many individual observations had to be digested into a coherent theory. Many of the famous astronomers and mathematicians of the 18th century such as Pierre Simon Laplace (1749–1827) in France and Karl Friedrich Gauss (1777–1855) in Germany were among the leaders in this field. The latter's lasting contribution to statistics is the development of the method of least squares, which will be encountered in later chapters of this book.

Perhaps the earliest important figure in biostatistic thought was Adolphe Quetelet (1796–1874), a Belgian astronomer and mathematician, who in his work combined the theory and practical methods of statistics and applied them to problems of biology, medicine, and sociology. Francis Galton (1822–1911), a cousin of Charles Darwin, has been called the father of biostatistics and eugenics, two subjects that he studied interrelatedly. The inadequacy of Darwin's genetic theories stimulated Galton to try to solve the problems of heredity. Galton's major contribution to biology is his application of statistical methodology to the analysis of biological variation, such as the analysis of variability and his study of regression and correlation in biological measurements. His hope of unraveling the laws of genetics through these procedures was in vain. He started with the most difficult material and with the wrong assumptions. However, his methodology has become the foundation for the application of statistics to biology. Karl Pearson (1857–1936) at University College, London, became interested in the application of statistical methods to biology, particularly in the demonstration of natural selection, through the influence of W. F. R. Weldon (1860–1906), a zoologist at the same institution. Weldon, incidentally, is credited with coining the term biometry for the type of studies pursued by him. Pearson continued in the tradition of Galton and laid the foundation for much of descriptive and correlational statistics. The dominant figure in statistics and biometry in this century has been Ronald A. Fisher (1890–1962). His many contributions to statistical theory will become obvious even to the cursory reader of this book.

Statistics today is a broad and extremely active field, whose applications touch almost every science and even the humanities. New applications for statistics are constantly being found, and no one can predict from what branch of statistics new applications to biology will be made.

1.3 The statistical frame of mind

The ever-increasing importance and application of statistics to biological data is evident even on cursory inspection of almost any biological journal. Why has there been such a marked increase in the use of statistics in biology? It has apparently come with the realization that in biology the interplay of causal and response variables obeys laws that are not in the classic mold of 19th century physical science. In that century, biologists such as Robert Mayer, Helmholtz, and others, in trying to demonstrate that biological processes were nothing but physicochemical phenomena, helped create the impression that the experimental methods and natural philosophy that had led to such dramatic progress in the physical sciences should be imitated fully in biology. Regrettably, opposition to this point of view was confounded with the vitalistic movement, which led to unproductive theorizing.

Thus, many biologists even to this day have retained the tradition of strictly mechanistic and deterministic concepts of thinking, while physicists, as their science became more refined and came to deal with ever more "elementary" particles, began to resort to statistical approaches. In biology most phenomena are affected by many causal factors, uncontrollable in their variation and often unidentifiable. Statistics is needed to measure such variable phenomena with a predictable error and to ascertain the reality of minute but important differences.

A misunderstanding of these principles and relationships has given rise to the attitude of some biologists that if differences induced by an experiment, or observed in nature, are not such as to be clear on plain inspection (and therefore not in need of statistical analysis), they are not worth investigating. There are few legitimate fields of inquiry in which, from the nature of the phenomena studied, statistical investigation is unnecessary.

It should be stressed that statistical thinking is not really different in kind from ordinary disciplined scientific thinking, in which we try to quantify our observations. In statistics we express our degree of belief or disbelief as a probability rather than as a vague, general statement. For example, the statements that species A is larger than species B or that females are more often found sitting on tree M rather than on tree N are of a kind commonly made by biological scientists. Such statements can and should be more precisely expressed in quantitative form. In many ways the human mind is a remarkable statistical machine, absorbing many facts from the outside world, digesting these, and regurgitating them in simple summary form. From our experience we know certain events to occur frequently, others rarely. "Man smoking cigarette" is a frequently observed event, "Man slipping on banana peel," rare. We know from experience that Japanese are on the average shorter than Englishmen and that Egyptians are on the average darker than Swedes. We associate thunder with lightning, almost always, flies with garbage cans frequently in the summer, but snow with the southern Californian desert extremely rarely. All such knowledge comes to us as a result of our lifetime experience, both directly and through that of others by direct communication or through reading. All these facts have been processed by that remarkable computer, the human brain, which furnishes an abstract. This summary is constantly under revision, and though occasionally faulty and biased, it is on the whole astonishingly sound; it is our knowledge of the moment.

Although statistics arose to satisfy the needs of scientific research, the development of its methodology in turn affected the sciences in which statistics is applied. Thus, in a positive feedback type of operation, statistics, created to serve the needs of natural science, has itself affected the philosophy of the biological sciences. To cite an example: analysis of variance has had a tremendous effect in influencing the types of experiments researchers carry out; the whole field of quantitative genetics, one of whose problems is the separation of environmental from genetic effects, depends upon the analysis of variance for its realization, and many of the concepts of quantitative genetics have been directly built around the analysis of variance.

2

Data in Biology

In Section 2.1 we explain the statistical meaning of the terms "sample" and "population," which we shall be using throughout this book. Then we come to the types of observations that we obtain from biological research material, with which we shall perform the various computations in the rest of this book (Section 2.2). The degree of accuracy necessary for recording data and the procedure for rounding off figures are discussed in Section 2.3. We shall then be ready to consider in Section 2.4 certain kinds of derived data, such as ratios and indices, frequently used in biological science, which present peculiar problems with relation to their accuracy and distribution. Knowing how to arrange data in frequency distributions is important, because such arrangements permit us to get an overall impression of their general appearance and to set them up for further computational procedures. Frequency distributions as well as the presentation of numerical data are discussed in the next section (2.5) of this chapter. In Section 2.6 we briefly describe the computational handling of data.

2.1 Samples and populations

We shall now define a number of important terms necessary for an understanding of biological data. The data in biostatistics are generally based on *individual observations. They are observations or measurements taken on the smallest sampling unit.* These smallest sampling units frequently, but not necessarily, are also individuals in the ordinary biological sense. If we measure weight in 100 rats, then the weight of each rat is an individual observation; the hundred rat weights together represent the *sample of observations,* defined as *a collection of individual observations selected by a specified procedure.* In this instance, one individual observation is based on one individual in a biological sense—that is, one rat. However, if we had studied weight in a single rat over a period of time, the sample of individual observations would be the weights recorded on one rat at successive times. If we wish to measure temperature in a study of ant colonies, where each colony is a basic sampling unit, each temperature reading for one colony is an individual observation, and the sample of observations is the temperatures for all the colonies considered. If we consider an estimate of the DNA content of a single mammalian sperm cell to be an individual observation, the sample of observations may be the estimates of DNA content of all the sperm cells studied in one individual mammal. A synonym for individual observation is *item.*

We have carefully avoided so far specifying what particular variable was being studied, because the terms "individual observation" and "sample of observations" as used above define only the structure but not the nature of the data in a study. *The actual property measured by the individual observations is the character or variable.* The more common term employed in general statistics is *variable.* However, in biology the word *character* is frequently used synonymously. More than one character can be measured on each smallest sampling unit. Thus, in a group of 25 mice we might measure the blood *p*H and the erythrocyte count. Each of the 25 mice (a biological individual) is the smallest sampling unit; blood *p*H and red cell count would be the two characters studied; the *p*H readings and cell counts are individual observations, and two samples of 25 observations on *p*H and erythrocyte count would result. Or we may speak of a *bivariate sample* of 25 observations, each referring to a *p*H reading paired with an erythrocyte count.

Next we define *population.* The biological definition of this term is well known. It refers to all the individuals of a given species (perhaps of a given life-history stage or sex) found in a circumscribed area at a given time. In statistics, population always means *the totality of individual observations about which inferences are to be made, existing anywhere in the world or at least within a definitely specified sampling area limited in space and time.* If you take five men and study the number of leucocytes in their peripheral blood and you are prepared to draw conclusions about all men from this sample of five, then the population from which the sample has been drawn represents the leucocyte counts of all extant males of the species *Homo sapiens*. If, on the other hand, you restrict yourself to a more narrowly specified sample, such as five male Chinese, aged 20, and you are restricting your conclusions to this particular group, then the population from which you are sampling will be leucocyte numbers of all Chinese males of age 20. The population in this statistical sense is sometimes referred to as the *universe*. A population may refer to variables of a concrete collection of objects or creatures, such as the tail lengths of all the white mice in the world, the leucocyte counts of all the Chinese men in the world of age 20, or the DNA content of all the hamster sperm cells in existence; or it may refer to the outcomes of experiments, such as all the heartbeat frequencies produced in guinea pigs by injections of adrenalin. In the first cases the population is generally finite. Although in practice it would be impossible to collect, count, and examine all hamster sperm cells, all Chinese men of age 20, or all white mice in the world, these populations are in fact finite. Certain smaller populations, such as all the whooping cranes in North America or all the pocket gophers in a given colony, may well lie within reach of a total census. By contrast, an experiment can be repeated an infinite number of times (at least in theory). A given experiment such as the administration of adrenalin to guinea pigs could be repeated as long as the experimenter could obtain material and his health and patience held out. The sample of experiments actually performed is a sample from an infinite number that *could* be performed. Some of the statistical methods to be developed later make a distinction between sampling from finite and from infinite populations. However, though populations are theoretically finite in most applications in biology, they are generally so much larger than samples drawn from them, that they can be considered as *de facto* infinitely sized populations.

2.2 Variables in biology

Each biological discipline has its own set of variables, which may include conventional morphological measurements, concentrations of chemicals in body fluids, rates of certain biological processes, frequencies of certain events as in genetics and radiation biology, physical readings of optical or electronic machinery used in biological research, and many more.

We have already referred to biological variables in a general way, but we have not yet defined them. We shall define a *variable as a property with respect to which individuals in a sample differ in some ascertainable way*. If the property does not differ within a sample at hand or at least among the samples being studied, it cannot be of statistical interest. Length, height, weight, number of teeth, vitamin C content, and genotypes are examples of variables in ordinary, genetically and phenotypically diverse, groups of organisms. Warm-bloodedness in a group of mammals is not, since they are all alike in this regard, although body temperature of individual mammals would, of course, be a variable.

We can divide biological variables as follows:

Variables
Measurement variables
Continuous variables
Discontinuous variables
Ranked variables
Attributes

Measurement variables are all those whose differing states can be expressed in a numerically ordered fashion. They are divisible into two kinds. The first of these are *continuous variables,* which at least theoretically can assume an infinite number of values between any two fixed points. For example, between the two length measurements 1.5 and 1.6 cm there is an infinite number of lengths that could be measured if one were so inclined and had a precise enough method of calibration to obtain such measurements. Any given reading of a continuous variable, such as a length of 1.57 mm, is therefore an approximation to the exact reading, which in practice is unknowable. Many of the variables studied in biology are continuous variables. Examples are lengths, areas, volumes, weights, angles, temperatures, periods of time, percentages, and rates.

Contrasted with continuous variables are the *discontinuous variables,* also known as *meristic* or *discrete variables.* These are variables that have only certain fixed numerical values, with no intermediate values possible in between. Thus the number of segments in a certain insect appendage may be 4 or 5 or 6 but never $5\frac{1}{2}$ or 4.3. Examples of discontinuous variables are numbers of a certain structure (such as segments, bristles, teeth, or glands), the numbers of offspring, the numbers of colonies of microorganisms or animals, or the numbers of plants in a given quadrat.

Some variables cannot be measured but at least can be ordered or ranked by their magnitude. Thus, in an experiment one might record the rank order of emergence of ten pupae without specifying the exact time at which each pupa emerged. In such cases we code the data as a *ranked variable,* the order of emergence. Special methods for dealing with such variables have been developed and several are furnished in this book. By expressing a variable as a series of ranks, such as 1, 2, 3, 4, 5, we do not imply that the difference in magnitude between, say, ranks 1 and 2 is identical to or even proportional to the difference between 2 and 3.

Variables that cannot be measured but must be expressed qualitatively are called *attributes.* These are all properties, such as black or white, pregnant or not pregnant, dead or alive, male or female. When such attributes are combined with frequencies, they can be treated statistically. Of 80 mice, we may, for instance, state that four were black, two agouti, and the rest gray. When attributes are combined with frequencies into tables suitable for statistical analysis, they are referred to as *enumeration data.* Thus the enumeration data on color in mice just mentioned would be arranged as follows:

Color	*Frequency*
Black	4
Agouti	2
Gray	74
Total number of mice	80

In some cases attributes can be changed into variables if this is desired. Thus colors can be changed into wavelengths or color chart values, which are measurement variables. Certain other attributes that can be ranked or ordered can be coded to become ranked variables. For example, three attributes referring to a structure as "poorly developed," "well developed," and "hypertrophied" could conveniently be coded 1, 2, and 3.

A term that has not yet been explained is *variate.* In this book we shall use it as a single reading, score, or observation of a given variable. Thus, if we have measurements of the length of the tails of five mice, tail length will be a continuous variable, and each of the five readings of length will be a variate. In this text we identify variables by capital letters, the most common symbol being Y. Thus Y may stand for tail length of mice. A variate will refer to a given length measurement; Y_i is the measurement of tail length of the ith mouse, and Y_4 is the measurement of tail length of the fourth mouse in our sample.

2.3 Accuracy and precision of data

"Accuracy" and "precision" are used synonymously in everyday speech, but in statistics we define them more rigorously. *Accuracy* is the closeness of a measured or computed value to its true value; *precision* is the closeness of repeated measurements of the same quantity. A biased but sensitive scale might yield inaccurate but precise weight. By chance an insensitive scale might result in an accurate reading, which would however be imprecise, since a repeated weighing would be unlikely to yield an equally accurate weight. Unless there is bias in a measuring instrument, precision will lead to accuracy. We need therefore mainly be concerned with the former.

Precise variates are usually but not necessarily whole numbers. Thus, when we count four eggs in a nest, there is no doubt about the exact number of eggs in the nest if we have counted correctly; it is four, not three nor five, and clearly it could not be four plus or minus a fractional part. Meristic variables are generally measured as exact numbers. Seemingly, continuous variables derived from meristic ones can under certain conditions also be exact numbers. For instance, ratios between exact numbers are themselves also exact. If in a colony of animals there are 18 females and 12 males, the ratio of females to males is 1.5, a continuous variate but also an exact number.

Most continuous variables, however, are approximate. We mean by this that the exact value of the single measurement, the variate, is unknown and probably unknowable. The last digit of the measurement stated should imply precision; that is, the limits on the measurement scale between which we believe the true measurement to lie. Thus a length measurement of 12.3 mm implies that the true length of the structure lies somewhere between 12.25 and 12.35 mm. Exactly where between these *implied limits* the real length is we do not know. But where would a true measurement of 12.25 fall? Would it not equally likely fall in either of the two classes 12.2 and 12.3, clearly an unsatisfactory state of affairs? Such an argument is correct, but when we record a number as either 12.2 or 12.3 we imply that the decision whether to put it into the higher or lower class has already been taken. This decision was not taken arbitrarily, but presumably was based on the best available measurement. If the scale of measurement is so precise that a value of 12.25 would clearly have been recognized, then the measurement should have been recorded originally to four significant figures. Implied limits therefore always carry one more figure beyond the last significant one measured by the observer.

Hence it follows that if we record the measurement as 12.32, we are implying that the true value lies between 12.315 and 12.325. Unless this is what we mean, there would be no point in adding the last decimal figure to our original measurements. If we do add another figure we must imply an increase in precision. We see therefore that accuracy and precision in numbers is not an absolute concept, but is relative. Assuming there is no bias, a number becomes increasingly more accurate as we are able to write more significant figures for it (increase its precision). To illustrate this concept of the relativity of accuracy consider the following three numbers.

	Implied limits
193	192.5 –193.5
192.8	192.75 –192.85
192.76	192.755–192.765

We may imagine these numbers to be the recorded measurements of the same structure. Let us assume that we had extramundane knowledge of the true length of the given structure as being 192.758 units. If that were so, the three measurements increase in accuracy from the top down. You will note that the implied limits of the topmost one are wider than those of the measurement below it, which in turn are wider than those of the third measurement.

Meristic variates, though ordinarily exact, may be recorded approximately when large numbers are involved. Thus when counts are reported to the nearest thousand, a count of 36,000 insects in a cubic meter of soil implies that the true number varied somewhere from 35,500 to 36,500 insects.

To how many significant figures should we record measurements? If we array the sample by order of magnitude from the smallest individual to the largest one, an easy rule to remember is that *the number of unit steps from the smallest to the largest measurement in an array should be between 30 and 300.* Thus, if we are measuring a series of shells to the nearest millimeter and the largest is 8 mm and the smallest is 4 mm wide, there are only four unit steps between the largest and the smallest measurement. Hence we should have measured our shells to one more significant decimal place. Then the two extreme measurements might have been 8.2 mm and 4.1 mm, with 41 unit steps between them (counting the last significant digit as the unit); this would have been an adequate number of unit steps. The reason for such a rule is that an error of 1 in the last significant digit of a reading of 4 mm would constitute an inadmissible error of 25%, but an error of 1 in the last digit of 4.1 is less than 2.5%. Similarly, if we had measured the height of the tallest of a series of plants as 173.2 cm and that of the shortest of these plants as 26.6 cm, the difference between these limits would comprise 1466 unit steps (of 0.1 cm), which are far too many. It would therefore have been advisable to record the heights to the nearest centimeter, as follows: 173 cm for the tallest and 27 cm for the shortest. This would yield 146 unit steps. Using the rule stated above, we shall record two or three digits for most measurements.

The last digit of an approximate number should always be significant; that is, it should imply a range for the true measurement of from half a "unit step" below to half a "unit step" above the recorded score, as illustrated earlier. This applies to all digits, zero included. Zeros should therefore not be written at the end of approximate numbers to the right of the decimal point unless they are meant to be significant digits. Thus 7.80 must imply the limits 7.795 to 7.805. If 7.75 to 7.85 is implied, the measurement should be recorded as 7.8.

When the number of significant digits is to be reduced, we carry out the process of *rounding off* numbers. The rules for rounding off are very simple. A digit to be rounded off is not changed if it is followed by a digit less than 5. If the digit to be rounded off is followed by a digit greater than 5 or by 5 followed by other nonzero digits, it is increased by one. When the digit to be rounded off is followed by a 5 standing alone or followed by zeros, it is unchanged if it is even but increased by one if it is odd. The reason for this last rule is that when such numbers are summed in a long series we should have as many digits raised as are being lowered on the average; these changes would therefore balance out. Practice the above rules by rounding off the following numbers to the indicated number of significant digits.

Number	*Significant digits desired*	*Answer*
26.58	2	27
133.7137	5	133.71
0.03725	3	0.0372
0.03715	3	0.0372
18,316	2	18,000
17.3476	3	17.3

2.4 Derived variables

The majority of variables in biometric work are observations recorded as direct measurements or counts of biological material or as readings that are the output of various types of instruments. However, there is an important class of variables in biological research that we may call the *derived* or *computed variables*, which are generally based on two or more independently measured variables whose relations are expressed in a certain way. We are referring to ratios, percentages, indices, rates, and the like.

A *ratio* expresses as a single value the relation that two variables have one to the other. In its simplest form it is expressed as in 64:24, which may represent the number of wild type versus mutant individuals or the number of males versus females, or the proportion of parasitized individuals versus those not parasitized and so on. The above examples implied ratios based on counts; a ratio based on a continuous variable might be similarly expressed as 1.2:1.8, which may represent the ratio of width to length in a sclerite of an insect or the ratio between the concentrations of two minerals contained in water or soil. Ratios may also be expressed as fractions; thus the two ratios above could be expressed as $\frac{64}{24}$ and $\frac{1.2}{1.8}$. However, for computational purposes it is most useful to express the ratio as a quotient. The two ratios cited above would therefore be 2.666 . . . and 0.666 . . . , respectively. These are pure numbers, not expressed in measurement units of any kind. It is this form for ratios that we shall consider further below. *Percentages* are also a type of ratio. Ratios and percentages are basic quantities in much biological research, widely used and generally familiar.

An *index* is the ratio of one anatomic variable divided by a larger, so-called standard one. A well-known example of an index in this sense is the cephalic index in physical anthropology. Conceived in the wide sense, an index could be the average of two measurements—either simply, such as $\frac{1}{2}$ (length of A + length of B), or in weighted fashion, such as $\frac{1}{3}$ [(2 × length of A) + length of B].

Rates will be important in many experimental fields of biology. The amount of a substance liberated per unit weight or volume of biological material, weight gain per unit time, reproductive rates per unit population size and time (birth rates), and death rates would fall in this category.

The use of ratios and percentages is deeply ingrained in scientific thought processes. Often ratios may be the only meaningful way to interpret and understand certain types of biological problems. If the biological process being investigated operates on the ratio of the variables studied, one must examine this ratio to understand the process. Thus, Sinnott and Hammond (1935) found that inheritance of the shapes of squashes, *Cucurbita pepo*, could be interpreted by a form index based on a length-width ratio, but not in terms of the independent dimensions of shape. Similarly, selection affecting body proportions must be found to exist in the evolution of almost any organism when properly investigated.

The disadvantages of using ratios are several: first is their relative inaccuracy. Let us return to the ratio $\frac{1.2}{1.8}$ mentioned above and recall from the previous section that a measurement of 1.2 indicates a true range of measurement of the variable from 1.15 to 1.25; similarly, a measurement of 1.8 implies a range from 1.75 to 1.85. We realize therefore that the true ratio may vary anywhere from $\frac{1.15}{1.85}$ to $\frac{1.25}{1.75}$, or 0.622 and 0.714, respectively. We note a possible maximal error of 4.2% if 1.2 were an original measurement: $[(1.25 - 1.2)/1.2]$; the corresponding maximal error for the ratio is 7.0%: $[(0.714 - 0.667)/0.667]$. Furthermore, the best estimate of a ratio is not usually the midpoint between its possible ranges. Thus in our example the midpoint between the implied limits is 0.668 and the true ratio is 0.666 . . . , only a slight difference, which may, however, be greater in other instances.

A second drawback to ratios and percentages is that their distributions may be rather unusual and they may therefore not be more or less normally distributed (see Chapter 5) as required by many statistical tests. This difficulty can frequently be overcome by transformation of the variable (as discussed in Chapter 10). Another disadvantage of ratios is that they do not provide information on the relationship between the two variables whose ratio is being taken. Often more may be learned by studying the variables singly and their relationships to each other.

2.5 Frequency distributions

If we were to sample, for instance, a population of birth weights of infants, we could represent each sampled measurement by a point along an axis denoting magnitude of birth weights. This is illustrated in Figure 2.1A, for a sample of 25 birth weights. If we sample repeatedly from the population and obtain 100 birth weights, we shall probably have to place some of these points on top of other points in order to record them all correctly (Figure 2.1B). As we continue sampling additional hundreds and thousands of birth

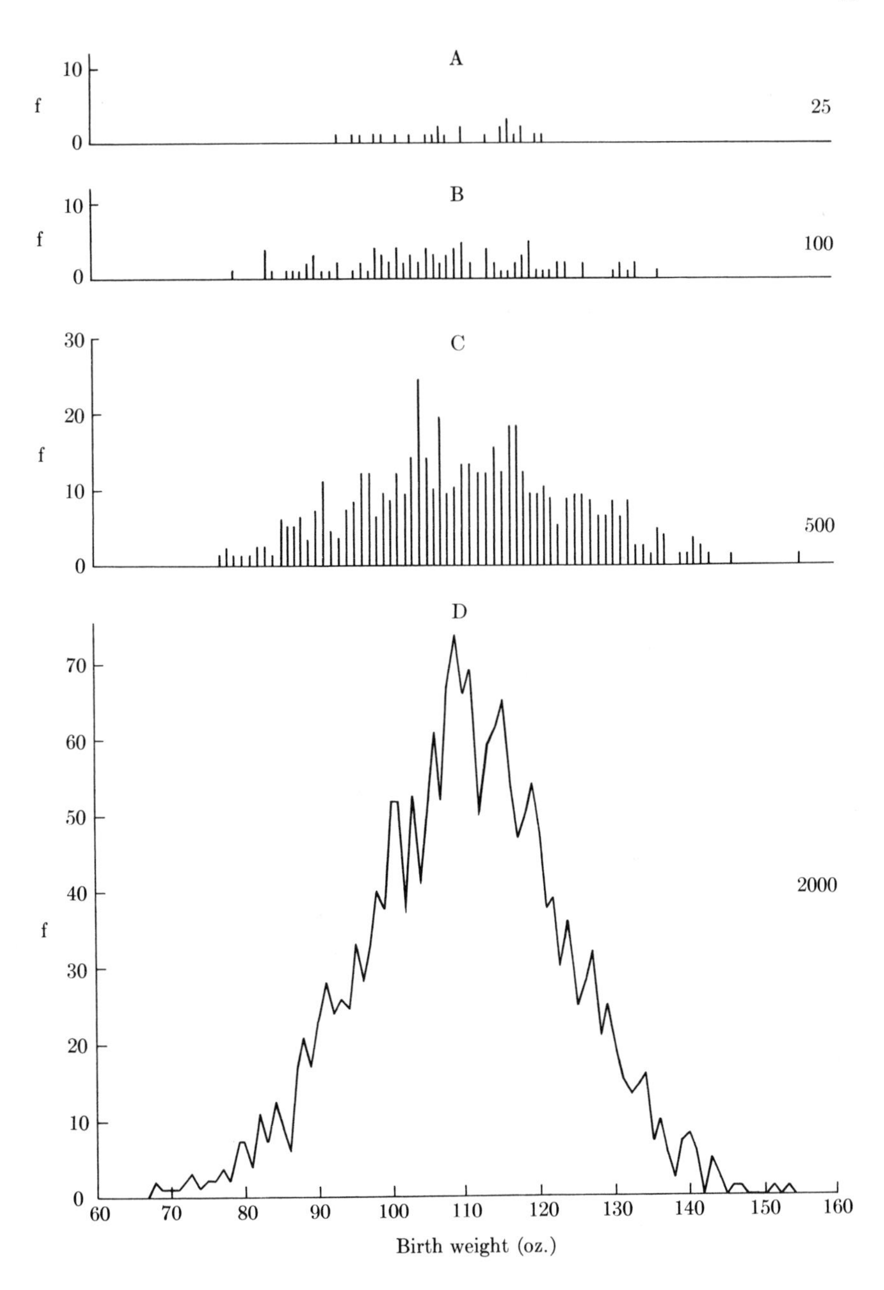

FIGURE 2.1
Sampling from a population of birth weights of infants (a continuous variable). A. A sample of 25. B. A sample of 100. C. A sample of 500. D. A sample of 2000.

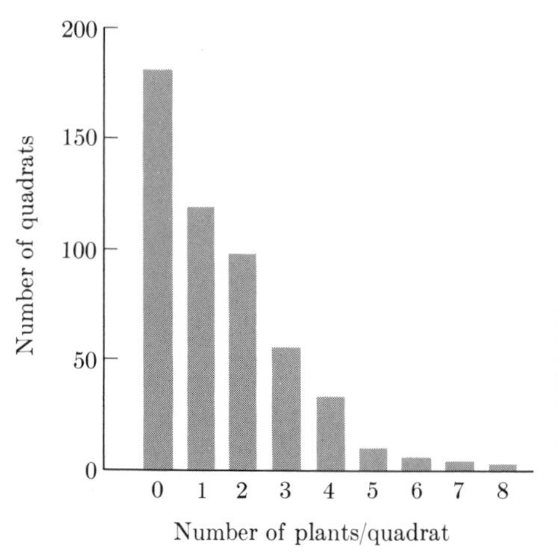

FIGURE 2.2
Bar diagram. Frequency of the sedge *Carex flacca* in 500 quadrats. Data from Table 2.2; originally from Archibald (1950).

weights (Figure 2.1C and D), the assemblage of points will continue to increase in size but will assume a fairly definite shape. The curve tracing the outline of the mound of points approximates the distribution of the variable. Remember that a continuous variable such as birth weight can assume an infinity of values between any two points on the abscissa. The refinement of our measurements will determine how fine the number of recorded divisions between any two points along the axis will be.

The distribution of a variable is of considerable biological interest. If we find that the distribution is asymmetrical and drawn out in one direction, it tells us that there is, perhaps, selection for or against organisms falling in one of the tails of the distribution, or possibly that the scale of measurement chosen is such as to bring about a distortion of the distribution. If, in a sample of immature insects, we discover that the measurements are bimodally distributed (with two peaks), this would indicate that the population is dimorphic; different species or races may have become intermingled in our sample, or the dimorphism could have arisen from the presence of both sexes or of different instars. There are several characteristic shapes of frequency distributions, the most common of which is the symmetrical bell-shaped distribution (approximated by the bottom graph in Figure 2.1), the normal frequency distribution discussed in Chapter 5. There are also skewed distributions (drawn out more at one tail than the other), L-shaped distributions as in Figure 2.2, U-shaped distributions, and others, which impart significant information about certain types of relationships. We shall have more to say about the implications of various types of distributions in later chapters and sections.

After data have been obtained in a given study they have to be arranged in a form suitable for computation and interpretation. We may assume that variates are randomly ordered initially or are in the order in which the measurements had been taken. A simple arrangement would be an *array* of the data by order of magnitude. Thus, for example, the variates 7, 6, 5, 7, 8, 9,

6, 7, 4, 6, 7 could be arrayed in order of decreasing magnitude as follows: 9, 8, 7, 7, 7, 7, 6, 6, 6, 5, 4. Where there are some variates of the same value, such as the 6's and 7's in this fictitious example, a time-saving device might immediately have occurred to you—namely, to list a frequency for each of the recurring variates, thus: 9, 8, 7(4 ×), 6(3 ×), 5, 4. Such a shorthand notation is one way to represent a *frequency distribution*, which is simply an arrangement of the classes of variates with the frequencies of each class indicated. Conventionally, a frequency distribution is stated in tabular form, as follows for the above example.

Variable	*Frequency*
Y	f
9	1
8	1
7	4
6	3
5	1
4	1

The above is an example of a *quantitative frequency distribution*, since Y is clearly a measurement variable. However, arrays and frequency distributions need not be limited to such variables. We can make frequency distributions of attributes, called *qualitative frequency distributions*. In these, the various classes are listed in some logical or arbitrary order. For example, in genetics we might have a qualitative frequency distribution as follows:

	f
$A-$	86
aa	32

This tells us that there are two classes of individuals, those identified by the $A-$ phenotype, of which 86 were found, and the homozygote recessive aa, of which 32 were seen in the sample. In ecology it is quite common to have a species list of inhabitants of a sampled ecological area. The arrangement of such tables is usually alphabetical. An example of an ecological list constituting a qualitative frequency distribution of families is shown in Table 2.1. In this instance the families of insects referred to are listed alphabetically; in other cases the sequence may be by convention, as for the families of flowering plants in botany.

A quantitative frequency distribution based on meristic variates is shown in Table 2.2. This is an example from plant ecology: the number of plants per quadrat sampled are listed at the left in the variable column; the observed frequency is shown at the right.

Quantitative frequency distributions based on a continuous variable are the most commonly employed frequency distributions and you should become thoroughly familiar with them. An example is shown in Box 2.1. It is based on 25 femur lengths of stem mothers (one of the life-history stages) of a

TABLE 2.1

A qualitative frequency distribution. Number of individuals of Heteroptera tabulated by family in total samples from a summer foliage insect community.

Family	*f*
Alydidae	2
Anthocoridae	37
Coreidae	2
Lygaeidae	318
Miridae	373
Nabidae	3
Neididae	5
Pentatomidae	25
Piesmidae	1
Reduviidae	3
Rhopalidae	2
Saldidae	1
Thyreocoridae	10
Tingidae	69
Total Heteroptera	851

Source: Data from Whittaker (1952).

TABLE 2.2

A meristic frequency distribution. Number of plants of the sedge *Carex flacca* found in 500 quadrats.

No. of plants per quadrat Y	*Observed frequency* f
0	181
1	118
2	97
3	54
4	32
5	9
6	5
7	3
8	1
Total	500

Source: Data from Archibald (1950).

species of aphids. The 25 readings (measured in coded micrometer units) are shown at the top of Box 2.1 in the order in which they were obtained as measurements. They could have been arrayed according to their magnitude. The data are next set up in a frequency distribution. The variates increase in magnitude by unit steps of 0.1. The frequency distribution is prepared by entering each variate in turn on the scale and indicating a count by a conventional tally mark. When all of the items have been tallied in the corresponding class, the tallies are converted into numerals indicating frequencies in the next column. Their sum is indicated by Σf.

What have we achieved in summarizing our data? The original 25 variates are now represented by only 15 classes. We find that variates 3.6, 3.8, and 4.3 have the highest frequencies However, we also note that there are several classes, such as 3.4 or 3.7, that are not represented by a single aphid. This gives the entire frequency distribution a rather drawn-out and scattered appearance. The reason for this is that we have only 25 aphids, rather too few to put into a frequency distribution with 15 classes. To obtain a more cohesive and smooth-looking distribution we have to condense our data into fewer classes. This process is known as *grouping of classes* of frequency distributions; it is illustrated in Box 2.1 and described in the following paragraphs.

We should realize that what we are doing when we group individual variates into classes of wider range is only an extension of the same process that took place when we obtained the initial measurement. Thus, as we have seen in Section 2.3, when we measure an aphid and record its femur length as 3.3 units, we imply thereby that the true measurement lies between 3.25 and 3.35 units, but that we were unable to measure to the second decimal place. In recording the measurement initially as 3.3 units, we estimated that it fell within this range. Had we estimated that it exceeded the value of 3.35, for example, we would have given it the next higher score, 3.4. Therefore all the measurements between 3.25 and 3.35 were in fact grouped into the class identified by the *class mark* 3.3. Our *class interval* was 0.1 units. If we now wish to make wider class intervals, we are doing nothing but extending the range within which measurements are placed into one class.

Reference to Box 2.1 will make this process clear. We group the data twice in order to impress upon the reader the flexibility of the process. In the first example of grouping, the class interval has been doubled in width; that is, it was made to equal 0.2 units. If we start at the lower end, the implied class limits will now be from 3.25 to 3.45, the limits for the next class from 3.45 to 3.65, and so forth.

Our next task is to find these class marks. This was quite simple in the frequency distribution shown at the left side of Box 2.1, in which the original measurements had been used as class marks. However, now we are using a class interval twice as wide as before, and the class marks are calculated by taking the midpoint of the new class intervals. Thus to find the class mark

BOX 2.1

Preparation of frequency distribution and grouping into fewer classes with wider class intervals.

Twenty-five femur lengths of stem mothers of the aphid *Pemphigus populi-transversus*. Measurements are in $mm \times 10^{-1}$.

Original measurements				
3.8	3.6	4.3	3.5	4.3
3.3	4.3	3.9	4.3	3.8
3.9	4.4	3.8	4.7	3.6
4.1	4.4	4.5	3.6	3.8
4.4	4.1	3.6	4.2	3.9

Original frequency distribution				*Grouping into 8 classes of interval 0.2*				*Grouping into 5 classes of interval 0.3*			
Implied limits	*Y*	*Tally marks*	*f*	*Implied limits*	*Class mark*	*Tally marks*	*f*	*Implied limits*	*Class mark*	*Tally marks*	*f*
3.25–3.35	3.3	|	1	3.25–3.45	3.35	|	1	3.25–3.55	3.4	||	2
3.35–3.45	3.4		0								
3.45–3.55	3.5	|	1	3.45–3.65	3.55	~~||||~~	5				
3.55–3.65	3.6	||||	4					3.55–3.85	3.7	~~||||~~ |||	8
3.65–3.75	3.7		0	3.65–3.85	3.75	||||	4				
3.75–3.85	3.8	||||	4								
3.85–3.95	3.9	|||	3	3.85–4.05	3.95	|||	3	3.85–4.15	4.0	~~||||~~	5
3.95–4.05	4.0		0								
4.05–4.15	4.1	||	2	4.05–4.25	4.15	|||	3				
4.15–4.25	4.2	|	1					4.15–4.45	4.3	~~||||~~ |||	8

4.25–4.35	4.3	||||	4	4.25–4.45	4.35	~~||||~~ ||	7				
4.35–4.45	4.4	|||	3								
4.45–4.55	4.5	|	1	4.45–4.65	4.55	|	1	4.45–4.75	4.6	||	2
4.55–4.65	4.6		0								
4.65–4.75	4.7	|	1	4.65–4.85	4.75	|	1				
Σf			25				25				25

Source: Data from R. R. Sokal.

Histogram of the original frequency distribution shown above and of the grouped distribution with 5 classes. Lower horizontal axis shows class marks for the grouped frequency distribution. Shaded bars represent original frequency distribution; hollow bars represent grouped distribution.

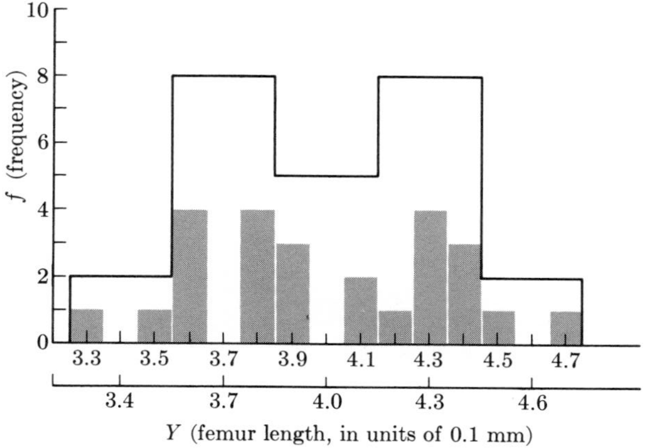

For a detailed account of the process of grouping, see Section 2.5

of the first class we take the midpoint between 3.25 and 3.45, which turns out to be 3.35. We note that the class mark has one more decimal place than the original measurements. We should not now be led to believe that we have suddenly achieved greater precision. Whenever we designate a class interval whose last *significant* digit is even (0.2 in this case), the class mark will carry one more decimal place than the original measurements. On the right side of the table in Box 2.1 the data are grouped once again, using a class interval of 0.3. Because of the odd last significant digit, the class mark now shows as many decimal places as the original variates, the midpoint between 3.25 and 3.55 being 3.4.

Once the implied class limits and the class mark for the first class have been correctly found, the others can be written down by inspection without any special computation. Simply add the class interval repeatedly to each of the values. Thus, starting with the lower limit 3.25, by adding 0.2 we obtain 3.45, 3.65, 3.85, and so forth; similarly for the class marks, we obtain 3.35, 3.55, 3.75, and so forth. It should be obvious that the wider the class intervals, the more compact the data become but also the less precise. However, looking at the frequency distribution of aphid femur lengths in Box 2.1, we notice that the initial rather chaotic structure is being simplified by grouping. When we group the frequency distribution into five classes with a class interval of 0.3 units, it becomes notably bimodal (that is, it possesses two peaks of frequencies).

In setting up frequency distributions, from 12 to 20 classes should be established. This rule need not be slavishly adhered to, but should be employed with some of the common sense that comes from experience in handling statistical data. The number of classes depends largely on the size of the sample studied. Samples of less than 40 or 50 should rarely be given as many as 12 classes, since that would provide too few frequencies per class. On the other hand, samples of several thousand may profitably be grouped into more than 20 classes. If the aphid data of Box 2.1 need to be grouped, they should probably not be grouped into more than 6 classes.

If the original data provide us with fewer classes than we think we should have, then nothing can be done if the variable is meristic, since this is the nature of the data in question. However, with a continuous variable a scarcity of classes would indicate that we probably have not made our measurements with sufficient precision. If we have followed the rules on number of significant digits for measurements stated in Section 2.3, this could not have happened. Often, regrettably, measurements have already been obtained before statistical advice is sought. Nothing can then be done if there are too few classes.

Whenever there are more than the desired number of classes, grouping should be undertaken. When the data are meristic, the implied limits of continuous variables are meaningless. Yet with many meristic variates, such as a bristle number varying from a low of 13 to a high of 81, it would probably be wise to group them into classes, each containing several counts. This can best be done by using an odd number as a class interval so that the class mark representing the data will be a whole rather than a fractional number. Thus if we were to group the bristle numbers 13, 14, 15, and 16 into one class, the class mark would have to be 14.5, a meaningless value in terms of bristle number. It would therefore be better to use a class ranging over 3 bristles or 5 bristles, giving the integral values 14 or 15 as a class mark.

When should data be grouped in frequency distributions? If a calculating machine is available and there are more than 100 or 150 observations, it is probably worthwhile to set up the data in a frequency distribution before carrying out the statistical computations. Such a decision is based on the fact that the computational time saved when a frequency distribution is employed has to be balanced against the time it takes to set up a frequency distribution. At the lower frequencies (that is, < 100) it would take longer to set up such a distribution than to compute from raw observations. If computer facilities are available the above rule is, of course, also invalid. Since a computer can handle hundreds and even thousands of observations in a very short time, it would generally not be necessary to group them unless we are dealing with extremely large samples, much in excess of 1000 observations per sample.

One further important reason for obtaining frequency distributions arises if the form of the distribution is a specific point of interest. Discovering in a sample of immature insects that some measurements are bimodally distributed (with two peaks) would indicate that the population is dimorphic. However, in certain cases, where repeated samples have been taken of a particular biological phenomenon or process and where the shape of the distribution is well known, we would not wish to set up a frequency distribution each time unless we needed to do so for computational reasons.

If the shape of a frequency distribution is of particular interest, we may often wish to present the distribution in graphic form when discussing the results. This is generally done by means of frequency diagrams, of which there are two common types. For a distribution of meristic data we employ a *bar diagram*, as shown in Figure 2.2 for the sedge data of Table 2.2. The abscissa represents the variable (in our case the number of plants per quadrat), and the ordinate represents the frequencies. The important point about such a diagram is that the bars do not touch each other, which indicates that the variable is not continuous. By contrast, continuous variables, such as the frequency distribution of the femur lengths of aphid stem mothers, are graphed as a *histogram*, in which the width of each bar along the abscissa represents a class interval of the frequency distribution and the bars touch each other to show that the actual limits of the classes are contiguous. The midpoint of the bar corresponds to the class mark. At the bottom of Box 2.1 are shown histograms of the frequency distribution of the aphid data, ungrouped and grouped. The height of the bars represents the frequency of each class. To illustrate that histograms are appropriate approximations to the continuous

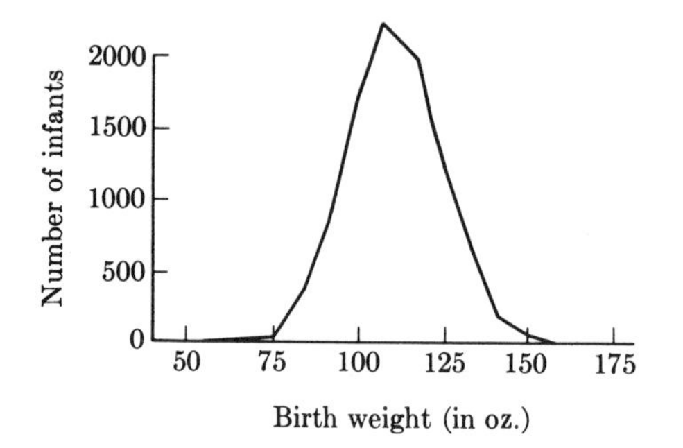

FIGURE 2.3
Frequency polygon. Birth weights of 9465 male infants. Chinese third-class patients in Singapore, 1950 and 1951. Data from Millis and Seng (1954).

distributions found in nature, we may take a histogram and make the class intervals more narrow, producing more classes. The histogram would then clearly fit more closely to a continuous distribution. We can continue this process until the class intervals approach the limit of infinitesimal width. At this point the histogram becomes the continuous distribution of the variable. Occasionally the class intervals of a grouped continuous frequency distribution are unequal. For instance, in a frequency distribution of ages we might have more detail on the different stages of young individuals and less accurate identification of the ages of old individuals. In such cases, the class intervals for the older age groups would be wider, those for the younger age groups, narrower. In representing such data, the bars of the histogram are drawn with different widths. Figure 2.3 shows another graphical mode of representation of a frequency distribution of a continuous variable—birth weight in infants. As we shall see later the shapes of distributions as seen in such frequency diagrams can reveal much about the biological situations affecting a given variable.

2.6 The handling of data

The skillful and expeditious handling of data is essential to the successful practice of statistics. Readers must become acquainted with the various techniques available for carrying out statistical computations. Although a desk calculator is well on its way to becoming as indispensable a research tool as a microscope, through one circumstance or another, you may find yourself without such a computational aid. In such instances, statistical computations must be carried out by so-called pencil and paper methods. As will be seen in the next chapter, data need to be coded in order to be rapidly processed, and the type of coding can be modified to suit pencil and paper methods.

Two other developments for the simplification of computation may be mentioned here. The first is the development of tables that provide answers for certain standard statistical problems without any computation whatsoever. The second is the development of some newer statistical techniques, which are so easy to carry out that no mechanical aids are needed. Some of these techniques are inherently simple (for example, some of the nonparametric methods, such as the sign test, Section 10.3); others are approximate and consequently less than fully efficient methods that can serve as a first summary of the desired results. An example of a less than fully efficient method is the employment of the median (Section 3.3) in lieu of the mean (Section 3.1).

Accuracy in computation will depend on the number of figures carried during computation. Different computing devices vary in their capacity for retaining the number of digits; therefore, the results obtained by solving the same problem on a desk calculator and a computer, for example, may differ. Even different computers will not necessarily give exactly the same answers.

Mechanical calculating machines are no longer being manufactured in most countries. However, they are still widely used and need brief mention. *Adding machines* are familiar to most persons from their extensive use in stores and offices. They consist of a keyboard for entering numbers into the machines and certain operational keys for adding, subtracting, subtotaling, totaling, and correcting entries, and for several other functions. The results are printed on a roll of paper tape attached to the machine. A modified version of an adding machine is the so-called *printing calculator*. This machine is a lineal descendant of the electric adding machine, to which automatic multiplication and division of numbers have been added. The results of these operations and some intermediate steps are printed on a paper tape.

A conventional *desk calculator* (also called rotary desk calculator) does not print the results of its calculations on paper tape but makes them visible to the operator in dials located in the carriage of the machine. This machine basically performs four operations: addition of numbers, subtraction of numbers, shifting of the carriage with relation to the keyboard, and clearing of the dials. Multiplication is performed by means of repeated addition; division is accomplished by means of repeated subtraction of the divisor from the dividend. Both multiplication and division are greatly simplified by shifting the carriage the appropriate number of decimal places. Some models possess a divided keyboard (with a separate multiplier keyboard); others feature a single keyboard.

In recent years *electronic desk calculators* have become widespread. These carry out computations by means of tubes, transistors, and circuitry, as in an electronic computer. These machines combine the convenience of accessibility and a relatively low price with the instantaneous speeds of computation available by electronic rather than mechanical means. Figures are entered by means of a keyboard and results are displayed on a screen or printed on a paper tape. Many of the models are programmable, permitting repetitive operations to be performed automatically as soon as a number is entered into the machine. Some of these models permit fairly sophisticated programs to be written (up to several hundred instructions), and these programs are recorded on tapes or cards that can be inserted in the machine, making it immediately ready to carry out a given set of computations.

Electronic computers, since the middle 40's, have revolutionized the handling and processing of data. A *digital computer* has the same basic arithmetical abilities as the desk calculators mentioned in the previous section (addition, subtraction, multiplication, and division), but it also has the following distinctive features. Data are entered and the results of computations displayed automatically under the control of a previously prepared set of detailed instructions stored within the computer. These instructions also control the exact sequence of arithmetic computations to be performed. Computers also have the ability to execute different sets of instructions, depending on whether a given quantity is negative, zero, or positive. Thus it is possible to develop sophisticated *programs* that can bring into play different commands, as a function of the outcomes of previous computations. The program may even change some of its own instructions because of intermediate results obtained during a computation.

A typical digital computer consists of three main components: the *memory*, which stores both the data and the instructions; the *processor*, which actually performs the arithmetic computations; and the *peripheral devices*, which handle the input, output, and intermediate storage of data and instructions.

The material presented in this book consists of relatively standard statistical computations, most of which have quite likely already been programmed in any given computer installation. Some standard programs covering topics in this book and written in the FORTRAN IV programming language are featured in the authors' more comprehensive textbook of biometry (Sokal and Rohlf, 1969).

Exercises 2

2.1 Differentiate between the following pairs of terms and give an example of each. (a) Statistical and biological populations. (b) Variate and individual. (c) Accuracy and precision (repeatability). (d) Class interval and class mark. (e) Bar diagram and histogram. (f) Abscissa and ordinate.

2.2 Round the following numbers to three significant figures: 106.55, 0.06819, 3.0495, 7815.01, 2.9149, and 20.1500. What are the implied limits before and after rounding? Round these same numbers to one decimal place.

2.3 Given 200 measurements ranging from 1.32 mm to 2.95 mm, how would you group them into a frequency distribution? Give class limits as well as class marks.

2.4 Group the following 40 measurements of interorbital width of a sample of domestic pigeons into a frequency distribution and draw its histogram (data from Olson and Miller, 1958). Measurements are in mm.

12.2	12.9	11.8	11.9	11.6	11.1	12.3	12.2	11.8	11.8
10.7	11.5	11.3	11.2	11.6	11.9	13.3	11.2	10.5	11.1
12.1	11.9	10.4	10.7	10.8	11.0	11.9	10.2	10.9	11.6
10.8	11.6	10.4	10.7	12.0	12.4	11.7	11.8	11.3	11.1

2.5 How precisely should you measure the wing length of a species of mosquitoes in a study of geographic variation, if the smallest specimen has a length of about 2.8 mm and the largest a length of about 3.5 mm?

2.6 Look up the common logarithms of the 40 measurements given in Exercise 2.4 and make a frequency distribution of these transformed variates. Comment on the resulting change in the pattern of the frequency distribution from that found before.

3

Descriptive Statistics

An early and fundamental stage in any science is the descriptive stage. Until the facts as they are can be accurately described, an analysis of their causes is premature. The question "What?" comes before "How?" Unless we know something about the usual distribution of the sugar content of blood in a population of guinea pigs, as well as its fluctuations from day to day and within days, we shall be unable to ascertain the effect of a given dose of a drug upon this variable. In a sizable sample it would obviously be tedious to obtain knowledge of the material by contemplating all the individual observations. We need some form of summary to permit us to deal with the data in manageable form, as well as to be able to share our findings with others in scientific talks and publications. A histogram or bar diagram of the frequency distribution would be one type of summary. However, for most purposes, a numerical summary is needed to describe concisely, yet accurately, the properties of the observed frequency distribution. Quantities providing such a summary may be called *descriptive statistics.* This chapter will introduce you to some of them and show how they are computed.

Two kinds of descriptive statistics will be discussed in this chapter: statistics of location and statistics of dispersion. The *statistics of location* describe the position of a sample along a given dimension representing a variable. Thus, when measuring the length of certain animals, we would like to know whether the sample measurements lie in the vicinity of 2 cm or 20 cm. A statistic of location, therefore, must yield a representative value for the sample of observations. However, such a statistic (sometimes also known as a measure of central tendency) will not describe the shape of a frequency distribution. This may be long or very narrow; it may be humped or U-shaped; it may contain two humps, or it may be markedly asymmetrical. Quantitative measures of such aspects of frequency distributions are required. To this end we need to define and study the *statistics of dispersion.*

The arithmetic mean described in Section 3.1 is undoubtedly the most important single statistic of location, but others (the geometric mean, the harmonic mean, the median, and the mode) are briefly mentioned in Sections 3.2, 3.3, and 3.4. A simple statistic of dispersion (the range) is briefly noted in Section 3.5, and the standard deviation, the most common statistic for describing dispersion, is explained in Section 3.6. Our first encounter with contrasts between sample statistics and population parameters occurs in Section 3.7, in connection with statistics of location and dispersion. In Section 3.8 there is a description of methods of coding data in order to simplify them for the machine computation of the mean and standard deviation, which is discussed in Section 3.9. The coefficient of variation (a statistic that permits us to compare the relative amount of dispersion in different samples) is explained in the last section (3.10).

The techniques that will be at your disposal after you have mastered this chapter will not be very powerful in solving biological problems, but they will be indispensable tools for any further work in biostatistics. Other descriptive statistics, of both location and dispersion, will be taken up in later chapters.

An important note: We shall first encounter the use of logarithms in this chapter. To avoid confusion here and in subsequent chapters, common logarithms have been consistently abbreviated as log, and natural logarithms as ln. Thus, $\log x$ means $\log_{10} x$ and $\ln x$ means $\log_e x$.

3.1 The arithmetic mean

The most common statistic of location is familiar to everyone. It is the *arithmetic mean,* commonly called the *mean* or *average.* The mean is calculated by summing all the individual observations or items of a sample and dividing this sum by the number of items in the sample. For instance, as the result of a gas analysis in a respirometer an investigator obtains the following four readings of oxygen percentages:

$$\begin{array}{r r} & 14.9 \\ & 10.8 \\ & 12.3 \\ & \underline{23.3} \\ \text{Sum} = & 61.3 \end{array}$$

He calculates the mean oxygen percentage as the sum of the four items divided by the number of items—here, by four. Thus the average oxygen percentage is

$$\text{Mean} = \frac{61.3}{4} = 15.325\%$$

Calculating a mean presents us with the opportunity for learning statistical symbolism. We have already seen (Section 2.2) that an individual observation is symbolized by Y_i, which stands for the ith observation in the sample. Four observations could be written symbolically as follows:

$$Y_1,\ Y_2,\ Y_3,\ Y_4$$

We shall define n, the *sample size*, as the number of items in a sample. In this particular instance, the sample size n is 4. Thus, in a large sample, we can symbolize the array from the first to the nth item as follows:

$$Y_1,\ Y_2,\ \ldots,\ Y_n$$

When we wish to sum items, we use the following notation:

$$\sum_{i=1}^{i=n} Y_i = Y_1 + Y_2 + \cdots + Y_n$$

The capital Greek sigma, Σ, simply means the sum of the items indicated. The $i = 1$ means that the items should be summed, starting with the first one, and ending with the nth one as indicated by the $i = n$ above the Σ. The subscript and superscript are necessary to indicate how many items should be summed. The "$i =$" in the superscript is usually omitted as superfluous. For instance, if we had wished to sum only the first three items, we would have written $\sum_{i=1}^{3} Y_i$. On the other hand, had we wished to sum all of them except the first one, we would have written $\sum_{i=2}^{n} Y_i$. With some exceptions (which will appear in later chapters), it is desirable to omit subscripts and superscripts, which generally add to the apparent complexity of the formula and, when they are unnecessary, distract the student's attention from the important relations expressed by the formula. Below are seen increasing simplifications of the complete summation notation shown at the extreme left:

$$\sum_{i=1}^{i=n} Y_i = \sum_{i=1}^{n} Y_i = \sum_{i} Y_i = \sum^{n} Y = \sum Y$$

The third symbol might be interpreted as meaning: sum the Y_i's over all available values of i. This is a frequently used notation, although we shall not employ it in this book. The next, with n as a superscript, tells us to sum n items of Y; note that the i subscript of the Y has been dropped as unnecessary. Finally, the simplest notation is shown at the right. It merely says sum the Y's. This will be the form we shall use most frequently, and if a summation sign precedes a variable the summation will be understood to be over n items (all the items in the sample) unless subscripts or superscripts specifically tell us otherwise.

We shall use the symbol $\bar{Y}$ for the arithmetic mean of the variable Y. Its formula is written as:

$$\bar{Y} = \frac{\sum Y}{n} = \frac{1}{n} \sum Y \qquad (3.1)$$

In effect, this formula tells us to sum all the (n) items and divide the sum by n. We shall postpone until Section 3.9 a discussion of the mechanics of efficiently computing a mean.

The mean of a sample represents the center of the observations in the sample. If you were to draw a histogram of an observed frequency distribution on a sheet of cardboard, then cut out the histogram and lay it flat against a blackboard, supporting it with a pencil beneath, chances are that it would be out of balance, toppling to either the left or the right. If you moved the supporting pencil point to a position about which the histogram would exactly balance, this point of balance would be the arithmetic mean. In fact, this would be an empirical method of finding the arithmetic mean of a frequency distribution.

3.2 Other means

We shall see in Chapters 10 and 11 that variables are sometimes transformed into their logarithms or reciprocals. If we calculate the means of such transformed variables and then change them back into the original scale, these means will not be the same as if we had computed the arithmetic means of the original variables. The resulting means have received special names in statistics. The back-transformed mean of the logarithmically transformed variables is called the *geometric mean*. It is computed as

$$G.M._Y = \text{antilog} \frac{1}{n} \sum \log Y \qquad (3.2)$$

which indicates that the geometric mean $G.M._Y$ is the antilogarithm of the mean of the logarithms of variable Y. Since addition of logarithms is equivalent to multiplication of their antilogarithms, another way of representing this quantity is

$$G.M._Y = \sqrt[n]{Y_1 Y_2 Y_3 \cdots Y_n} \tag{3.3}$$

The geometric mean permits us to become familiar with another operator symbol: capital pi, Π, which may be read as product. Just as Σ symbolizes summation of the items that follow it, so Π symbolizes the multiplication of the items that follow it. The subscripts and superscripts have exactly the same meaning as in the summation case. Thus Expression (3.3) for the geometric mean can be rewritten more compactly as follows:

$$G.M._Y = \sqrt[n]{\prod_{i=1}^{n} Y_i} \tag{3.3a}$$

The computation of the geometric mean by Expression (3.3a) is quite tedious. In practice the geometric mean has to be computed by transforming the variates into logarithms.

The reciprocal of the arithmetic mean of reciprocals is called the *harmonic mean*. If we symbolize it by H_Y, the formula for the harmonic mean can be written in concise form (without subscripts and superscripts) as

$$\frac{1}{H_Y} = \frac{1}{n} \sum \frac{1}{Y} \tag{3.4}$$

The reader may wish to convince himself that the geometric mean and the harmonic mean of the four oxygen percentages are 14.65% and 14.09%, respectively. Unless the individual items do not vary, the geometric mean is always less than the arithmetic mean, and the harmonic mean is always less than the geometric mean.

Some beginners in statistics have difficulty in accepting the fact that measures of location or central tendency other than the arithmetic mean are permissible or even desirable. They feel that the arithmetic mean is the "logical" average, and that any other mean would distort the data. This whole problem relates to the proper scale of measurement for representing data; this scale is not always the linear scale familiar to everyone, but is sometimes by preference a logarithmic or reciprocal scale. If you have doubts about this question, we shall try to allay these in Chapter 10, where we discuss the reasons for transforming variables.

3.3 The median

The *median* M is a statistic of location occasionally useful in biological research. It is defined as that value of the variable (in an ordered array) that has an equal number of items on either side of it. Thus, the median divides a frequency distribution into two halves. In the following sample of five measurements,

$$14, 15, 16, 19, 23$$

$M = 16$, since the third observation has an equal number of observations on both sides of it. We can visualize the median easily if we think of an array from largest to smallest—for example, a row of men lined up by their heights. The median individual will then be that person having an equal number of men on his right and left sides. His height will be the median height of the sample considered. This quantity is easily evaluated from a sample array with an odd number of individuals. When the number in the sample is even, the median is conventionally calculated as the midpoint between the $(n/2)$th and the $[(n/2) + 1]$th variate. Thus, for the sample of four measurements

$$14, 15, 16, 19$$

the median would be the midpoint between the second and third items, or 15.5.

Whenever any one value of a variate occurs more than once, problems may develop in locating the median. Computation of the median item becomes more involved, because all the members of a given class in which the median item is located will have the same class mark. The median then is the $(n/2)$th variate in the frequency distribution. It is usually computed as that point between the class limits of the median class where the median individual would be located (assuming the individuals in the class were evenly distributed).

The median is just one of a family of statistics dividing a frequency distribution into equal areas. It divides the distribution into two halves. The three *quartiles* cut the distribution at the 25, 50, and 75% points—that is, at points dividing the distribution into first, second, third, and fourth quarters by area (and frequencies). The second quartile is, of course, the median. (There are also quintiles, deciles, and percentiles, dividing the distribution into 5, 10, and 100 equal portions, respectively.)

From the point of view of applying it in later, more advanced statistical work, the median is not a useful statistic (except for "nonparametric" methods; see Chapter 10). However, in certain special cases it is a more representative measure of location than the arithmetic mean. Such instances almost always involve asymmetric distributions. An often quoted example from economics would be a suitable measure of location for the "typical" salary of an employee of a corporation. The very high salaries of the few senior executives would shift the arithmetic mean, the center of gravity, toward a completely unrepresentative value. The median, on the other hand, would be little affected by a few high salaries; it would give the particular point on the salary scale above which lie 50% of the salaries in the corporation, the other half being lower than this figure.

In biology an example of the preferred application of a median over the arithmetic mean may be in populations showing skewed distribution, such as weights. Thus a median weight of American males 50 years old may be a more meaningful statistic than the average weight. The median is also of importance in cases where it may be difficult or impossible to obtain and measure all the items of a sample necessary to obtain a mean. An example will clarify this situation. An animal behaviorist is studying the time it takes for a sample of animals to perform a certain behavioral step. The variable that he is meas-using is the time from the beginning of the experiment until each individual has performed. What he wants to obtain is an average time of performance. Such an average time, however, could only be calculated after records have been obtained on all the individuals. It may take a long time for the slowest animals to complete their performance, longer than the observer wishes to spend looking at them. Moreover, some of them may never respond appropriately, making the computation of a mean impossible. Therefore, a convenient statistic of location to describe these animals may be the median time of performance, or a related statistic such as the 75th or 90th percentile. Thus, so long as the observer knows what the total sample size is, he need not have measurements for the right-hand tail of his distribution. Similar examples would be the responses to a drug or poison in a group of individuals (the median lethal or effective dose, LD_{50} or ED_{50}) or the median time for a mutation to appear in a number of lines of a species.

3.4 The mode

The *mode* refers to the most "fashionable" value of the variable in a frequency distribution, or the value represented by the greatest number of individuals. When seen on a frequency distribution it is the value of the variable at which the curve peaks. In grouped frequency distributions the mode as a point has not much meaning. It usually suffices to identify the modal class. In biology, the mode does not have many applications.

Distributions having two peaks (equal or unequal in height) are called *bimodal;* those with more than two peaks are *multimodal.* In those rare distributions that are **U**-shaped, we refer to the low point at the middle of the distribution as an *antimode.*

In evaluating the relative merits of the arithmetic mean, the median, and the mode, a number of considerations have to be kept in mind. The mean is generally preferred in statistics since it has a smaller standard error than other statistics of location (explained in Section 6.2), it is easier to work with mathematically, and it has an additional desirable property (explained in Section 6.1)—it will tend to be normally distributed even if the original data are not. The mean is markedly affected by outlying observations; the median and mode are not. The mean is generally more sensitive to changes in the shape of a frequency distribution and, if it is desired to have a statistic re-

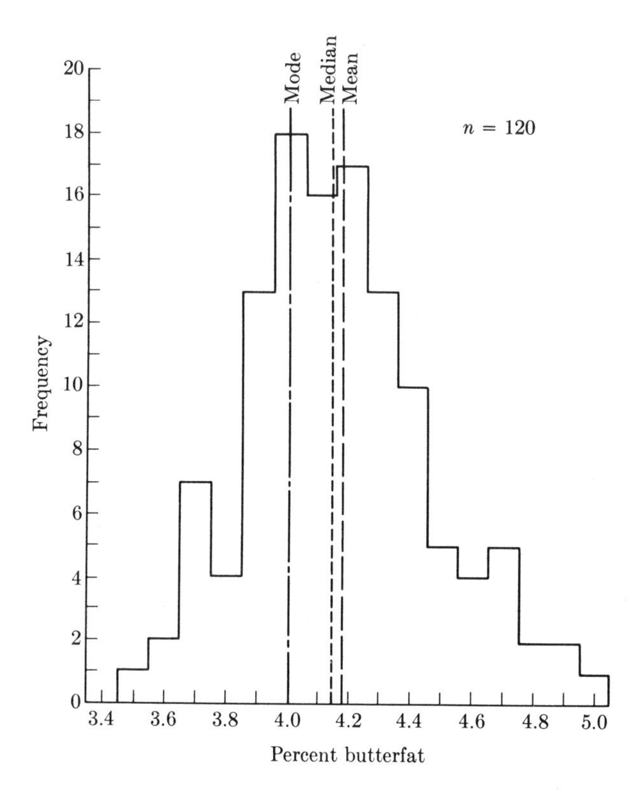

FIGURE 3.1

An asymmetrical frequency distribution (skewed to the right) showing location of the mean, median, and mode. Percent butterfat in 120 samples of milk (from a Canadian cattle breeders' record book).

flecting such changes, the mean may be preferred. In unimodal, symmetrical distributions the mean, the median, and the mode are all identical. A prime example of this is the well-known normal distribution of Chapter 5. In a typical asymmetrical distribution such as the one shown in Figure 3.1, the relative positions of the mode, median, and mean are generally these: the mean is closest to the drawn-out tail of the distribution, the mode is farthest, and the median is in between these. An easy way to remember this sequence is to recall that they occur in alphabetical order from the longer tail of the distribution.

3.5 The range

We now turn to measures of dispersion. Figure 3.2 demonstrates that radically different looking distributions may possess the identical arithmetic mean. It is therefore obvious that other ways of characterizing distributions must be found.

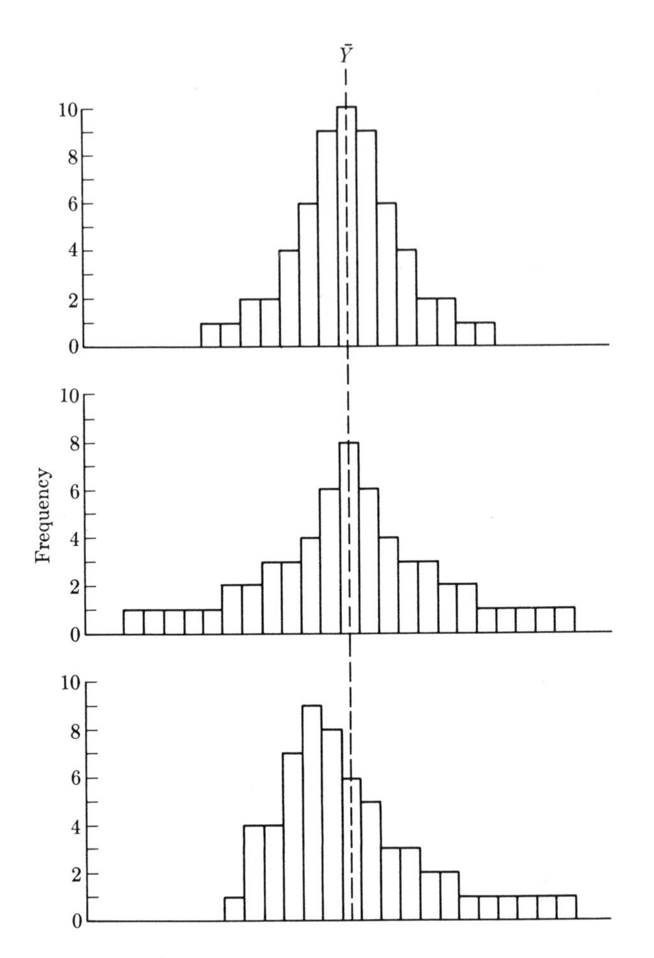

FIGURE 3.2

Three frequency distributions with identical means and sample sizes but differing in dispersion pattern.

One simple measure of dispersion is the *range*. It is the difference between the largest and the smallest items in a sample. Thus the range of the four oxygen percentages listed earlier (Section 3.1) is

$$\text{Range} = 23.3 - 10.8 = 12.5\%$$

and the range of the aphid femur lengths (Box 2.1) is

$$\text{Range} = 4.7 - 3.3 = 1.4 \text{ units of } 0.1 \text{ mm}$$

Since the range is a measure of the span of the variates along the scale of the variable, it is in the same units as the original measurements. The range is clearly affected by even a single outlying value and for this reason is only a rough estimate of the dispersion of all the items in the sample.

3.6 The standard deviation

A desirable measure of dispersion will take all items of a distribution into consideration, weighting each item by its distance from the center of the distribution. We shall now try to construct such a statistic. In Table 3.1 we

TABLE 3.1

Deviations from the mean. Femur lengths of aphids (from Box 2.1).

(1)	(2)	(3)	(4)		(5)	
Y	f	fY	$y = Y - \bar{Y}$		fy	
3.4	2	6.8	−0.6		−1.2	
3.7	8	29.6	−0.3		−2.4	
4.0	5	20.0	0.0		0.0	
4.3	8	34.4		0.3		2.4
4.6	2	9.2		0.6		1.2
Total	25	100.0			−3.6	3.6

$$\bar{Y} = \frac{\Sigma fY}{\Sigma f} = \frac{100.0}{25} = 4.0$$

show the grouped frequency distribution of the femur lengths of aphid stem mothers from Box 2.1. The first two columns show the class marks and the frequencies. The third column is needed for the computation of the mean of the frequency distribution (see Section 3.9 below). It is the class mark Y multiplied by the frequency f. The computation of the mean is shown below the table. The mean femur length turns out to be 4.0 units.

The distance of each class mark from the mean is computed as the following deviation:

$$y = Y - \bar{Y}$$

Each individual deviation or *deviate* is by convention computed as the individual observation minus the mean, $Y - \bar{Y}$, rather than the reverse, $\bar{Y} - Y$. Deviates are symbolized by lowercase letters corresponding to the capital letters of the variables. Column (4) in Table 3.1 gives the deviates computed in this manner. The column has been split into negative and positive deviates for convenience in computation. The deviates now have to be multiplied by their respective frequencies, so that those deviates that occur frequently contribute more to our measure of dispersion than those present only rarely. We therefore multiply the deviates by their frequencies f. The results of these computations are shown in column (5), which retains the separation between negative and positive deviates.

We now propose to calculate an average deviation that would sum all the deviates and divide them by the number of deviates in the sample. However, when we sum our deviates we note that the negative and positive deviates cancel out, as is shown by the sums at the bottom of column (5). This is always true of the sum of deviations from the arithmetic mean and is related to the fact that the mean is the center of gravity. Consequently the average deviation would also always equal zero. You are urged to study Appendix A1.1, which demonstrates that the sum of deviations around the mean of a sample is always equal to zero.

Squaring the deviates overcomes the fact that the sum of the deviations around the mean is always zero and results in other desirable mathematical properties, which we shall take up in a later section. In Table 3.2 the data on femur length of aphids are again presented. Columns (1), (2), and (3) are the class marks, frequencies, and deviates as given previously. The deviates are now not separated into negative and positive columns. Column (4) lists their squares, all of which are, of course, positive. Finally, column (5) shows the squared deviates multiplied by their frequencies—that is, column (4) multiplied by column (2). The sum of these squared deviates is 2.88. This is a very important quantity in statistics, called for short the *sum of squares* and identified symbolically as Σy^2. In Table 3.2 the sum of squares is symbolized as $\Sigma f y^2$, but the f is usually omitted since the Σ indicates summation over all the items possible. Another common symbol for the sum of squares is SS.

The next step is to obtain the average of the n squared deviations. The resulting quantity is known as the *variance* or the *mean square:*

$$\text{Variance} = \frac{\sum y^2}{n} = \frac{2.88}{25} = 0.1152$$

The variance is a measure of fundamental importance in statistics and we shall employ it throughout this book. At the moment we need only remember that because of the squaring of the deviations the variance is expressed in squared units. To undo the effect of the squaring we now take the positive square root of the variance and obtain the *standard deviation:*

$$\text{Standard deviation} = +\sqrt{\frac{\sum y^2}{n}} = 0.3394$$

A standard deviation is again expressed in the original units of measurement, since it is a square root of the squared units of the variance.

An important note: Do not use the technique just learned and illustrated in Table 3.2 for hand computation of a variance and standard deviation. This technique is far too tedious.

The observant reader may have noticed that we have avoided assigning any symbol to either the variance or the standard deviation. We shall explain why in the next section.

TABLE 3.2

The standard deviation. Long method, not recommended for actual computations but shown here to illustrate meaning of standard deviation. Same data as in Table 3.1.

(1) Y	(2) f	(3) $y = Y - \overline{Y}$	(4) y^2	(5) fy^2
3.4	2	−0.6	0.36	0.72
3.7	8	−0.3	0.09	0.72
4.0	5	0.0	0.00	0.00
4.3	8	0.3	0.09	0.72
4.6	2	0.6	0.36	0.72
Total	25			2.88

$\overline{Y} = 4.0$ $\Sigma fy^2 = 2.88$

$$\text{Variance} = \frac{\Sigma fy^2}{n} = \frac{2.88}{25} = 0.1152 \qquad \text{Standard deviation} = \sqrt{0.1152} = 0.3394$$

3.7 Sample statistics and parameters

Up to now we have calculated statistics from samples without giving too much thought to what these statistics represent. When correctly calculated, a mean and standard deviation will always be absolutely true measures of location and dispersion for the samples on which they are based. Thus the true mean of the four oxygen percentage readings in Section 3.1 is in fact 15.325%. The standard deviation of the 25 femur lengths is 0.3394 units when the items are grouped as shown. However, rarely in biology (or in statistics in general for that matter) are we interested in measures of location and dispersion only as descriptive summaries of the samples we have studied. Almost always we are interested in the *populations* from which the samples were taken. We therefore would like to know not the mean of the particular four oxygen percentages but rather the true oxygen percentage of the universe of readings from which the four readings had been sampled. Similarly, we would like to know the true mean femur length of the population of aphid stem mothers, not merely the mean of the 25 individuals we measured. When studying dispersion we generally wish to learn the true standard deviations of the populations and not those of the samples. These population statistics, however, are unknown and (generally speaking) are unknowable. Who would be able to collect all the stem mothers of this particular aphid population and measure them? Thus we need to use *sample statistics* as estimators of *population statistics* or *parameters.*

It is conventional in statistics to use Greek letters for population parameters and Roman letters for sample statistics. Thus the sample mean $\overline{Y}$ estimates μ, the parametric mean of the population. Similarly, a sample variance, symbolized by s^2, estimates a parametric variance, symbolized by σ^2. Such estimators should be *unbiased*. By this we mean that samples (regardless of the sample size), taken from a population with a known parameter, should give sample statistics that, when averaged, will give the parametric value. An estimator that does not do so is called *biased*. The sample mean $\bar{Y}$ is an unbiased estimator of the parametric mean μ. However, the sample variance as computed above (in Section 3.6) is not unbiased. On the average it will underestimate the magnitude of the population variance σ^2. To overcome this bias, mathematical statisticians have shown that when sums of squares are divided by $n - 1$ rather than by n the resulting sample variances will be unbiased estimators of the population variance. For this reason it is customary to compute variances by dividing the sum of squares by $n - 1$. The formula for the standard deviation is therefore customarily given as follows:

$$s = +\sqrt{\frac{\sum y^2}{n - 1}} \tag{3.5}$$

In the aphid stem mother data the standard deviation would thus be computed as

$$s = \sqrt{\frac{2.88}{24}} = 0.3464$$

We note that this value is only slightly larger than our previous estimate of 0.3394. Of course, the greater the sample size, the less difference there will be between division by n and by $n - 1$. However, regardless of sample size, it is good practice to divide a sum of squares by $n - 1$ when computing a variance or standard deviation. It may be assumed that when the symbol s^2 is encountered, it refers to a variance obtained by division of the sum of squares by the *degrees of freedom*, as the quantity $n - 1$ is generally referred to. The only time when division of the sum of squares by n is appropriate is when the interest of the investigator is truly limited to the sample at hand and to its variance and standard deviation as descriptive statistics of the sample, in contrast to using these as estimates of the population parameters. There are also the rare cases in which the investigator possesses data on the entire population; in such cases division by n is perfectly justified, because then he is not estimating a parameter but is in fact evaluating it. Thus the variance of the wing lengths of all adult whooping cranes would be a parametric value; similarly, if the I.Q.'s of all winners of the Nobel Prize in physics had been measured, their variance would be a parameter since it is based on the entire population.

3.8 Coding of data before computation

Before we can discuss how to compute means and standard deviations in a practical manner on a desk calculator, one more important procedure must be learned—the coding of the original data. By *coding* we mean the addition or subtraction of a constant number to the original data and/or the multiplication or division of these data by a constant. Data frequently need to be coded because they are originally expressed in too many digits or are very large numbers that may cause difficulties and errors during data handling. The importance of correct coding of data cannot be overestimated. Successful statistical computation—obtaining the correct result and not becoming bogged down in a morass of figures—is often facilitated by correct coding of the original data.

We shall call *additive coding* the addition or subtraction of a constant (since subtraction is only addition of a negative number). We shall similarly call *multiplicative coding* the multiplication or division by a constant (since division is multiplication by the reciprocal of the divisor). We shall call *combination coding* the application of both additive and multiplicative coding to the same set of data.

In Appendix A1.2 we examine the consequences of the three types of coding in the computation of means, variances, and standard deviations. The results *for means* can be summarized as follows: when a constant has been added to each variate, the coded mean $\bar{Y}_c$ is decoded by subtracting the constant.

When variates have been coded by being multiplied by a constant, $\bar{Y}_c$ can be decoded by being divided by the same constant; similarly, when variates have been coded by division, the mean is decoded by multiplication by this same constant.

Means coded by combination coding can be decoded by performing *the inverse operation, in reverse sequence.* This can be made clear by a little mnemonic diagram, which, incidentally, works equally well for simple multiplicative or additive coding. Thus if we have coded a set of variates by subtracting 5 and multiplying by 10, we decode the mean by first dividing by 10, then adding 5:

$$\text{code} \xrightarrow{-5 \qquad\qquad \times 10} \bar{Y}_c$$
$$\bar{Y} \xleftarrow[+5 \qquad\qquad \div 10]{} \text{decode}$$

Similarly, a coded mean based on variates that had been divided by 100 and to which 1.2 had then been added could be decoded by first subtracting 1.2, then multiplying $\times$ 100:

$$\text{code} \xrightarrow{\div 100 \qquad\qquad +1.2} \bar{Y}_c$$
$$\bar{Y} \xleftarrow[\times 100 \qquad\qquad -1.2]{} \text{decode}$$

The following diagrams illustrate simple coding:

$$\text{code} \xrightarrow{+10} \quad \xleftarrow[-10]{} \text{decode} \qquad \text{or} \qquad \text{code} \xrightarrow{\times 4} \quad \xleftarrow[\div 4]{} \text{decode}$$

On considering the effects of coding variates on the values of *variances and standard deviations*, we find first of all that additive codes have no effect on the sums of squares, variances, or standard deviations. The mathematical proof is given in Appendix A1.2, but this can be seen intuitively because an additive code has no effect on the distance of an item from its mean. The distance from an item of 15 to its mean of 10 would be 5. If we were to code the variates by subtracting a constant of 10, the item would now be 5 and the mean zero. The difference between them would still be 5. Thus if only additive coding is employed, the only statistic in need of decoding is the mean. But multiplicative coding does have an effect on sums of squares, variances, and standard deviations. The coded standard deviations s_c have to be divided by the multiplicative code, just as had to be done for the mean. However, the sums of squares or variances have to be divided by the multiplicative codes squared because they are squared terms, and the multiplicative factor became squared during the operations. In combination coding the additive code can be ignored. The following mnemonic diagrams can be used to summarize decoding for standard deviations:

$$\text{code} \xrightarrow{+5 \qquad\qquad \times 10} s_c$$
$$s \xleftarrow[\div 10]{} \text{decode}$$

and for sums of squares or variances:

$$\text{code} \xrightarrow{\div 100 \qquad\qquad +1.2} \sum y_c^2 \text{ or } s_c^2$$
$$\sum y^2 \text{ or } s^2 \xleftarrow[\times 100^2]{} \text{decode}$$

Examples of coding and decoding data are shown in Boxes 3.1 and 3.2, which are discussed in the next section.

3.9 Practical methods for computing mean and standard deviation

The procedure used to compute the sum of squares in Section 3.6 can be expressed by the following formula:

$$\sum y^2 = \sum (Y - \bar{Y})^2 \tag{3.6}$$

This formulation explains most clearly the nature of the sum of squares, but except on digital computers it is the most awkward one. To carry out the computation of Σy^2 by Expression (3.6), we would first have to compute the mean, then calculate the deviation of each item from the mean. Next, each deviation would have to be squared and finally these squares accumulated.

The most tiresome procedure would be the calculation and squaring of the deviations, which on most desk calculators cannot be done automatically without transcribing the deviations or at least entering them manually. In our experience many students once having learned some particular technique will repeat it by rote. Therefore, having learned to calculate a sum of squares as shown in Table 3.2 they might tend to repeat this particular procedure. We cannot therefore emphasize strongly enough that Table 3.2 is presented for pedagogic reasons only; the sum of squares is ordinarily not calculated as shown there. Let us now develop a computational formula for the standard deviation. In review, there are three steps necessary for computing this statistic: (1) find Σy^2, the sum of squares, (2) divide by $n - 1$ to give the variance, and (3) take the square root of the variance to obtain the standard deviation. Steps (2) and (3) are single operations and no time saving is possible for them. Our efforts at simplifying the computation must center on step (1), the computation of the sum of squares. The customary computational formula for this quantity is

$$\sum y^2 = \sum Y^2 - \frac{(\sum Y)^2}{n} \tag{3.7}$$

Let us be clear about what this formula represents. The first term on the right side of the equation, ΣY^2, is the sum of all the Y's squared as follows:

$$\sum Y^2 = Y_1^2 + Y_2^2 + Y_3^2 + \cdots + Y_n^2$$

When referred to by name, ΣY^2 should be called the "sum of Y squared" and should be carefully distinguished from Σy^2, "the sum of squares of Y." These names are unfortunate, but they are too well established to think of amending them. The other quantity in Expression (3.7) is $(\Sigma Y)^2/n$. It is often called the *correction term*, CT. The numerator of this term is the square of the sum of the Y's; that is, all the Y's are first summed and this sum is then squared. In general, this quantity is different from ΣY^2, which first squares the Y's and then sums them. These two terms are identical only if all the Y's are equal. If you are not certain about this, you can convince yourself of this fact by calculating these two quantities for a few numbers.

Why is Expression (3.7) identical with Expression (3.6)? The proof of this identity is very simple and is given in Appendix A1.3.

From Expressions (3.1) and (3.7) we see that it is necessary to have three quantities—n, ΣY, and ΣY^2—for computing the mean and standard deviation of a sample. The detailed operations depend on whether the data are unordered or arrayed in a frequency distribution. We have recommended leaving the data unordered when the number of variates is less than 150 because the time needed to set up a frequency distribution would be about equivalent to the time saved during computation if the data were in a frequency distribution form. When the data are unordered the computation

BOX 3.1

Calculation of $\overline{Y}$ and s from unordered data.

Aphid femur length data, unordered as shown at the head of Box 2.1.

The data were coded by multiplying by 10 to avoid decimal points during the computation. Coded variables and statistics are identified by the subscript c.

Computation	*Coding and decoding*
$n = 25$	Code: $Y_c = 10Y$
$\sum Y_c = 1001$	
$\overline{Y}_c = \frac{1}{n} \sum Y_c = 40.04$	To decode $\overline{Y}_c$: $\overline{Y} = \frac{\overline{Y}_c}{10} = \frac{40.04}{10} = 4.004$
$\sum Y_c^2 = 40{,}401$	
$\sum y_c^2 = \sum Y_c^2 - \frac{(\sum Y_c)^2}{n}$	
$= 40{,}401 - \frac{(1001)^2}{25}$	
$= 320.960$	
$s_c^2 = \frac{\sum y_c^2}{n-1} = \frac{320.960}{24}$	
$= 13.37$	To decode s_c^2: $s^2 = \frac{s_c^2}{10^2} = \frac{13.37}{100} = 0.1337$
$s_c = \sqrt{13.37} = 3.656$	To decode s_c: $s = \frac{s_c}{10} = \frac{3.656}{10} = 0.3656$

proceeds as in Box 3.1, which is based on the unordered aphid femur length data shown at the head of Box 2.1. The data were coded for convenience by multiplying variates $\times 10$ to remove the decimal point during the computation.

When the data are arrayed in a frequency distribution the computations are considerably simpler. An example is shown in Box 3.2. These are the birth weights of male Chinese children, first encountered in Figure 2.3. First of all they have to be coded, to remove the rather awkward class marks. This is done by subtracting 59.5, the lowest class mark of the array. The resulting class marks are values such as 0, 8, 16, 24, 32, etc. They are then divided by 8, which changes them to 0, 1, 2, 3, 4, etc., which is the desired format. The computation is now carried out quite simply and elegantly. We accumulate the products of fY_c and fY_c^2 to yield ΣfY_c and ΣfY_c^2. Summing the frequencies gives $\Sigma f = n$.

BOX 3.2

Calculation of $\bar{Y}$, s, and CV from a frequency distribution.

Birth weights of male Chinese in ounces.

(1) Class mark Y	(2) f	(3) Coded class mark Y_c
59.5	2	0
67.5	6	1
75.5	39	2
83.5	385	3
91.5	888	4
99.5	1729	5
107.5	2240	6
115.5	2007	7
123.5	1233	8
131.5	641	9
139.5	201	10
147.5	74	11
155.5	14	12
163.5	5	13
171.5	1	14
	9465 = n	

Source: Millis and Seng (1954).

Computation

$\sum fY_c = 59{,}629$

$\bar{Y}_c = \sum f\bar{Y}_c / \sum f = 6.300$

$\sum fY_c^2 = 402{,}987$

$$CT = \frac{(\sum fY_c)^2}{n} = 375{,}659.550$$

$\sum fy_c^2 = \sum fY_c^2 - CT = 27{,}327.450$

$$s_c^2 = \frac{\sum fy_c^2}{n-1} = 2.888$$

$s_c = 1.6991$

Coding and decoding

$$\text{Code: } Y_c = \frac{Y - 59.5}{8}$$

To decode $\bar{Y}_c$: $\bar{Y} = 8\bar{Y}_c + 59.5$

$= 50.4 + 59.5$

$= 109.9$ oz.

To decode s_c: $s = 8s_c = 13.593$ oz.

$$CV = \frac{s}{\bar{Y}} \times 100 = \frac{13.593}{109.9} \times 100 = 12.369\%$$

A common rule of thumb for estimating the mean is to average the largest and smallest observation to obtain the so-called *midrange*. For the aphid stem mother data of Box 2.1, this value is $(4.7 + 3.3)/2 = 4.0$, which happens to fall exactly on the computed sample mean. Standard deviations can be estimated from ranges by appropriate division of the range.

For samples of	*divide the range by*
10	3
30	4
100	5
500	6
1000	$6\frac{1}{2}$

The range of the aphid data is 1.4. When this value is divided by 4 we get an estimate of the standard deviation of 0.35, which compares not too badly with the calculated value of 0.3656 in Box 3.1. These approximate methods are very useful for detecting gross errors in computation.

3.10 The coefficient of variation

Having obtained the standard deviation as a measure of the amount of variation in the data, you may be led to ask, "Now what?" At this stage in our comprehension of statistical theory, nothing really useful comes of the computations we have carried out, although the skills just learned are basic to all later statistical work. So far, the only use that we might have for the standard deviation is as an estimate of the amount of variation in a population. Thus we may wish to compare the magnitudes of the standard deviations of similar populations and see whether population A is more or less variable than population B. However, when populations differ appreciably in their means, the comparison of their variances or standard deviations would be quite risky. For instance, the standard deviation of the tail lengths of elephants is obviously much greater than the entire tail length of a mouse. In order to compare the amount of variation in populations having different means, the *coefficient of variation* has been developed. This is simply the standard deviation expressed as a percentage of the mean. Its formula is

$$CV = \frac{s \times 100}{\bar{Y}} \tag{3.8}$$

For example, the coefficient of variation of the birth weights in Box 3.2 is 12.37%, as shown at the bottom of that box. The coefficient of variation is independent of the unit of measurement and is expressed as a percentage.

Coefficients of variation are extensively used when one wishes to compare the variation of two populations independent of the magnitude of their means. It is probably of little interest to discover whether the birth weights of the Chinese children are more or less variable than the femur lengths of the aphid stem mothers. We can calculate the latter as $0.3656 \times 100/4.004 = 9.13\%$, which would suggest that the birth weights are more variable. More commonly, we shall wish to test whether a given biological sample is more variable for one character than for another. Thus, for a sample of rats, is body weight more variable than blood sugar content? A second, frequent type of comparison, especially in systematics, is among different populations for the same character. Thus, we may have measured wing length in samples of birds from several localities. We wish to know whether any one of these populations is more variable than the others. An answer to this question can be obtained by examining the coefficients of variation of wing length in these samples.

Exercises 3

3.1 Find the mean, standard deviation, and coefficient of variation for the pigeon data given in Exercise 2.4. Group the data into ten classes, recompute $\bar{Y}$ and s, and compare them with the results obtained from the ungrouped data. Compute the median for the grouped data.

3.2 Find $\bar{Y}$, s, CV, and the median for the following data (mg of glycine per mg of creatinine in the urine of 37 chimpanzees; from Gartler, Firschein, and Dobzhansky, 1956). ANS. $\bar{Y} = 0.115$, $s = 0.10404$.

.008	.018	.056	.055	.135	.052	.077	.026	.440	.300
.025	.036	.043	.100	.120	.110	.100	.350	.100	.300
.011	.060	.070	.050	.080	.110	.110	.120	.133	.100
.100	.155	.370	.019	.100	.100	.116			

3.3 The following are percentages of butterfat from 120 registered three-year-old Ayrshire cows selected at random from a Canadian stock record book.

(a) Calculate $\bar{Y}$, s, and CV directly from the data. You will find it advantageous to calculate ΣY and ΣY^2 separately for each of the four columns, since in case an error occurs you will have to recompute only one column instead of all four. Save the data on each column. They will be used later.

(b) Group the data in a frequency distribution and again calculate $\bar{Y}$, s, and CV. Compare the results with those of (a). How much precision has been lost by grouping? Also calculate the median.

4.32	4.24	4.29	4.00
3.96	4.48	3.89	4.02
3.74	4.42	4.20	3.87
4.10	4.00	4.33	3.81
4.33	4.16	3.88	4.81
4.23	4.67	3.74	4.25
4.28	4.03	4.42	4.09
4.15	4.29	4.27	4.38
4.49	4.05	3.97	4.32
4.67	4.11	4.24	5.00

(columns continued on next page.)

4.60	4.38	3.72	3.99
4.00	4.46	4.82	3.91
4.71	3.96	3.66	4.10
4.38	4.16	3.77	4.40
4.06	4.08	3.66	4.70
3.97	3.97	4.20	4.41
4.31	3.70	3.83	4.24
4.30	4.17	3.97	4.20
4.51	3.86	4.36	4.18
4.24	4.05	4.05	3.56
3.94	3.89	4.58	3.99
4.17	3.82	3.70	4.33
4.06	3.89	4.07	3.58
3.93	4.20	3.89	4.60
4.38	4.14	4.66	3.97
4.22	3.47	3.92	4.91
3.95	4.38	4.12	4.52
4.35	3.91	4.10	4.09
4.09	4.34	4.09	4.88
4.28	3.98	3.86	4.58

3.4 What effect would adding a constant 5.2 to all observations have upon the numerical values of the following statistics? $\bar{Y}$, s, CV, average deviation, median, mode, range. What would be the effect of adding 5.2 and then multiplying the sums by 8.0? Would it make any difference in the above statistics if we multiplied by 8.0 first and then added 5.2?

3.5 Show that the equation for the variance can also be written as

$$s^2 = \frac{\sum Y^2 - n\bar{Y}^2}{n - 1}$$

(In actual practice this expression is not used much because it may lead to serious rounding errors unless $\bar{Y}$ is calculated to many significant figures.)

3.6 Estimate μ and σ using the midrange and the range (see Section 3.9) for the data in Exercises 3.1, 3.2, and 3.3. How well do these estimates agree with the estimates given by $\bar{Y}$ and s? ANS. Estimates of μ and σ for Exercise 3.2 are 0.224 and 0.1014.

4

Introduction to Probability Distributions: Binomial and Poisson

In Section 2.5 we first encountered frequency distributions. For example, Table 2.2 shows a distribution for a meristic, or discrete (discontinuous), variable, the number of sedge plants per quadrat. Examples of distributions for continuous variables are the femur lengths of aphid stem mothers in Box 2.1 or the human birth weights in Box 3.2. Each of these distributions informs us about the absolute frequency of any given class and permits computation of the relative frequencies of any class of variable. Thus, most of the quadrats contained either no sedges, or one or two plants. In the 139.5-oz. class of birth weights we find only 201 out of the total of 9465 babies recorded; that is, approximately only 2.1% of the infants are in that birth weight class. We realize, of course, that these frequency distributions are only samples from given populations. The birth weights represent a population of male Chinese infants from a given geographical area. But if we knew our sample to be representative of that population, we could make all sorts of predictions based upon the sample frequency distribution. For instance, we could say that approximately 2.1% of male Chinese babies born in this population should weigh between 135.5 and 143.5 oz. at birth. Similarly, we might say that the probability of any one baby in this population weighing 139.5 oz. at birth is quite low. If each of the 9465 weights were given a registration number, the numbers mixed up in a hat, and a single one pulled out, the probability that we would pull out one of the 201 in the 139.5-oz. class would be very low indeed—only 0.021. It would be much more probable that we would sample an infant of 107.5 or 115.5 oz., since the infants in these classes are represented by frequencies 2240 and 2007, respectively. Finally, if we were to sample from an unknown population of babies and find that the very first individual sampled had a birth weight of 170 oz., we would probably reject any hypothesis that the unknown population was the same as that sampled in Box 3.2. We would arrive at this conclusion because in the distribution in Box 3.2 only one out of almost 10,000 infants had a birth weight that high. Though it is possible that we could have sampled from the population of male Chinese babies and obtained a birth weight of 170 oz., the probability that the first individual sampled would have such a value is very low indeed. It seems much more reasonable to suppose that the unknown population from which we are sampling is different in mean and possibly in variance from the one in Box 3.2.

We have used this empirical frequency distribution to make certain predictions (with what frequency a given event will occur) or to make judgments and decisions (is it likely that an infant of a given birth weight belongs to this population?). In many cases in biology, however, we shall make such predictions not from empirical distributions, but on the basis of theoretical considerations that in our judgment are pertinent. We may feel that the data should be distributed in a certain way because of basic assumptions about the nature of the forces acting on the example at hand. If our actually observed data do not conform to the values expected on the basis of these assumptions, we shall have serious doubts about our assumptions. This is a common use of frequency distributions in biology. The assumptions being tested generally lead to a theoretical frequency distribution known also as a *probability distribution*. This may be a simple two-valued distribution such as the 3:1 ratio in a Mendelian cross, or it may be a more complicated function trying to predict the number of plants in a quadrat. If we find that the observed data do not fit the expectations on the basis of theory, we are often led to the discovery of some biological mechanism causing this deviation from expectation. The phenomena of linkage in genetics, of preferential mating between different phenotypes in animal behavior, of congregation of animals at certain favored places or, conversely, their territorial dispersion are cases in point. We shall thus make use of probability theory to test our assumptions about the laws of occurrence of certain biological phenomena. We should point out to the reader, however, that probability theory underlies the entire structure of statistics, since, owing to the nonmathematical orientation of this book, this may not be entirely obvious.

In the sections that follow we shall first of all present a brief discussion of probability (Section 4.1), limited to the amount necessary for comprehension of the sections that follow at the intended level of mathematical sophistication. Next we shall take up the binomial frequency distribution (Section 4.2), which not only is important in certain types of studies, such as genetics, but is fundamental to an understanding of the various kinds of probability distributions to be discussed in this book.

The Poisson distribution, which follows in Section 4.3, is of wide applicability in biology, especially for tests of randomness of occurrence of certain events. Both the binomial and Poisson distributions are discrete probability distributions. The most common continuous probability distribution is the normal frequency distribution, discussed in the following chapter.

4.1 Probability, random sampling, and hypothesis testing

We shall start this discussion with an example that is not biometrical or biological in the strict sense. We have often found it pedagogically effective to introduce new concepts through situations thoroughly familiar to the student, even if the example is not relevant to the general subject matter of biostatistics.

Let us betake ourselves to Matchless University, a state institution somewhere between the Appalachians and the Rockies. Looking at its enrollment figures we notice the following breakdown of the student body: 70% of the students are American undergraduates (AU) and 26% are American graduate students (AG); the remaining 4% are from abroad. Of these, 1% are foreign undergraduates (FU) and 3% are foreign graduate students (FG). In much of our work we shall use proportions rather than percentages as a useful convention. Thus the enrollment consists of 0.70 AU's, 0.26 AG's, 0.01 FU's, and 0.03 FG's. The total student body, corresponding to 100%, is therefore represented by the figure 1.0.

If we were to assemble all the students and sample 100 of them at random, we would intuitively expect that, on the average, 3 will be foreign graduate students. The actual outcome might vary. There might not be a single FG student among the 100 sampled or there may be quite a few more than 3. The ratio of the number of foreign graduate students sampled divided by the total number of students sampled might therefore vary from zero to considerably greater than 0.03. If we increased our sample size to 500 or 1000, it is less likely that the ratio will fluctuate widely around 0.03. The greater the sample taken, the closer the ratio of FG students sampled to the total students sampled will approach 0.03. In fact, the *probability* of sampling a foreign student is defined as the limit reached by the ratio of foreign students to the total number of students sampled, as sample size keeps increasing. Thus, we may formally summarize the situation by stating that the probability of a student at Matchless University being a foreign graduate student is $P[FG] = 0.03$. Similarly, the probability of sampling a foreign undergraduate is $P[FU] = 0.01$, that of sampling an American undergraduate is $P[AU] = .70$, and that for American graduate students, $P[AG] = 0.26$.

Now let us imagine the following experiment: we try to sample a student at random from among the student body at Matchless University. This is not as easy a task as might be imagined. If we wish to do this operation physically, we would have to set up a collection or trapping station somewhere on campus. And to make certain that the sample is truly random with respect to the entire student population, we would have to know the ecology of students on campus very thoroughly. We should try to locate our trap at some station where each student has an equal probability of passing. Few, if any, such places can be found in a university. The student union facilities are likely to be frequented more by independent and foreign students, less by those living in organized houses and dormitories. Fewer foreign and graduate students might be found along fraternity row. Clearly we would not wish to place our trap near the International Club or House because our probability of sampling a foreign student would be greatly enhanced. In front of the bursar's window we might sample students paying tuition. But those on scholarships might not be found there. We do not know whether the proportion of scholarships among foreign or graduate students is the same or different from that among the American or undergraduate students. Athletic events, political rallies, dances, and the like would all draw a differential spectrum of the student body; indeed, no easy solution seems in sight. The time of sampling is equally important, in the seasonal as well as the diurnal cycle.

Those among the readers who are interested in sampling of organisms from nature will already have perceived parallel problems in their work. If we were to sample only students wearing turbans or saris, their probability of being foreign students would be almost 1. We could no longer speak of a random sample. In the familiar ecosystem of the university these violations of proper sampling procedure are obvious to all of us, but they are not nearly so obvious in real biological instances where we are unfamiliar with the true nature of the environment. How should we proceed to obtain a random sample of leaves from a tree, of insects from a field, or of mutations in a culture? In sampling at random we are attempting to permit the frequencies of various events occurring in nature to be reproduced unalteredly in our records; that is, we hope that on the average the frequencies of these events in our sample will be the same as they are in the natural situation. Another way of saying this is that in a random sample every individual in the population being sampled has an equal probability of being included in the sample.

We might go about obtaining a random sample by using records representing the student body, such as the student directory, selecting a page from it at random and a name at random from the page. Or we could assign an arbitrary number to each student, write each on a chip or disk, put these in a large container, stir well, and then pull out a number.

Imagine now that we sample a single student physically by the trapping method, after carefully planning the placement of the trap in such a way as to make sampling random. What are the possible outcomes? Clearly the student could be either an AU, AG, FU, or FG. This *set* of four possible outcomes exhausts the possibilities of this experiment. This set, which we can represent as {AU, AG, FU, FG} is called the *sample space*. Any single experiment would result in only one of the four possible outcomes (elements) in the set. Such an element in a sample space is called a *simple event*. It is distinguished from an *event*, which is any subset of the sample space. Thus in the sample space defined above {AU}, {AG}, {FU}, or {FG} are each simple events. The following sampling results are some of the possible events: {AU, AG, FU}, {AU, AG, FG}, {AG, FG}, {AU, FG}, . . . By the definition of event, simple events as well as the entire sample space are also events. The meaning of these events should be clarified. Thus {AU, AG, FU} implies being either an American or an undergraduate, or both.

Given the sampling space described above the event $\mathbf{A} = \{AU, AG\}$ encompasses all possible outcomes in the space yielding an American student. Similarly, the event $\mathbf{B} = \{AG, FG\}$ summarizes the possibilities for obtaining a graduate student. The *intersection* of events **A** and **B**, written $\mathbf{A} \cap \mathbf{B}$, describes only those events that are shared by **A** and **B**. Clearly only AG qualifies, as can be seen below:

{AU, AG}

{AG, FG}

Thus $\mathbf{A} \cap \mathbf{B}$ is that event in the sample space giving rise to the sampling of an American graduate student. When the intersection of two events is empty, as in $\mathbf{B} \cap \mathbf{C}$, where $\mathbf{C} = \{AU, FU\}$, the events **B** and **C** are mutually exclusive. Thus there is no common element in these two events in the sampling space.

We may also define events that are *unions* of two other events in the sample space. Thus $\mathbf{A} \cup \mathbf{B}$ indicates that **A** or **B** or both **A** and **B** occur. As defined above $\mathbf{A} \cup \mathbf{B}$ would describe all students who are either American or graduate students, or both—namely, American graduate students.

Why are we concerned with defining sample spaces and events? Because these concepts lead us to useful definitions and operations regarding the probability of various outcomes. If we can assign a number $0 \leq p \leq 1$ to each simple event in a sample space such that the sum of these p's over all simple events in the space equals unity, then the space becomes a (finite) *probability space*. In our example above, the following numbers were associated with the appropriate simple events in the sample space:

$$\{AU, AG, FU, FG\}$$
$$\{0.70, 0.26, 0.01, 0.03\}$$

Given this probability space we are now able to make statements regarding the probability of given events. For example, what is the probability of a student sampled at random being an American graduate student? Clearly it is $P[\{AG\}] = 0.26$. What is the probability that a student is either American or a graduate student? In terms of the events defined earlier this is a $P[\mathbf{A}] \cup P[\mathbf{B}] = P[\{AU, AG\}] + P[\{AG, FG\}] - P[\{AG\}] = 0.96 + 0.29 - 0.26 = 0.99$. We subtract $P[\{AG\}]$ because if we did not do so it would be included twice, once in $P[\mathbf{A}]$ and once in $P[\mathbf{B}]$ and lead to the absurd result of a probability >1.

Now let us assume that we have sampled our single student from the student body of Matchless University. He turns out to be a foreign graduate student. What can we conclude from this? By chance alone, this result would happen 0.03 or 3% of the time, or not very frequently. The assumption that we have sampled at random should probably be rejected, since if we accept the hypothesis of random sampling the outcome of the experiment is improbable. Please note that we said *improbable,* not *impossible.* It is obvious that we could have chanced upon an FG as the very first one to be sampled. However, it is not very likely. The probability is 0.97 that a single student sampled would be a non-FG. If we could be certain that our sampling method was random (as when drawing student numbers out of a container), we would of course have to decide that an improbable event has occurred. The decisions of this paragraph are all based on our definite knowledge that the proportion of students at Matchless University is indeed as specified by the probability space. If we were uncertain about this, we would be led to assume a higher proportion of foreign graduate students as a consequence of the outcome of our sampling experiment.

We shall now extend our experiment and sample two students rather than just one. What are the possible outcomes of this sampling experiment? The new sampling space can best be depicted by a diagram (Figure 4.1) that shows the set of the sixteen possible simple events as points in a lattice. The possible combinations, ignoring which student was sampled first, are {AU, AU}, {AU, AG}, {AU, FU}, {AU, FG}, {AG, AG}, {AG, FU}, {AG, FG}, {FU, FU}, {FU, FG}, and {FG, FG}.

What would be the expected probabilities of these outcomes? We know the expected outcomes for sampling one student from the former probability space, but what will be the probability space corresponding to the new sampling space of 16 elements? Now the nature of the sampling procedure becomes quite important. We may sample with or without *replacement;* that is, we may return the first student sampled to the population or may keep

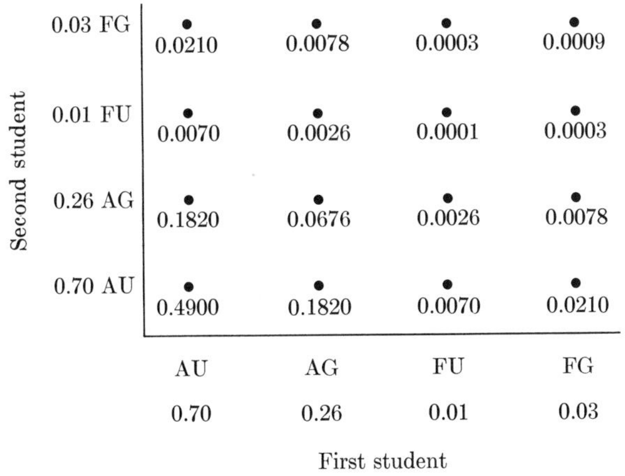

Second student	AU 0.70	AG 0.26	FU 0.01	FG 0.03
0.03 FG	0.0210	0.0078	0.0003	0.0009
0.01 FU	0.0070	0.0026	0.0001	0.0003
0.26 AG	0.1820	0.0676	0.0026	0.0078
0.70 AU	0.4900	0.1820	0.0070	0.0210

First student

FIGURE 4.1

Sample space for sampling two students from Matchless University. (For further explanation see text.)

him out of the pool of the individuals to be sampled. If we do not replace the first individual sampled, the probability of sampling a foreign graduate student will no longer be exactly 0.03. This is easily seen. Let us assume that Matchless University has 10,000 students. Then, since there are 3% of foreign students, there must be 300 FG students at the university. After sampling a foreign graduate student first, this number is reduced to 299 out of 9999 students. Consequently, the probability of sampling an FG student now becomes 299/9999 = 0.0299, a slightly lower probability than the value of 0.03 for sampling the first FG student. If, on the other hand, we return the original foreign student to the student population and make certain that the population is thoroughly randomized before being sampled again (that is, give him a chance to lose himself among the campus crowd or, in drawing student numbers out of a container, mix up the disks with the numbers on them), the probability of sampling a second FG student is the same as before—0.03. In fact, if we keep on replacing the sampled individuals in the original population, we can sample from it as though it were an infinitely sized population.

Biological populations are, of course, finite, but they are frequently so large that for purposes of sampling experiments we can consider them effectively infinite whether we replace sampled individuals or not. After all, even in this relatively small population of 10,000 students, the probability of sampling a second foreign graduate student (without replacement) is only minutely different from 0.03. For the rest of this section we shall consider sampling to be with replacement, so that the probability level of obtaining a foreign student does not change.

There is a second potential source of difficulty in this design. We not only have to assume that the probability of sampling a second foreign student is equal to that of the first, but also that it is *independent* of it. By independence of events we mean that if one event occurs with a certain probability, a second such event will occur with the same probability, not affected by whether the first event has already occurred or not. In the case of the students, having sampled one foreign student, is it more or less likely that a second student sampled in the same manner will also be a foreign student? Independence of the events may depend on where we sample the students or on the method of sampling. If we sampled students on campus it is quite likely that the events are not independent; that is, having sampled one foreign student, the probability that the second student is foreign is increased, since foreign students tend to congregate. Thus, at Matchless University the probability that a student walking with a foreign graduate student is also an FG would be greater than 0.03.

Events **D** and **E** in a sample space will be defined as independent whenever $P[\mathbf{D} \cap \mathbf{E}] = P[\mathbf{D}]P[\mathbf{E}]$. The probability values assigned to the sixteen points in the sample-space lattice of Figure 4.1 have been computed to satisfy the above condition. Thus letting $P[\mathbf{D}]$ equal the probability of the first student being an AU, that is, $P[\{AU_1AU_2, AU_1AG_2, AU_1FU_2, AU_1FG_2\}]$, and $P[\mathbf{E}]$ equal the probability of the second student being an FG, that is, $P[\{AU_1FG_2, AG_1FG_2, FU_1FG_2, FG_1FG_2\}]$, we note that the intersection $\mathbf{D} \cap \mathbf{E}$ is $\{AU_1FG_2\}$. This has a value of 0.0210 in the probability space of Figure 4.1. We find that this value is the product of $P[\{AU\}]P[\{FG\}] = 0.70 \times 0.03 = 0.0210$. These mutually independent relations have been deliberately imposed upon all points in the probability space. Therefore, if the sampling probabilities for the second student are independent of the type of student sampled first, we can compute the probabilities of the outcomes simply as the product of the independent probabilities. Thus the probability of obtaining two FG students is $P[\{FG\}]P[\{FG\}] = 0.03 \times 0.03 = 0.0009$.

The probability of obtaining one AU and one FG student in the sample should be the product 0.70×0.03. However, it is in fact twice that probability. It is easy to see why. There is only one way of obtaining two FG students, namely by sampling first one FG and then again another FG. Similarly, there is only one way to sample two AU students. However, sampling one of each type of student can be done by first sampling an AU followed by an FG or by first sampling an FG followed by an AU. Thus the probability is $2P[\{AU\}]P[\{FG\}] = 2 \times (0.70) \times (0.03) = 0.0420$.

If we conduct such an experiment and obtain a sample of two FG students we would be led to the following conclusions. Only 0.0009 of the samples (9/100th of 1% or 9 out of 10,000 cases) would be expected to consist of two foreign graduate students. It is quite improbable to obtain such a result by chance alone. Given $P[\{FG\}] = 0.03$ as a fact, we would therefore suspect that sampling was not random or that the events were not independent (or that both assumptions—random sampling and independence of events—were incorrect).

Random sampling is sometimes confused with randomness in nature. The former is the faithful representation in the sample of the distribution of the events in nature; the latter is the independence of the events in nature. The first of these generally is or should be under the control of the experimenter and is related to the strategy of good sampling. The second generally describes an innate property of the objects being sampled and thus is of greater biological interest. The confusion between random sampling and independence of events arises because lack of either can yield observed frequencies of events differing from expectation. We have already seen how lack of independence in samples of foreign students can be interpreted from both points of view in our illustrative example from Matchless University.

The above account of probability is adequate for our present purposes but far too sketchy to convey an understanding of the field. Readers interested in extending their knowledge of the subject are referred to Mosimann (1968) for a simple introduction.

4.2 The binomial distribution

For purposes of the discussion to follow we shall simplify our sample space to consist of only two elements, foreign and American students, represented by {F, A}, and ignore whether they are undergraduates or graduates. Let us symbolize the probability space by $\{p, q\}$ where $p = P[\text{F}]$, the probability of being a foreign student and $q = P[\text{A}]$, the probability of being an American student. As before, we can compute the probability space of samples of two students as follows:

$$\{\text{FF, FA, AA}\}$$
$$\{p^2,\ 2pq,\ q^2\}$$

If we were to sample three students, the probability space of samples of three students is as follows:

$$\{\text{FFF, FFA, FAA, AAA}\}$$
$$\{p^3,\ 3p^2q,\ 3pq^2,\ q^3\}$$

Samples of three foreign or three American students can again be obtained in only one way, and their probabilities are p^3 and q^3, respectively. However, in samples of three there are three ways of obtaining two students of one kind and one student of the other. As before, if A stands for American and F stands for foreign, then the sampling sequence could be AFF, FAF, and FFA for two foreign students and one American. Thus the probability of this outcome will be $3p^2q$. Similarly, the probability for two Americans and one foreign student is $3pq^2$.

A convenient way to summarize these results is by means of the binomial expansion, which is applicable to samples of any size from populations in which objects occur only in two classes—students who may be foreign or American, individuals who may be dead or alive, male or female, black or white, rough or smooth, and so forth. This is accomplished by expanding the binomial term $(p + q)^k$, where k equals sample size, p equals the probability of occurrence of the first class, and q equals the probability of occurrence of the second class. By definition, $p + q = 1$; hence q is a function of p: $q = 1 - p$. We shall expand the expression for samples of k from 1 to 3:

For samples of 1, $(p + q)^1 = p + q$
For samples of 2, $(p + q)^2 = p^2 + 2pq + q^2$
For samples of 3, $(p + q)^3 = p^3 + 3p^2q + 3pq^2 + q^3$

It will be seen that these expressions yield the same probability spaces discussed previously. The coefficients (the numbers before the powers of p and q) express the number of ways a particular outcome is obtained. An easy method for evaluating the coefficients of the expanded terms of the binomial expression is through the use of Pascal's triangle, which is shown below.

```
k
1            1   1
2          1   2   1
3        1   3   3   1
4      1   4   6   4   1
5    1   5  10  10   5   1
...  .   .   .   .   .   .   .
```

Pascal's triangle provides the coefficients of the binomial expression—that is, the number of possible outcomes of the various combinations of events. For $k = 1$ the coefficients are 1, 1 respectively; for the second line ($k = 2$), write 1 at the left-hand margin of the line. The 2 in the middle of this line is the sum of the values to the left and right of it in the line above. The line is concluded with a 1. Similarly, the values at the beginning and end of the third line are 1, the other numbers are sums of the values to their left and right in the line above; thus 3 is the sum of 1 and 2. This principle continues for every line. You can work out the coefficients for any size sample in this manner. The line for $k = 6$ would consist of the following coefficients: 1, 6, 15, 20, 15, 6, 1. The p and q values receive powers in a consistent pattern, which should be easy to imitate for any value of k. We give it here for $k = 4$:

$$p^4q^0 + p^3q^1 + p^2q^2 + p^1q^3 + p^0q^4$$

The power of p decreases from 4 to 0 (k to 0 in the general case) as the power of q increases from 0 to 4 (0 to k in the general case). Since any value to the power 0 is 1 and any term to the power 1 is simply itself, we can simplify this expression as shown below and at the same time provide it with the coefficients from Pascal's triangle for the case $k = 4$:

$$p^4 + 4p^3q + 6p^2q^2 + 4pq^3 + q^4$$

Thus we are able to write down almost by inspection the expansion of the binomial to any reasonable power.

Suppose we have a population of insects, exactly 40% of which are infected with a given virus X. If we take samples of $k = 5$ insects each and examine each insect separately for presence of virus, what distribution of samples could we expect if the probability of infection of each insect in a sample were independent from that of other insects in the sample? In this case $p = 0.4$, the proportion infected, and $q = 0.6$, the proportion not infected. It is assumed that the population is so large that the question of whether sampling is with or without replacement is irrelevant for practical purposes. The expected proportions would be the expansion of the binomial:

$$(p + q)^k = (0.4 + 0.6)^5$$

With the aid of Pascal's triangle this expansion is the probability space

$$\{p^5 + 5p^4q + 10p^3q^2 + 10p^2q^3 + 5pq^4 + q^5\}$$

or

$$(0.4)^5 + 5(0.4)^4(0.6) + 10(0.4)^3(0.6)^2 + 10(0.4)^2(0.6)^3 + 5(0.4)(0.6)^4 + (0.6)^5$$

representing the expected proportions of samples of five infected insects, four infected and one noninfected insects, three infected and two noninfected insects, and so on. The reader has probably realized by now that the terms of the binomial expansion actually yield a type of frequency distribution for these different outcomes. Associated with each outcome, such as "five infected insects," there is a probability of occurrence—in this case $(0.4)^5 = 0.01024$. This is a theoretical frequency distribution or *probability distribution* of events that can occur in two classes. It describes the expected distribution of outcomes in random samples of five insects, 40% of which are infected. The probability distribution described here is known as the *binomial distribution*, and the binomial expansion yields the expected frequencies of the classes of the binomial distribution.

A convenient layout for presentation and computation of a binomial distribution is shown in Table 4.1. The firsts column lists the number of infected insects per sample, the second column shows decreasing powers of p from p^5 to p^0, and the third column shows increasing powers of q from q^0 to q^5. The binomial coefficients from Pascal's triangle are shown in column (4). The *relative expected frequencies*, which are the probabilities of the various outcomes, are shown in column (5). We label such expected frequencies $\hat{f}_{rel}$.

They are simply the product of columns (2), (3), and (4). Their sum is equal to 1.0, since the events listed in column (1) exhaust the possible outcomes. We see from column (5) in Table 4.1 that only about 1% of samples are expected to consist of 5 infected insects, and 25.9% are expected to contain 1 infected and 4 noninfected insects. We shall test whether these predictions hold in an actual experiment.

Experiment 4.1. Simulate the sampling of infected insects by using a table of random numbers such as Table **I**. These are randomly chosen one-digit numbers in which each digit 0 through 9 has an equal probability of appearing. The numbers are grouped in blocks of twenty for convenience. Since there is an equal probability for any one digit to appear, you can let any four digits (say 0, 1, 2, 3) stand for the infected insects and the remaining digits (4, 5, 6, 7, 8, 9) stand for the noninfected insects. The probability of any one digit selected from the table representing an infected insect (that is, being a 0, 1, 2, or 3) is therefore 40% or 0.4, since these are four of the ten possible digits. Also, successive digits are assumed to be independent of the values of previous digits. Thus the assumptions of the binomial distribution should be met in this experiment. Enter the table of random numbers at an arbitrary point (not always at the beginning!) and look at successive groups of five digits, noting in each group how many of the digits were 0, 1, 2, or 3. Take as many groups of five as you can find time to do, but no less than 100 groups.

Column (7) in Table 4.1 shows the results of one such experiment during one year by a biostatistics class. A total of 2423 samples of five numbers were obtained from the table of random numbers and the distribution of the four

TABLE 4.1

Expected frequencies of infected insects in samples of 5 insects sampled from an infinitely large population with an assumed infection rate of 40 percent.

(1)	*(2)*	*(3)*	*(4)*	*(5)*	*(6)*	*(7)*
Number of infected insects per sample Y	*Powers of* $p = 0.4$	*Powers of* $q = 0.6$	*Binomial coefficients*	*Relative expected frequencies* $\hat{f}_{rel}$	*Absolute expected frequencies* $\hat{f}$	*Observed frequencies* f
5	0.01024	1.00000	1	0.01024	24.8	29
4	0.02560	.60000	5	0.07680	186.1	197
3	0.06400	.36000	10	0.23040	558.3	535
2	0.16000	.21600	10	0.34560	837.4	817
1	0.40000	.12960	5	0.25920	628.0	643
0	1.00000	.07776	1	0.07776	188.4	202
			$\Sigma\hat{f}$ or Σf $(= n)$	1.00000	2423.0	2423
			ΣY	2.00000	4846.1	4815
			Mean	2.00000	2.00004	1.98721
			Standard deviation	1.09545	1.09543	1.11934

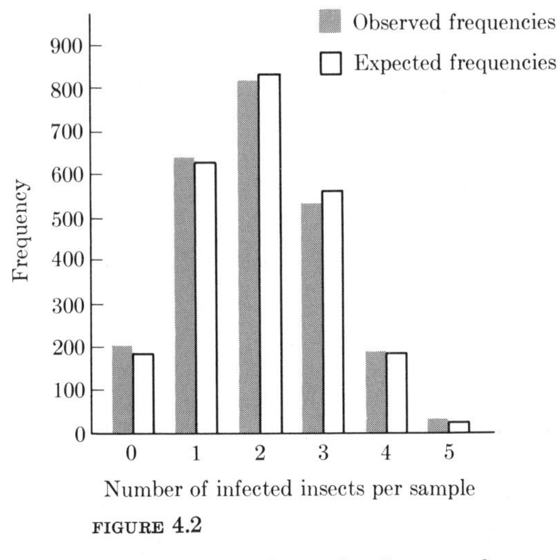

FIGURE 4.2

Bar diagram of observed and expected frequencies given in Table 4.1.

digits simulating the percentage of infection is shown in this column. The observed frequencies are labeled f. To calculate the expected frequencies for this actual example we multiplied the relative expected frequencies $\hat{f}_{\text{rel}}$ of column (5) times $n = 2423$, the number of samples taken. This results in *absolute expected frequencies*, labeled $\hat{f}$, shown in column (6). When we compare the observed frequencies in column (7) with the expected frequencies in column (6) we note general agreement between the two columns of figures. The two distributions are also illustrated in Figure 4.2. If the observed frequencies would not fit expected frequencies, we might believe that the lack of fit was due to chance alone. Or we might be led to reject one or more of the following hypotheses: (1) that the true proportion of digits 0, 1, 2, and 3 is 0.4 (rejection of this hypothesis would normally not be reasonable, for we may rely on the fact that the proportion of digits 0, 1, 2, and 3 in a table of random numbers *is* 0.4 or very close to it); (2) that sampling was at random; and (3) that events are independent.

These statements can be reinterpreted in terms of the original infection model with which we started this discussion. If, instead of a sampling experiment of digits by a biostatistics class, this had been a real sampling experiment of insects, we would conclude that the insects had indeed been randomly sampled and that we had no evidence to reject the hypothesis that the proportion of infected insects was 40%. If the observed frequencies had not fitted the expected frequencies, the lack of fit might be attributed to chance, or to the conclusion that the true proportion of infection is not 0.4, or we would have to reject one or both of the following assumptions: (1) that sampling was at random, and (2) that the occurrence of infected insects in these samples was independent.

Experiment 4.1 was designed to yield random samples and independent events. How could we simulate a sampling procedure in which the occurrences of the digits 0, 1, 2, and 3 were not independent? We could, for example, instruct the sampler to sample as indicated previously, but every time he found a 3, to search through the succeeding digits until he found another one of the four digits standing for infected individuals and to incorporate this in the sample. Thus, once a 3 was found, the probability would be 1.0 that another one of the indicated digits would be included in the sample. After repeated samples, this would result in higher frequencies of classes of two or more indicated digits and in lower frequencies than expected (on the basis of the binomial distribution) of classes of one event. A variety of such different sampling schemes could be devised. It should be quite clear to the reader that the probability of the second event occurring would be different from that of the first and dependent on it.

How would we interpret a large departure of the observed frequencies from expectation in another example? We have not as yet learned techniques for testing whether observed frequencies differ from those expected by more than can be attributed to chance alone. This will be taken up in Chapter 13. Assume that such a test has been carried out and that it has shown us that our observed frequencies are significantly different from expectation. Two main types of departure from expectation are likely: (1) *clumping* and (2) *repulsion*, shown in fictitious examples in Table 4.2. In actual examples we

TABLE 4.2

Artificial distributions to illustrate clumping and repulsion. Expected frequencies from Table 4.1.

(1) Number of infected insects per sample Y	*(2)* Absolute expected frequencies $\hat{f}$	*(3)* Clumped (contagious) frequencies f	*(4)* Deviation from expectation	*(5)* Repulsed frequencies f	*(6)* Deviation from expectation
5	24.8	47	+	14	−
4	186.1	227	+	157	−
3	558.3	558	0	548	−
2	837.4	663	−	943	+
1	628.0	703	+	618	−
0	188.4	225	+	143	−
$\Sigma\hat{f}$ or n	2423.0	2423		2423.0	
ΣY	4846.1	4846		4846	
Mean	2.00004	2.00000		2.00000	
Standard deviation	1.09543	1.20074		1.01435	

would have no a priori notions about the magnitude of p, the probability of one of the two possible outcomes. In such cases it is customary to obtain p from the observed sample and to calculate the expected frequencies, using the sample p. This would mean that the hypothesis that p is a given value cannot be tested, since by design the expected frequencies will have the same p value as the observed frequencies. Therefore the hypotheses tested are whether the samples are random and the events independent.

The clumped frequencies in Table 4.2 have an excess of observations at the tails of the frequency distribution and consequently a shortage of observations at the center. Such a distribution is also called *contagious*. Remember that the total number of items must be the same in both observed and expected frequencies in order to make them comparable. In the repulsed frequency distribution there are more observations than expected at the center of the distribution and fewer at the tails. These discrepancies are easiest seen in columns (4) and (6) of Table 4.2, where the deviations of observed from expected frequencies are shown as plus or minus signs.

What do these phenomena imply? In the clumped frequencies more samples were entirely infected (or largely infected) and similarly more samples were entirely noninfected (or largely noninfected) than you would expect if probabilities of infection were independent. This could be due to poor sampling design. If, for example, the investigator in collecting his samples of five insects always tended to pick out like ones—that is, infected ones or noninfected ones—then such a result would likely appear. But if the sampling design is sound, the results become more interesting. Clumping would then mean that the samples of five are in some way related, so that if one insect is infected, others in the same sample are more likely to be infected. This could be true if they come from adjacent locations in a situation in which neighbors are easily infected. Or they could be siblings jointly exposed to a source of infection. Or possibly the infection might spread among members of a sample between the time that the insects have been sampled and the time they are examined.

The opposite phenomenon, repulsion, is more difficult to interpret biologically. There are fewer homogeneous groups and more mixed groups in such a distribution. This involves the idea of a compensatory phenomenon: if some of the insects in a sample are infected, the others in the sample are less likely to be. If the infected insects in the sample could in some way transmit immunity to their associates in the sample, such a situation could arise logically, but it is biologically improbable. A more reasonable interpretation of such a finding is that for each sampling unit there were only a limited number of pathogens available and that once several of the insects have become infected, the others go free of infection simply because there is no more infectious agent. This is an unlikely situation in microbial infections, but in situations in which a limited number of parasites enter the body of the host, repulsion may be more reasonable.

From the expected and observed frequencies in Table 4.1, we may calculate the mean and standard deviation of the number of infected insects per sample. These values are given at the bottom of columns (5), (6), and (7) in Table 4.1. We note that the means and standard deviations in columns (5) and (6) are almost identical and differ only trivially because of rounding errors. Column (7), being a sample from a population whose parameters are the same as those of the expected frequency distribution in columns (5) or (6), differs somewhat. The mean is slightly smaller and the standard deviation is slightly greater than in the expected frequencies. If we wish to know the mean and standard deviation of expected binomial frequency distributions we need not go through the computations shown in Table 4.1. The mean and standard deviation of a binomial frequency distribution are, respectively,

$$\mu = kp, \qquad \sigma = \sqrt{kpq}$$

Substituting the values $k = 5$, $p = 0.4$, and $q = 0.6$ of the above example, we obtain $\mu = 2.0$ and $\sigma = 1.09545$, which are identical to the values computed from column (5) in Table 4.1. Note that we use the Greek parametric notation here because μ and σ are parameters of an expected frequency distribution, not sample statistics, as are the mean and standard deviation in column (7). The proportions p and q are parametric values also and strictly speaking should be distinguished from sample proportions. In fact, in later chapters we resort to $\hat{p}$ and $\hat{q}$ for parametric proportions (rather than π, which conventionally is used as the ratio of the circumference to the diameter of a circle). Here, however, we prefer to keep our notation simple. It is interesting to look at the standard deviations of the clumped and repulsed frequency distributions of Table 4.2. We note that the clumped distribution has a standard deviation greater than expected, and that of the repulsed one is less than expected. Comparison of sample standard deviations with their expected values is a useful measure of dispersion in such instances. If we wish to express our variable as a proportion rather than as a count—that is, to indicate mean incidence of infection in the insects as 0.4, rather than as 2 per sample of 5—we can use other formulas for the mean and standard deviation in a binomial distribution:

$$\mu = p, \qquad \sigma = \sqrt{pq/k}$$

We shall now employ the binomial distribution to solve a biological problem. On the basis of our knowledge of the cytology and biology of species A, we expect the sex ratio among its offspring to be 1:1. The study of a litter in nature reveals that of 17 offspring 14 were females and 3 were males. What conclusions can we draw from this evidence? Assuming that $p_{♀}$ (the probability of being a female offspring) $= 0.5$ and that this probability is independent among the members of the sample, the pertinent probability distribution is the binomial for sample size $k = 17$. Expanding the binomial to the power 17 is a formidable task, which, as we shall see, fortunately need not be done in its entirety. However, we must have the binomial coefficients, which can be obtained either from an expansion of Pascal's triangle (fairly tedious unless once obtained and stored for future use) or by working out the expected frequencies for any given class of Y from the general formula for any term of the binomial distribution

$$C(k, Y)p^{k-Y}q^{Y} \tag{4.1}$$

The expression $C(k, Y)$ stands for the number of combinations that can be formed from k items taken Y at a time. This can be evaluated as $k!/[Y!(k - Y)!]$, where ! means "factorial." In mathematics k factorial is the product of all the integers from 1 up to and including k. Thus: $5! = 1 \times 2 \times 3 \times 4 \times 5 = 120$. By convention, $0! = 1$. In working out fractions containing factorials, note that any factorial will always cancel against a higher factorial. Thus $5!/3! = (5 \times 4 \times 3!)/3! = 5 \times 4$. For example, the binomial coefficient for the expected frequency of samples of 5 items containing 2 infected insects is $C(5, 2) = 5!/2!3! = (5 \times 4)/2 = 10$.

TABLE 4.3

Some expected frequencies of males and females for samples of 17 offspring on the assumption that the sex ratio is 1:1 $[p_{♀} = 0.5, q_{♂} = 0.5; (p_{♀} + q_{♂})^{k} = (0.5 + 0.5)^{17}]$.

(1) ♀♀	(2) ♂♂	(3) $p_{♀}$	(4) $q_{♂}$	(5) *Binomial coefficients*	(6) *Relative expected frequencies* $\hat{f}_{rel}$	
17	—	0.000,007,63	1	1	0.000,007,63	} 0.006,363,42
16	1	0.000,015,26	.5	17	0.000,129,71	
15	2	0.000,030,52	.25	136	0.001,037,68	
14	3	0.000,061,04	.125	680	0.005,188,40	
13	4	0.000,122,07	.0625	2380	0.018,157,91	

The setup of the example is shown in Table 4.3. Decreasing powers of $p_{♀}$ from $p_{♀}^{17}$ down, and increasing powers of $q_{♂}$ are computed (from power 0 to power 4). Note that for the purposes of our problem we need not proceed beyond the term for 13 females and 4 males. Calculating the relative expected frequencies in column (6), we note that the probability of 14 females and 3 males is 0.005,188,40, a very small value. If we add to this value all "worse" outcomes—that is, all outcomes that are even more unlikely than 14 females and 3 males on the assumption of a 1:1 hypothesis—we obtain a probability of 0.006,363,42, still a very small value. It is a common practice in statistics to calculate the probability of a deviation as large or larger than a given value.

On the basis of these findings one or more of the following assumptions is unlikely: (1) that the true sex ratio in species A is 1:1; (2) that we have sampled at random in the sense of obtaining an unbiased sample; or (3) that the sexes of the offspring are independent of one another. Lack of independence of events may mean that although the average sex ratio is 1:1, the individual sibships or litters are largely unisexual, so that the offspring from a given mating would tend to be all (or largely) females or all (or largely) males. To confirm this hypothesis we would need to have more samples and then examine the distribution of samples for clumping, which would indicate a tendency for unisexual sibships.

We must be very precise about the questions we ask of our data. There are really two questions we could ask about the sex ratio. One, are the sexes unequal in frequency so that females will appear more often than males? Two, are the sexes unequal in frequency? We may be only concerned with the first of these questions, since we know from past experience that in this particular group of organisms the males are never more frequent than females. In such a case the reasoning followed above is appropriate. However, if we know very little about this group of organisms and if our question is simply whether the sexes among the offspring are unequal in frequency, then we have to consider both tails of the binomial frequency distribution; departures from the 1:1 ratio could occur in either direction. We should then not only consider the probabilities of samples with 14 females and 3 males (and all worse cases) but also the probability of samples of 14 males and 3 females (and all worse cases in that direction). Since this probability distribution is symmetrical (because $p_♀ = q_♂ = 0.5$), we simply double the cumulative probability of 0.006,363,42 obtained previously, which results in 0.012,726,84. This new value is still very small, making it quite unlikely that the true sex ratio is 1:1. This is your first experience with one of the most important applications of statistics—hypothesis testing. A formal introduction to this field will be deferred until Section 6.8. We may simply point out here that the two approaches followed above are known appropriately as *one-tailed tests* and *two-tailed tests*, respectively. Students sometimes have difficulty knowing which of the two tests to apply. In future examples we shall try to point out in each case why a one-tailed or a two-tailed test is being used.

We have said that a tendency for unisexual sibships would result in a clumped distribution of observed frequencies. An actual case of this nature is a classic in the literature, the sex ratio data obtained by Geissler (1889) from hospital records in Saxony. Table 4.4 reproduces sex ratios of 6115 sibships of 12 children each from the more extensive study by Geissler. All columns of the table should by now be familiar. The expected frequencies were not calculated on the basis of a 1:1 hypothesis, since it is known that in human populations the sex ratio at birth is not 1:1. As the sex ratio varies in different human populations, the best estimate of it for the population in Saxony was simply obtained using the mean proportion of males in these data. This can be obtained by calculating the average number of males per sibship ($\bar{Y}$ = 6.23058) for the 6115 sibships and converting this into a proportion. This value turns out to be 0.519,215. Consequently, the proportion of females = 0.480,785. In the deviations of the observed frequencies from the absolute expected frequencies shown in column (9) of Table 4.4, we notice considerable clumping. There are many more instances of families with all male or all female children (or nearly so) than independent probabilities would indicate. The genetic basis for this is not clear, but it is evident that there are some families which "run to girls" and similarly those which "run to boys." Evidence of clumping can also be seen from the fact that s^2 is much larger than we would expect on the basis of the binomial distribution [$\sigma^2 = kpq = 12(0.519215)0.480785 = 2.99557$].

There is a distinct contrast between the data in Table 4.1 and those in Table 4.4. In the insect infection data of Table 4.1 we had a hypothetical proportion of infection based on outside knowledge. In the sex ratio data of Table 4.4 we had no such knowledge; we used an *empirical value of p obtained from the data,* rather than a *hypothetical value external to the data.* This is a distinction whose importance will become apparent later. In the sex ratio data of Table 4.3, as in much work in Mendelian genetics, a hypothetical value of p is used.

TABLE 4.4

Sex ratios in 6115 sibships of twelve in Saxony.

(1)	*(2)*	*(3)*	*(4)*	*(5)*	*(6)*	*(7)*	*(8)*	*(9)*
♂♂ Y	♀♀	$p_{♂}$	$q_{♀}$	*Binomial coefficients*	*Relative expected frequencies* $\hat{f}_{rel}$	*Absolute expected frequencies* $\hat{f}$	*Observed frequencies* f	*Deviation from expectation* $(f - \hat{f})$
12	—	0.000,384	1	1	0.000,384	2.3	7	+
11	1	0.000,739	0.480,785	12	0.004,264	26.1	45	+
10	2	0.001,424	0.231,154	66	0.021,725	132.8	181	+
9	3	0.002,742	0.111,135	220	0.067,041	410.0	478	+
8	4	0.005,282	0.053,432	495	0.139,703	854.3	829	−
7	5	0.010,173	0.025,689	792	0.206,973	1265.6	1112	−
6	6	0.019,592	0.012,351	924	0.223,590	1367.3	1343	−
5	7	0.037,734	0.005,938	792	0.177,459	1085.2	1033	−
4	8	0.072,676	0.002,855	495	0.102,708	628.1	670	+
3	9	0.139,972	0.001,373	220	0.042,280	258.5	286	+
2	10	0.269,584	0.000,660	66	0.011,743	71.8	104	+
1	11	0.519,215	0.000,317	12	0.001,975	12.1	24	+
—	12	1	0.000,153	1	0.000,153	0.9	3	+
Total					0.999,998	6115.0	6115	
			$\bar{Y}$ = 6.23058		s^2 = 3.48985			

Source: Geissler (1889).

4.3 The Poisson distribution

In the typical application of the binomial we had relatively small samples (2 students, 5 insects, 17 offspring, 12 siblings) in which two alternative states occurred at varying frequencies (American and foreign, infected and non-infected, male and female). Quite frequently, however, we study cases in which sample size k is very large. These would present a considerable computational problem. We have seen that the expansion of the binomial $(p + q)^k$ is quite tiresome when k is large. Suppose you had to expand the expression $(0.001 + 0.999)^{1000}$. Note that not only is the sample size large in this expression, but one of the events (represented by probability q) is very much more frequent than the other (represented by probability p). Expanding this binomial by the methods you have learned so far requires a very accurate table of logarithms to as many as 10 decimal places. Yet such expressions are often of great biological importance and are commonly encountered. In such cases we are generally interested in one tail of the distribution only. This is the tail represented by the terms

$$p^0q^k,\ C(k, 1)p^1q^{k-1},\ C(k, 2)p^2q^{k-2},\ C(k, 3)p^3q^{k-3}, \ldots$$

The first term represents no rare events and k frequent events in a sample of k events, the second term represents one rare event and $k - 1$ frequent events, the third term 2 rare events and $k - 2$ frequent events, and so forth. The expressions of the form $C(k, i)$ are the binomial coefficients, represented by the combinatorial terms discussed in the previous section. Although this expression would permit the computation of the desired tail of the curve, it would still be a very complicated procedure in view of the magnitude of k. The reader might try computing one or more of these terms for $(0.001 + 0.999)^{1000}$ to convince himself of this. Fortunately, it is much easier to compute another distribution, the Poisson distribution, which closely approximates the desired results.

The *Poisson distribution* is also a discrete frequency distribution of the number of times a rare event occurs. But, in contrast to the binomial distribution, the number of times that an event does not occur is infinitely large. For purposes of our treatment here, a Poisson variable will be studied in either a spatial or temporal sample. An example of the first would be the number of moss plants in a sampling quadrat on a hillside or the number of parasites on an individual host; an example of a temporal sample is the number of mutations occurring in a genetic strain in the time interval of one month or the reported cases of influenza in one town during one week. The Poisson variable, Y, will be the number of events per sample. It can assume discrete values from 0 on up. To be distributed in Poisson fashion the variable must have two properties: (1) Its mean must be small relative to the maximum possible number of events per sampling unit. Thus the event should be "rare." For example, a quadrat in which moss plants are counted must be large enough that a substantial number of moss plants could occur there physically if the biological conditions were such as to favor the development of numerous moss plants in the quadrat. A quadrat of 1-cm square would be far too small for mosses to be distributed in Poisson fashion. Similarly, a time span of 1 minute would be unrealistic for reporting new influenza cases in a town, but within 1 week a great many such cases could occur. (2) An occurrence of the event must be independent of prior occurrences within the sampling unit. Thus, the presence of one moss plant in a quadrat must not enhance or diminish the probability of other moss plants developing in the quadrat. Similarly, the fact that one influenza case has been reported must not affect the probability of reporting subsequent influenza cases. Events that meet these conditions ("rare and random events") should be distributed in Poisson fashion.

The purpose of fitting a Poisson distribution to numbers of rare events in nature is to test whether the events occur independently with respect to each other. If they do, they will follow the Poisson distribution. If the occurrence of one event enhances the probability of a second such event, we obtain a clumped or contagious distribution. If the occurrence of one event impedes that of a second such event in the sampling unit, we obtain a repulsed or spatially uniform distribution. The Poisson can be used as a test for randomness or independence of distribution not only spatially but also in time.

The distribution is named after the French mathematician Poisson, who described it in 1837. It is an infinite series whose terms add to one (as must be true for any probability distribution). The series can be represented as

$$\frac{1}{e^{\mu}}, \frac{\mu}{1!e^{\mu}}, \frac{\mu^2}{2!e^{\mu}}, \frac{\mu^3}{3!e^{\mu}}, \frac{\mu^4}{4!e^{\mu}}, \ldots, \frac{\mu^r}{r!e^{\mu}}, \ldots \tag{4.2}$$

which are the relative expected frequencies corresponding to the following counts of the rare event Y:

$$0, \quad 1, \quad 2, \quad 3, \quad 4, \quad \ldots, \quad r, \quad \ldots$$

Thus, the first of these terms represents the relative expected frequency of samples containing no rare event; the second term, one rare event; the third term, two rare events; and so on. The denominator of each term contains e^{μ}, where e is the base of the natural or Naperian logarithm, a constant whose value, accurate to 5 decimal places, is 2.71828. We recognize μ as the parametric mean of the distribution; it is a constant for any given problem. The exclamation mark after the coefficient in the denominator means "factorial" and has been explained in the previous section.

One way to learn more about the Poisson distribution is to apply it to an actual case. At the top of Box 4.1 is a well-known result from the early statistical literature. It tests the distribution of yeast cells in 400 squares of a hemacytometer, a counting chamber such as is used in making counts of blood cells and other microscopic structures suspended in liquid. Column (1 lists the number of yeast cells observed in each hemacytometer square and column (2) gives the observed frequency—the number of squares containing

BOX 4.1

Calculation of expected Poisson frequencies.

Yeast cells in 400 squares of a hemacytometer: $\bar{Y} = 1.8$ cells per square; $n = 400$ squares sampled.

(1) Number of cells per square Y	(2) Observed frequencies f		(3) Absolute expected frequencies $\hat{f}$		(4) Deviation from expectation $f - \hat{f}$	
0	75		66.1		+	
1	103		119.0		−	
2	121		107.1		+	
3	54		64.3		−	
4	30		28.9		+	
5	13		10.4		+	
6	2		3.1		−	
7	1	17	0.8	14.5	+	+
8	0		0.2		−	
9	1		0.0		+	
	400		399.9			

Source: "Student" (1907).

Computational steps

Flow of computation based on Expression (4.3) multiplied by n, since we wish to obtain absolute expected frequencies, $\hat{f}$.

1. Calculate $e^{\bar{Y}}$.

$$\log e^{\bar{Y}} = \bar{Y} \log e = \bar{Y}(0.43429) = (1.8)(0.43429)$$

$$= 0.78172$$

$$\text{antilog}\,(0.78172) = 6.0495$$

2. $\hat{f}_0 = \dfrac{n}{e^{\bar{Y}}} = ne^{-\bar{Y}} \qquad = \dfrac{400}{6.0495} = 400(0.16530) = 66.12$

3. $\hat{f}_1 = \left(\dfrac{n}{e^{\bar{Y}}}\right)\bar{Y} \qquad = 66.12(1.8) \qquad = 119.02$

4. $\hat{f}_2 = \left(\dfrac{n\bar{Y}}{e^{\bar{Y}}}\right)\dfrac{\bar{Y}}{2} \qquad = 119.02\left(\dfrac{1.8}{2}\right) \qquad = 107.12$

5. $\hat{f}_3 = \left(\dfrac{n\bar{Y}^2}{2e^{\bar{Y}}}\right)\dfrac{\bar{Y}}{3} \qquad = 107.12\left(\dfrac{1.8}{3}\right) \qquad = 64.27$

6. $\hat{f}_4 = \left(\dfrac{n\bar{Y}^3}{2 \times 3e^{\bar{Y}}}\right)\dfrac{\bar{Y}}{4} \qquad = 64.27\left(\dfrac{1.8}{4}\right) \qquad = 28.92$

BOX 4.1 continued

7. $\hat{f}_5 = \left(\frac{n\bar{Y}^4}{2 \times 3 \times 4e^{\bar{Y}}}\right)\frac{\bar{Y}}{5} \quad = 28.92\left(\frac{1.8}{5}\right) \quad = \quad 10.41$

8. $\hat{f}_6 = \left(\frac{n\bar{Y}^5}{2 \times 3 \times 4 \times 5e^{\bar{Y}}}\right)\frac{\bar{Y}}{6} \quad = 10.41\left(\frac{1.8}{6}\right) \quad = \quad 3.12$

9. $\hat{f}_7 = \left(\frac{n\bar{Y}^6}{2 \times 3 \times 4 \times 5 \times 6e^{\bar{Y}}}\right)\frac{\bar{Y}}{7} \quad = 3.12\left(\frac{1.8}{7}\right) \quad = \quad 0.80$

10. $\hat{f}_8 = \left(\frac{n\bar{Y}^7}{2 \times 3 \times 4 \times 5 \times 6 \times 7e^{\bar{Y}}}\right)\frac{\bar{Y}}{8} = 0.80\left(\frac{1.8}{8}\right) \quad = \quad 0.18$

Total 399.96

$\hat{f}_9$ and beyond 0.04

a given number of yeast cells. We note that 75 squares contained no yeast cells, but that most squares held either 1 or 2 cells. Only 17 squares contained 5 or more yeast cells.

Why would we expect this frequency distribution to be distributed in Poisson fashion? We have here a relatively rare event, the frequency of yeast cells per hemacytometer square, the mean of which has been calculated and found to be 1.8. That is, on the average there are 1.8 cells per square. Relative to the amount of space provided in each square and the number of cells that could have come to rest in any one square, the actual number found is low indeed. We might also expect that the occurrence of individual yeast cells in a square is independent of the occurrence of other yeast cells. This is a commonly encountered class of application of the Poisson distribution.

The mean of the rare event is the only quantity that we need to know to calculate the relative expected frequencies of a Poisson distribution. Since we do not know the parametric mean of the yeast cells in this problem, we employ an estimate (the sample mean) and calculate expected frequencies of a Poisson distribution whose μ equals the sample mean of the observed frequency distribution of Box 4.1. It is convenient for the purpose of computation to rewrite Expression (4.2) as

$$\frac{1}{e^{\bar{Y}}}, \frac{1}{e^{\bar{Y}}}\left(\frac{\bar{Y}}{1}\right), \frac{\bar{Y}}{e^{\bar{Y}}}\left(\frac{\bar{Y}}{2}\right), \frac{\bar{Y}^2}{2e^{\bar{Y}}}\left(\frac{\bar{Y}}{3}\right), \frac{\bar{Y}^3}{2 \times 3e^{\bar{Y}}}\left(\frac{\bar{Y}}{4}\right), \ldots \tag{4.3}$$

Note first of all that the parametric mean μ has been replaced by the sample mean $\bar{Y}$. Each term is mathematically exactly the same as its corresponding term in Expression (4.2), but it has been factored into a convenient form for computation. After the first term in Expression (4.3), all subsequent terms consist of the previous term multiplied by the mean over an integer that increases by 1 for each succeeding term. Thus, we need only compute the expression $1/e^{\bar{Y}}$ once to obtain the frequency of the first term, multiply this by $\bar{Y}/1$ to get the second term, multiply the second term by $\bar{Y}/2$ for the third term, and so forth. It is important to make no computational error, since in such a chain multiplication the correctness of each term depends on the accuracy of the term before it. Expression (4.3) yields relative expected frequencies. If, as is more usual, absolute expected frequencies are desired, simply multiply the first term by n, the number of samples, and then proceed with the computational steps as before. By this process of chain multiplication the n continues as a factor in every term. The actual computation is illustrated in Box 4.1 and the expected frequencies so obtained are listed in column (3).

What have we learned from this computation? When we compare the observed with the expected frequencies we notice quite a good fit of our observed frequencies to a Poisson distribution of mean 1.8, although we have not as yet learned a statistical test for goodness of fit (Chapter 13). No clear pattern of deviations from expectation is shown. We cannot test a hypothesis about the mean because the mean of the expected distribution was taken from the sample mean of the observed variates. As in the binomial distribution, clumping or aggregation would mean that the probability that a second yeast cell will be found in a square is not independent of the presence of the first one, but is higher. This would result in a clumping of the items in the classes at the tails of the distribution so that there would be some squares with large numbers of cells.

The biological interpretation of the dispersion pattern varies with the problem. The yeast cells seem to be randomly distributed in the counting chamber, indicating thorough mixing of the suspension. However, unless the proper suspension fluid is used, red blood cells will often stick together because of an electrical charge. This so-called rouleaux effect would be indicated by clumping of the observed frequencies.

Note that in Box 4.1, as in the subsequent tables giving examples of the application of the Poisson distribution, we group the low frequencies at one tail of the curve, uniting them by means of a bracket. This tends to simplify the patterns of distribution somewhat. However, the main reason for this grouping is related to the G-test for goodness of fit (of observed to expected frequencies), which is discussed in Section 13.2. For purposes of this test, no expected frequency $\hat{f}$ should be less than 5.

Before we turn to other examples we need to learn a few more facts about the Poisson distribution. You probably noticed that in computing expected frequencies we needed to know only one parameter—the mean of the distribution. By comparison, in the binomial we needed two parameters, p and k. Thus the mean completely defines the shape of a given Poisson distribution. From this it follows that the variance is some function of the mean and in fact in a Poisson distribution we have a very simple relationship between the two: $\mu = \sigma^2$, the variance being equal to the mean. The variance

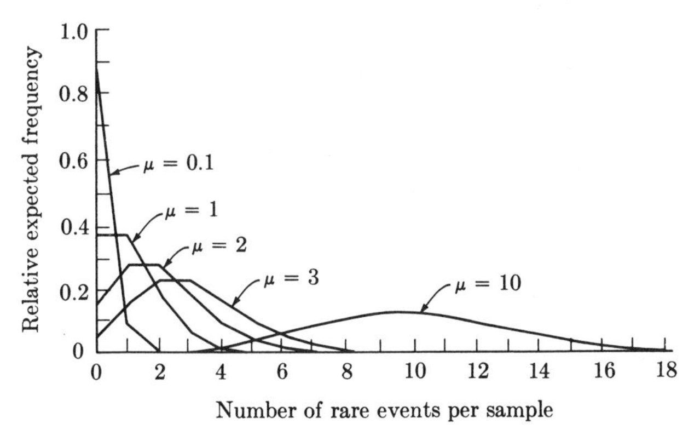

FIGURE 4.3

Frequency polygons of the Poisson distribution for various values of the mean.

of the number of yeast cells per square based on the observed frequencies in Box 4.1 equals 1.965, not much larger than the mean of 1.8, indicating again that the yeast cells are distributed in Poisson fashion, hence randomly. This relationship between variance and mean suggests a rapid test of whether an observed frequency distribution is distributed in Poisson fashion even without fitting expected frequencies to the data. We simply compute a *coefficient of dispersion*

$$C.D. = \frac{s^2}{\bar{Y}}$$

This value will be near 1 in distributions that are essentially Poisson, will be >1 in clumped samples, and <1 in cases of repulsion. In the yeast cell example, $C.D. = 1.092$.

The shapes of five Poisson distributions of different means are shown in Figure 4.3 as frequency polygons (line connecting the midpoints of a bar diagram. We notice that for the low value of $\mu = 0.1$ the frequency polygon is extremely L-shaped, but with an increase in the value of μ the distributions become humped and eventually nearly symmetrical.

We conclude our study of the Poisson distribution with a consideration of two examples. The first example shows the distribution of a water mite on adults of a chironomid fly (Table 4.5). This example is similar to the distribution of yeast cells except that here the sampling unit is a fly instead of a hemacytometer square. The rare event is a mite infesting the fly. The coefficient of dispersion is 2.225 and this is clearly reflected by the observed frequencies, which are greater than expected in the tails and less than expected in the center. This relationship is easily seen in the last column of the frequency distribution, which shows the signs of the deviations (observed minus expected frequencies) and shows a characteristic clumped pattern. A possible

TABLE 4.5

Mites (*Arrenurus* sp.) infesting 589 chironomid flies (*Calopsectra akrina*).

(1) Number of mites per fly Y	(2) Observed frequencies f		(3) Poisson expected frequencies $\hat{f}$		(4) Deviation from expectation $f - \hat{f}$	
0	442		380.7		+	
1	91		166.1		−	
2	29		36.2		−	
3	14	27	5.3	5.7	+	+
4	4		0.6		+	
5	6		0.1		+	
6	2		0.0		+	
7	0		0.0		0	
8	1		0.0		+	
Total	589		589.0			

$\bar{Y} = 0.4363$ $s^2 = 0.9709$ $C.D. = 2.225$

Source: Data by F. J. Rohlf.

explanation is that the density of mites in the several ponds from which the chironomid flies emerged differed considerably. Those chironomids that emerged from ponds with many mites would be parasitized by more than one mite, but those from ponds where mites were scarce would show little or no infestation.

The second distribution (Table 4.6) is extracted from an experimental study of the effects of different densities of parents of the Azuki bean weevil. We are studying the number of weevils emerging per bean. Larvae of these weevils enter the beans, feed and pupate inside them, and then emerge through an emergence hole. Thus the number of holes per bean is a good measure of the number of adults that have emerged. The rare event in this case is the weevil present in the bean. We note that the distribution is strongly repulsed. There are many more beans containing one weevil than the Poisson distribution would predict. A statistical finding of this sort leads to the biological question, "Why?" In this case it was found that the adult female weevils tended to deposit their eggs evenly rather than randomly over the available beans. This prevented too many eggs being placed on any one bean and precluded heavy competition among the developing larvae on any one bean. A contributing factor was competition between remaining larvae feeding on the same bean, resulting in all but one generally being killed or driven out. Thus it is easily understood how the above biological phenomena would give rise to a repulsed distribution.

TABLE 4.6

Azuki bean weevils (*Callosobruchus chinensis*) emerging from 112 azuki beans (*Phaseolus radiatus*).

(1)	(2)	(3)	(4)
Number of weevils emerging per bean Y	*Observed frequencies* f	*Poisson expected frequencies* $\hat{f}$	*Deviation from expectation* $f - \hat{f}$
0	61	70.4	−
1	50	32.7	+
2	1 } 1	7.6 } 8.9	− } −
3	0	1.2	−
4	0	0.1	−
Total	112	112.0	
$\bar{Y} = 0.4643$	$s^2 = 0.269$	$C.D. = 0.579$	

Source: Utida (1943).

Exercises 4

4.1 In man the sex ratio of newborn infants is about 100 ♀♀ : 105 ♂♂. Were we to take 10,000 random samples of 6 newborn infants from the total population of such infants for one year, what would be the expected frequency of groups of 6 males, 5 males, 4 males, and so on?

4.2 The two columns below give fertility of eggs of the CP strain of *Drosophila melanogaster* raised in 100 vials of 10 eggs each (data from R. R. Sokal). Find the expected frequencies on the assumption of independence of mortality for each egg in a vial. Use the observed mean. Calculate the expected variance and compare it with the observed variance. Interpret results, knowing that the eggs of each vial are siblings and that the different vials contain descendants from different parent pairs. ANS. $\sigma^2 = 2.417$, $s^2 = 6.636$.

Number of eggs hatched	*Number of vials*
Y	f
0	1
1	3
2	8
3	10
4	6
5	15
6	14
7	12
8	13
9	9
10	9

4.3 Calculate Poisson expected frequencies for the frequency distribution given in Table 2.2 (number of plants of the sedge *Carex flacca* found in 500 quadrats).

4.4 The Army Medical Corps is concerned over the intestinal disease X. From previous experience they know that soldiers suffering from the disease invariably harbor the pathogenic organism in their feces and that to all practical purposes every diseased stool specimen contains these organisms. However, the organisms are never abundant and thus only 20% of all slides prepared by the standard procedure will contain some (we assume that if an organism is present on a slide it will be seen). How many slides should they direct their laboratory technicians to prepare and examine per stool specimen, so that in case a specimen is positive it will be erroneously diagnosed negative in less than 1% of the cases (on the average)? On the basis of your answer would you recommend that the Corps attempt to improve their diagnostic methods? ANS. 21 slides.

4.5 A cross is made in a genetic experiment in drosophila in which it is expected that $\frac{1}{4}$ of the progeny will have white eyes and $\frac{1}{2}$ will have the trait called "singed bristles." Assume that the two gene loci segregate independently. (a) What proportion of the progeny should exhibit both traits simultaneously? (b) If 4 flies are sampled at random, what is the probability that they would all be white-eyed? (c) What is the probability that none of the 4 flies would have either white eyes or "singed bristles?" (d) If two flies are sampled, what is the probability that at least one of the flies has either white eyes or "singed bristles" or both traits?

4.6 Those readers who have had a semester or two of calculus may wish to try to prove that Expression (4.1) tends to Expression (4.2) as k becomes indefinitely large (and p becomes infinitesimal, so that $\mu = kp$ remains constant).

$$\text{HINT: } \left(1 - \frac{x}{n}\right)^n \longrightarrow e^{-x} \text{ as } n \longrightarrow \infty.$$

4.7 Summarize and compare the assumptions and parameters on which the binomial and Poisson distributions are based.

4.8 If the frequency of the gene A is p and the frequency of a is q, what are the expected frequencies of the zygotes AA, Aa, or aa (assuming a diploid zygote represents a random sample of size 2)? What would the expected frequency be for an autotetraploid (for a locus close to the centromere a zygote can be thought of as a random sample of size 4)?

4.9 A population consists of three types of individuals A_1, A_2, and A_3 with relative frequencies of 0.5, 0.2, and 0.3, respectively. (a) What is the probability of obtaining only individuals of type A_1 in samples of size 1, 2, 3, . . . , n? (b) What would be the probabilities of obtaining only individuals that were not of types A_1 or A_2 in a sample of size n? (c) What is the probability of obtaining a sample containing at least one representation of each type in samples of size 1, 2, 3, 4, 5 . . . , n?

5

The Normal Probability Distribution

The theoretical frequency distributions in the last chapter were all discrete. Their variables assumed values that changed in integral steps (meristic variables). Thus the number of infected insects per sample was 0 or 1 or 2 but could not assume an intermediate value between these. Similarly, the number of yeast cells per hemacytometer square is a meristic variable and requires a discrete probability function to describe it. However, most variables encountered in biology are continuous (such as the infant birth weights or the aphid femur lengths used as examples in Chapters 2 and 3). This chapter deals more extensively with the distributions of continuous variables.

The first section (5.1) introduces frequency distributions of continuous variables. In Section 5.2 we show one way of deriving the most common such distribution, the normal probability distribution, and we examine its properties in Section 5.3. A few applications of the normal distribution are illustrated in Section 5.4. A graphic technique for pointing out departures from normality and for estimating mean and standard deviation in approximately normal distributions is given in Section 5.5, as are some of the reasons for departure from normality in observed frequency distributions.

5.1 Frequency distributions of continuous variables

For continuous variables the theoretical probability distribution or *probability density function* can be represented by a continuous curve, as shown in Figure 5.1. The height of the curve gives the density for a given value of the variable. By *density* we mean the relative concentration of variates along the Y-axis (as indicated in Figure 2.1). Note that the Y-axis is the abscissa in the figure, with the frequency f or the density being the ordinate. In order to compare the theoretical with the observed frequency distribution it is necessary to divide the two into corresponding classes, as shown by the vertical lines in Figure 5.1. Probability density functions are defined so that the expected frequency of observations between two class limits (vertical lines) is represented by the area between these limits under the curve. The total area under the curve is therefore equal to the sum of the expected frequencies (1.0 or n, depending on whether relative or absolute expected frequencies were calculated).

When forming a frequency distribution of observations of a continuous variable, the choice of class limits is arbitrary because all values of a variable are theoretically possible. In a continuous distribution one cannot evaluate the probability of the variable being exactly equal to a given value such as 3 or 3.5. One can only estimate the frequency of observations falling between two limits. This is so because the area of the curve corresponding to any point along the curve is an infinitesimal. Thus to calculate expected frequencies for a continuous distribution we shall have to calculate the proportions of the area under the curve between the class limits. We shall see how this is done in the normal frequency distribution in Sections 5.3 and 5.4.

Continuous frequency distributions may start and terminate at finite points along the Y-axis, as shown in Figure 5.1, or one or both ends of the curve may extend indefinitely, as in Figures 6.11 and 5.3. The idea of an area

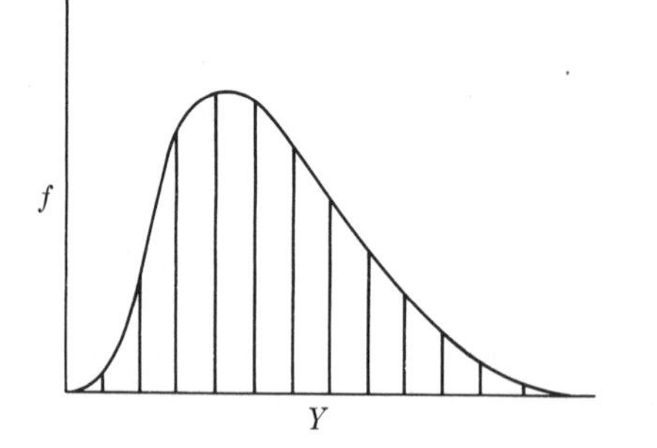

FIGURE 5.1

A probability distribution of a continuous variable. (See text for explanation.)

under a curve, when one or both ends go to infinity, may trouble those of you not acquainted with the calculus. Fortunately, however, this is not a great conceptual stumbling block, since in all the cases that we shall encounter, the tail of the curve will approach the Y-axis rapidly enough, so that the portion of the area beyond a certain point is for all practical purposes zero and the frequencies it represents are infinitesimal.

We may fit continuous frequency distributions to some sets of meristic data as, for example, the number of teeth in an organism. In such cases we have reason to believe that underlying biological variables causing differences in numbers of the structure are really continuous, even though expressed as a discrete variable.

We shall now proceed to discuss the most important probability density function in statistics, the normal frequency distribution.

5.2 Derivation of the normal distribution

There are several ways of deriving the normal frequency distribution from elementary assumptions. Most of these require more mathematics than we expect of our readers. We shall therefore use a largely intuitive approach, which we have found of heuristic value.

Let us consider a binomial distribution of the familiar form $(p + q)^k$ in which k becomes indefinitely large. What type of biological situation could give rise to such a binomial distribution? An example might be one in which many factors cooperate additively in producing a biological result. The following hypothetical case is possibly not too far removed from reality. The intensity of skin pigmentation in an animal will be due to the summation of many factors, some genetic, others environmental. As a simplifying assumption let us state that every factor can occur in two states only: present or absent. When the factor is present it contributes one unit of pigmentation to skin color, but it contributes nothing to pigmentation when it is absent. Each factor, regardless of its nature or origin, has the identical effect and the effects are additive; if three out of five possible factors are present in an individual, the pigmentation intensity would be three units, being the sum of three contributions of one unit each. One final assumption: each factor has an equal probability of being present or absent in a given individual. Thus, $p = P[F] = 0.5$, the probability of the factor being present, while $q = P[f] = 0.5$, the probability of the factor being absent.

With only one factor ($k = 1$), expansion of the binomial $(p + q)^1$ would yield two pigmentation classes among the animals as follows:

$\{F,$	$f\ \}$	pigmentation classes (probability space)
$\{0.5,$	$0.5\}$	expected frequency
$\{1,$	$0\ \}$	pigmentation intensity

Half the animals would have intensity 1, the other half 0. With $k = 2$ factors

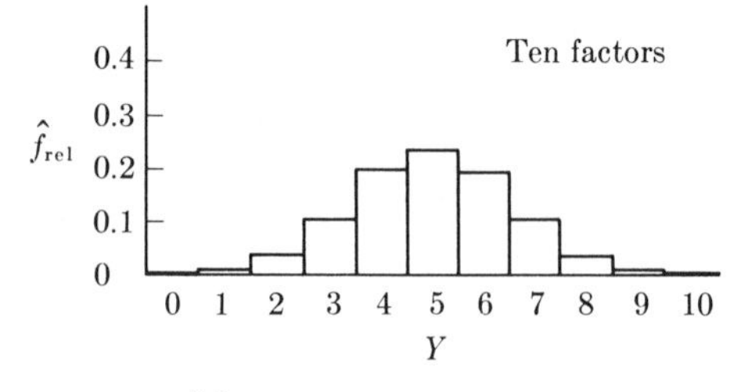

FIGURE 5.2

Histogram based on relative expected frequencies resulting from expansion of binomial $(0.5 + 0.5)^{10}$. The Y-axis measures the number of pigmentation factors F. (For further explanation see text.)

present in the population (the factors are assumed to occur independently of each other), the distribution of pigmentation intensities would be represented by the expansion of the binomial $(p + q)^2$:

{FF,	Ff,	ff }	pigmentation classes (probability space)
{0.25,	0.50,	0.25}	expected frequency
{2,	1,	0 }	pigmentation intensity

One-fourth of the individuals would have pigmentation intensity 2, half 1, and the remaining fourth 0.

The number of classes in the binomial increases with the number of factors. The frequency distributions are symmetrical and the expected frequencies at the tails become progressively less as k increases. The binomial distribution for $k = 10$ is graphed as a histogram in Figure 5.2 (rather than as a bar diagram as it should be drawn). We note that it approaches the familiar bell-shaped outline of the normal frequency distribution (Figures 5.3 and 5.4). Were we to expand the expression for $k = 20$, our histogram would be so close to a normal frequency distribution that we could not show the difference between them on a graph the size of this page. At the beginning of this procedure we made a number of severe limiting assumptions for the sake of simplicity. What happens when these are removed? When $p \neq q$, the distribution also approaches normality as k approaches infinity. This is intuitively difficult to see because when $p \neq q$ the histogram is at first asymmetrical. However, it can be shown that when k, p, and q are such that $kpq \geq 3$, the normal distribution will be closely approximated. In a more realistic situation factors would be permitted to occur in more than 2 states, one state making a large contribution, a second state a smaller contribution, and so forth. However, it can also be shown that the multinomial $(p + q + r + \cdots + z)^k$ approaches the normal frequency distribution as k approaches infinity. Different factors may be present in different frequencies and may have different quantitative effects. As long as these are additive and independent, normality is still approached as k approaches infinity.

Lifting these restrictions makes the assumptions leading to a normal distribution compatible with innumerable biological situations. It is therefore not surprising that so many biological variables are approximately normally distributed.

Let us review the conditions that would tend to produce normal frequency distributions: (1) if there are many factors that are single or composite; (2) if these factors are independent in occurrence; (3) if the factors are independent in effect—that is, if their effects are additive; and (4) if they make equal contributions to the variance. The fourth condition we are not yet in a position to discuss and mention here only for completeness. We shall return to it in Chapter 7.

5.3 Properties of the normal distribution

Formally the *normal probability density function* can be represented by the expression

$$Z = \frac{1}{\sigma\sqrt{2\pi}} e^{-(Y-\mu)^2/2\sigma^2} \tag{5.1}$$

Here Z indicates the height of the ordinate of the curve, which represents the density of the items. It is the dependent variable in the expression, being a function of the variable Y. There are two constants in the equation: π, well known to be approximately 3.14159, making $1/\sqrt{2\pi}$ equal 0.39894, and e, the base of the Naperian or natural logarithms, whose value approximates 2.71828. There are two parameters in a normal probability density function. These are the parametric mean μ and the parametric standard deviation σ, which determine the location and shape of the distribution. Thus there is not just one normal distribution, as might appear to the uninitiated who keep encountering the same bell-shaped image in elementary textbooks; rather, there are an infinity of such curves, since these parameters can assume an infinity of values. This is illustrated by the three normal curves in Figure 5.3, representing the same total frequencies. Curves A and B differ in their means, and hence are at different locations. Curves B and C have identical means but different standard deviations. Since the standard deviation of curve C is only half that of curve B, it presents a much narrower appearance.

In theory a normal frequency distribution extends from negative infinity to positive infinity along the axis of the variable (labeled Y, although it is frequently the abscissa). This means that a normally distributed variable can assume any value, however large or small, although values farther from the mean than plus or minus three standard deviations are quite improbable, their relative expected frequencies being very rare. This can be seen from Expression (5.1). When Y is very large or very small, the term $(Y - \mu)^2/2\sigma^2$ will necessarily become very large. Hence e raised to the negative power of that term will be very small and Z will therefore be very small.

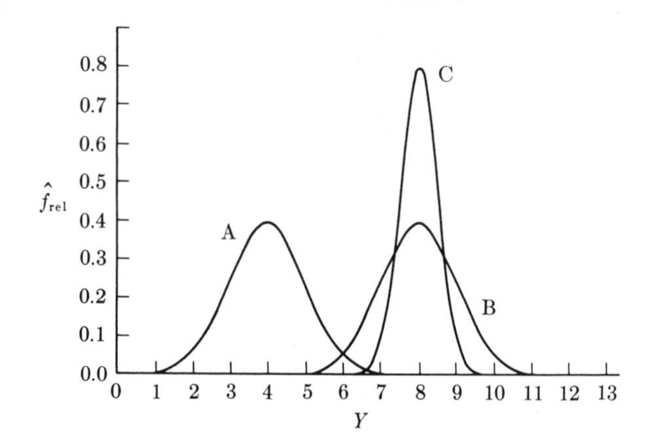

FIGURE 5.3

Illustration of how changes in the two parameters of the normal distribution affect the shape and position of curves. A. $\mu = 4$, $\sigma = 1$; B. $\mu = 8$, $\sigma = 1$; C. $\mu = 8$, $\sigma = 0.5$.

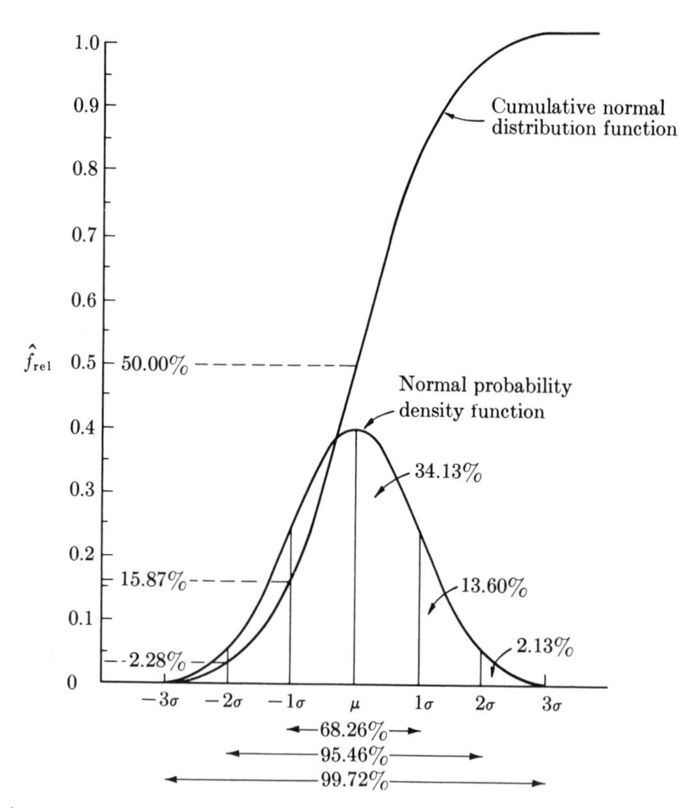

FIGURE 5.4

Areas under the normal probability density function and the cumulative normal distribution function.

The curve is symmetrical around the mean. Therefore the mean, median, and mode of the normal distribution are all at the same point. The following percentages of items in a normal frequency distribution lie within the indicated limits:

$\mu \pm \sigma$ contains 68.26% of the items
$\mu \pm 2\sigma$ contains 95.46% of the items
$\mu \pm 3\sigma$ contains 99.73% of the items

Conversely,

50% of the items fall between $\mu \pm 0.674\sigma$
95% of the items fall between $\mu \pm 1.960\sigma$
99% of the items fall between $\mu \pm 2.576\sigma$

These relations are shown in Figure 5.4. How have these percentages been calculated? The direct calculation of any portion of the area under the normal curve requires an integration of the function shown as Expression (5.1). Fortunately, for those of you who do not know calculus (and even for those of you who do) the integration has already been carried out in an alternative form of the normal distribution: the *normal distribution function* (the theoretical *cumulative* distribution function of the normal probability density function), shown in Figure 5.4. It gives the total frequency from negative infinity up to any point along the abscissa. We can therefore look up directly the probability that an observation will be less than a specified value of Y. For example, Figure 5.4 shows that the total frequency up to the mean is 50.00% and the frequency up to a point one standard deviation below the mean is 15.87%. These frequencies are found, graphically, by raising a vertical line from a point, such as $-\sigma$, until it intersects the cumulative distribution curve, and then reading the frequency (15.87%) off the ordinate. The probability of an observation falling between two arbitrary points can be found by subtracting the probability of an observation falling below the lower point from the probability of an observation falling below the upper point. For example, we can see from Figure 5.4 that the probability of an observation falling between the mean and a point one standard deviation below the mean is $0.5000 - 0.1587 = 0.3413$.

The normal distribution function is tabulated in Table **II**, Areas of the Normal Curve, and, for convenience in later calculations, 0.5 has been subtracted from all of the entries. This table therefore lists the proportion of the area between the mean and any point a given number of standard deviations above it. Thus, for example, the area between the mean and the point 0.50 standard deviations above the mean is 0.1915 of the total area of the curve. Similarly, the area between the mean and the point 2.64 standard deviations above the mean is 0.4959 of the curve. A point 4.0 standard deviations from the mean includes 0.499968 of the area between it and the mean.

TABLE 5.1

Populations of wing lengths and milk yields. *Column 1.* Rank number. *Column 2.* Lengths (in mm $\times 10^{-1}$) of 100 wings of houseflies arrayed in order of magnitude; $\mu = 45.5$, $\sigma^2 = 15.21$, $\sigma = 3.90$; distribution approximately normal. *Column 3.* Total annual milk yield (in hundreds of pounds) of 100 two-year-old registered Jersey cows arrayed in order of magnitude; $\mu = 66.61$, $\sigma^2 = 124.4779$, $\sigma = 11.1597$; distribution departs strongly from normality.

(1)	*(2)*	*(3)*	*(1)*	*(2)*	*(3)*	*(1)*	*(2)*	*(3)*	*(1)*	*(2)*	*(3)*	*(1)*	*(2)*	*(3)*
01	36	51	21	42	58	41	45	61	61	47	67	81	49	76
02	37	51	22	42	58	42	45	61	62	47	67	82	49	76
03	38	51	23	42	58	43	45	61	63	47	68	83	49	79
04	38	53	24	43	58	44	45	61	64	47	68	84	49	80
05	39	53	25	43	58	45	45	61	65	47	69	85	50	80
06	39	53	26	43	58	46	45	62	66	47	69	86	50	81
07	40	54	27	43	58	47	45	62	67	47	69	87	50	82
08	40	55	28	43	58	48	45	62	68	47	69	88	50	82
09	40	55	29	43	58	49	45	62	69	47	69	89	50	82
10	40	56	30	43	58	50	45	63	70	48	69	90	50	82
11	41	56	31	43	58	51	46	63	71	48	70	91	51	83
12	41	56	32	44	59	52	46	63	72	48	72	92	51	85
13	41	57	33	44	59	53	46	64	73	48	73	93	51	87
14	41	57	34	44	59	54	46	65	74	48	73	94	51	88
15	41	57	35	44	60	55	46	65	75	48	74	95	52	88
16	41	57	36	44	60	56	46	65	76	48	74	96	52	89
17	42	57	37	44	60	57	46	65	77	48	74	97	53	93
18	42	57	38	44	60	58	46	65	78	49	74	98	53	94
19	42	57	39	44	60	59	46	67	79	49	75	99	54	96
20	42	57	40	44	61	60	46	67	80	49	76	00	55	98

Source: Column 2—Data adapted from Sokal and Hunter (1955). Column 3—Data from Canadian government records.

However, since the normal distribution extends from negative to positive infinity, one needs to go an infinite distance from the mean to reach fully half of the area under the curve. The use of the table of areas of the normal curve will be illustrated in the next section.

A sampling experiment will give you a "feel" for the distribution of items under a normal curve.

Experiment 5.1. You are asked to sample from two populations. The first one is an approximately normal frequency distribution of 100 wing lengths of houseflies. The second population deviates strongly from normality. It is a frequency distribution of the total annual milk yield of 100 Jersey cows. Both populations are shown in Table 5.1. You are asked to sample from them repeatedly in order to simulate sampling from an infinite population. Obtain samples of 35 items from each of the two populations. This is done by obtaining two sets of 35 two digit random numbers from the table of random numbers (Table **I**) with which you became familiar in Experiment 4.1. Write down the random numbers in blocks of five, and copy next to them the value of Y (for either wing length or milk yield) corresponding to the random number. An example of such a block of five numbers and the computations required for it are shown below, using the housefly wing lengths as an example.

Random number	*Wing length* Y
16	41
59	46
99	54
36	44
21	42
$\Sigma Y =$	227
$\Sigma Y^2 =$	10,413
$\bar{Y} =$	45.4

These samples and the computations carried out for each sample will be used in subsequent chapters. Therefore, preserve your data carefully!

In this experiment consider the 35 variates for each variable as a single sample, without breaking them down into groups of five. Since the true mean and standard deviation (μ and σ) of the two distributions are known, you can calculate the expression $(Y_i - \mu)/\sigma$ for each variate Y_i. Thus for the first housefly wing length sampled above, you compute

$$\frac{(41 - 45.5)}{3.90} = -1.1538$$

This means that the first wing length is 1.1538 standard deviations below the true mean of the population. The deviation from the mean measured in standard deviation units is called a *standardized deviate* or *standard deviate.* The arguments of Table **II**, expressing distance from the mean in units of σ are called *standard normal deviates.* Group all 35 variates in a frequency distribution, then do the same for milk yields. Since you know the parametric mean and standard deviation, you need not compute each deviate separately, but can simply write down class limits in terms of the actual variable as well as in standard deviation form. The class limits for such a frequency distribution are shown in Table 5.2. Combine the results of your sampling with those of your classmates and study the percentage of the items in the distribution one, two, and three standard deviations to each side of the mean. Note the marked differences in distribution between the housefly wing lengths and the milk yields.

TABLE 5.2

Table for recording frequency distributions of standard deviates $(Y_i - \mu)/\sigma$ for samples of Experiment 5.1.

	Wing lengths			*Milk yields*	
	Variates falling between these limits	f		*Variates falling between these limits*	f
$-\infty$			$-\infty$		
-3σ			-3σ		
$-2\frac{1}{2}\sigma$			$-2\frac{1}{2}\sigma$		
-2σ	36, 37		-2σ		
$-1\frac{1}{2}\sigma$	38, 39		$-1\frac{1}{2}\sigma$		
$-\sigma$	40, 41		$-\sigma$	51–55	
$-\frac{1}{2}\sigma$	42, 43		$-\frac{1}{2}\sigma$	56–61	
$\mu = 45.5$	44, 45		$\mu = 66.61$	62–66	
$\frac{1}{2}\sigma$	46, 47		$\frac{1}{2}\sigma$	67–72	
σ	48, 49		σ	73–77	
$1\frac{1}{2}\sigma$	50, 51		$1\frac{1}{2}\sigma$	78–83	
2σ	52, 53		2σ	84–88	
$2\frac{1}{2}\sigma$	54, 55		$2\frac{1}{2}\sigma$	89–94	
3σ			3σ	95–98	
$+\infty$			$+\infty$		

5.4 Applications of the normal distribution

The normal frequency distribution is the most widely used distribution in statistics, and time and again we shall have recourse to it in a variety of situations. For the moment we may subdivide its applications as follows.

1. We sometimes have to know whether a given sample is normally distributed before we can apply a certain test to it. To test whether a given sample is normally distributed we have to calculate expected frequencies for a normal curve of the same mean and standard deviation using the table

of areas of the normal curve. In this book we shall employ only approximate graphic methods for testing normality. These are featured in the next section.

2. Knowing whether a sample is normally distributed may confirm or reject certain underlying hypotheses about the nature of the factors affecting the phenomenon studied. This is related to the conditions making for normality in a frequency distribution, discussed in Section 5.2. Thus if we find a given variable to be normally distributed we have no reason for rejecting the hypothesis that the causal factors affecting the variable are additive and independent and of equal variance. On the other hand, when we find departure from normality, this may indicate certain forces, such as selection, affecting the variable under study. For instance, bimodality may indicate a mixture of samples from two populations. Skewness of milk yield data may reflect the fact that these were records of selected cows and substandard milk cows were not included in the record.

3. If we assume a given distribution to be normal, we may make predictions and tests of given hypotheses based upon this assumption. An example of such an application is shown below.

You will recall the birth weights of male Chinese children, illustrated in Box 3.2. The mean of this sample of 9,465 birth weights is 109.9 oz. and its standard deviation is 13.593 oz. Sampling at random from the birth records of this population, what is the chance of obtaining a birth weight of 151 oz. or heavier? Such a birth weight is considerably above the mean of our sample, the difference being $151 - 109.9 = 41.1$ oz. However, we cannot consult the table of areas of the normal curve with a difference in ounces. We must *standardize* the difference—that is, divide it by the standard deviation to convert it into a standard deviate. When we divide the difference by the standard deviation we obtain $41.1/13.593 = 3.02$. This means that a birth weight of 151 oz. is 3.02 standard deviation units greater than the mean. Assuming that the birth weights are normally distributed we may consult the table of areas of the normal curve (Table **II**), where we find a value of 0.4987 for 3.02 standard deviations. This means that 49.87% of the area of the curve lies between the mean and a point 3.02 standard deviations from it. Conversely, 0.0013 or 0.13% of the area lies beyond 3.02 standard deviation units above the mean. Thus, assuming a normal distribution of birth weights, only 0.13% or 13 out of 10,000 of the infants would have a birth weight of 151 oz. *or farther* from the mean. It is quite improbable that a single sampled item from that population would deviate by so much from the mean, and if a random sample of one weight was obtained from the records of an unspecified population we might therefore be justified in doubting whether the observation did in fact come from the population known to us.

The above probability was calculated from one tail of the distribution. We found the probability of an individual being 3.02 standard deviations *greater* than the mean. If we have no prior knowledge that the individual will be either heavier or lighter than the mean but are merely concerned with how different it is from the population mean, an appropriate question would be: assuming that the individual belongs to the population, what is the probability of observing a birth weight of an individual as deviant from the mean in either direction? That probability must be computed by using both tails of the distribution. The previous probability can be simply doubled, since the normal curve is symmetrical. Thus, $2 \times 0.0013 = 0.0026$. This, too, is so small that we would conclude that a birth weight as deviant as 151 oz. is unlikely to have come from the population represented by our sample of male Chinese children.

We can learn one more important point from this example. Our assumption has been that the birth weights are normally distributed. Inspection of the frequency distribution in Box 3.2, however, shows clearly that the distribution is asymmetrical, tapering to the right. Though there are eight classes above the mean class, there are only six classes below the mean class. In view of this asymmetry, conclusions about one tail of the distribution would not necessarily pertain to the second tail. We calculated that 0.13% of the items would be found beyond 3.02 standard deviations above the mean, which corresponds to 151 oz. In fact our sample contains 20 items $(14 + 5 + 1)$ beyond the 147.5-oz. class, the upper limit of which is 151.5 oz., almost the same as the single birth weight. However, 20 items of the 9465 of the sample is approximately 0.21%, more than the 0.13% expected from the normal frequency distribution. Although it would still be improbable to find a single birth weight as heavy as 151 oz. in the sample, conclusions based on the assumption of normality might be in error if the exact probability were critical for a given test. Our statistical conclusions are only as valid as our assumptions about the data.

5.5 Departures from normality and graphic methods

In many cases an observed frequency distribution will depart obviously from normality. We shall emphasize two types of departure from normality. One is *skewness*, which is another name for asymmetry; skewness means that one tail of the curve is drawn out more than the other. In such curves the mean and the median will not coincide. Curves are called skewed to the right or left, depending upon whether the right or left tails are drawn out. The other type of departure from normality is *kurtosis*, or "peakedness" of a curve. A *leptokurtic* curve has more items near the mean and at the tails, with fewer items in the intermediate regions relative to a normal distribution with the same mean and variance. A *platykurtic* curve has fewer items at the mean and at the tails than the normal curve but has more items in intermediate regions. A bimodal distribution is an extreme platykurtic distribution.

Graphic methods have been developed that examine the shape of an observed distribution for departures from normality. These methods also permit estimates of the mean and standard deviation of the distribution without computation.

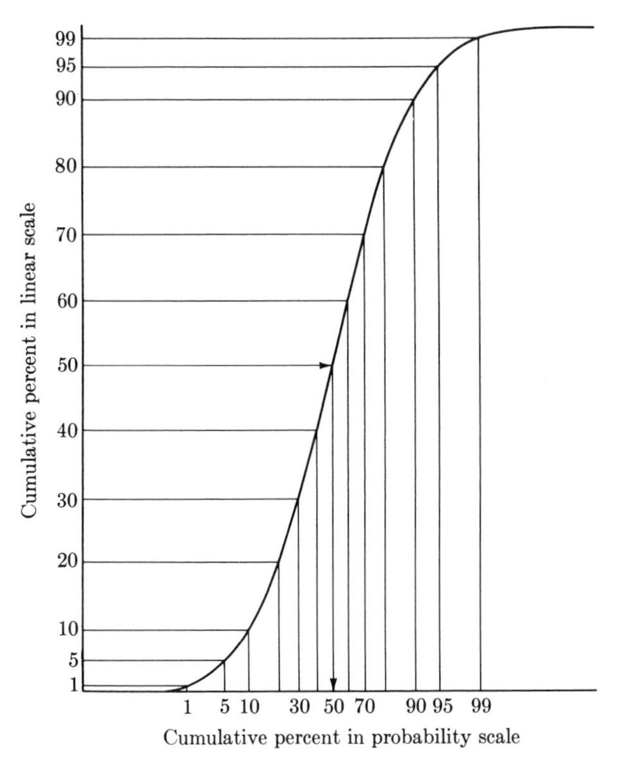

FIGURE 5.5

Transformation of cumulative percentages into normal probability scale.

The graphic methods are based on a cumulative frequency distribution. In Figure 5.4 we saw that a normal frequency distribution graphed in cumulative fashion describes an **S**-shaped curve, called a sigmoid curve. In Figure 5.5 the ordinate of the sigmoid curve is given as relative frequencies expressed as percentages. The slope of the cumulative curve reflects changes in height of the frequency distribution on which it is based. Thus the steep middle segment of the cumulative normal curve corresponds to the relatively greater height of the normal curve around its mean. The ordinate in Figures 5.4 and 5.5 is in linear scale. Another possible scale is the *normal probability scale* (often simply called *probability scale*), which can be generated by dropping perpendiculars from the cumulative normal curve, corresponding to given percentages on the ordinate, to the abscissa (as shown in Figure 5.5). The scale represented by the abscissa compensates for the nonlinearity of the cumulative normal curve. It contracts the scale around the median and expands it at the low and high cumulative percentages. This scale can be found on *arithmetic* or *normal probability graph paper* (or simply *probability graph paper*), which is generally available. Such paper generally has the long

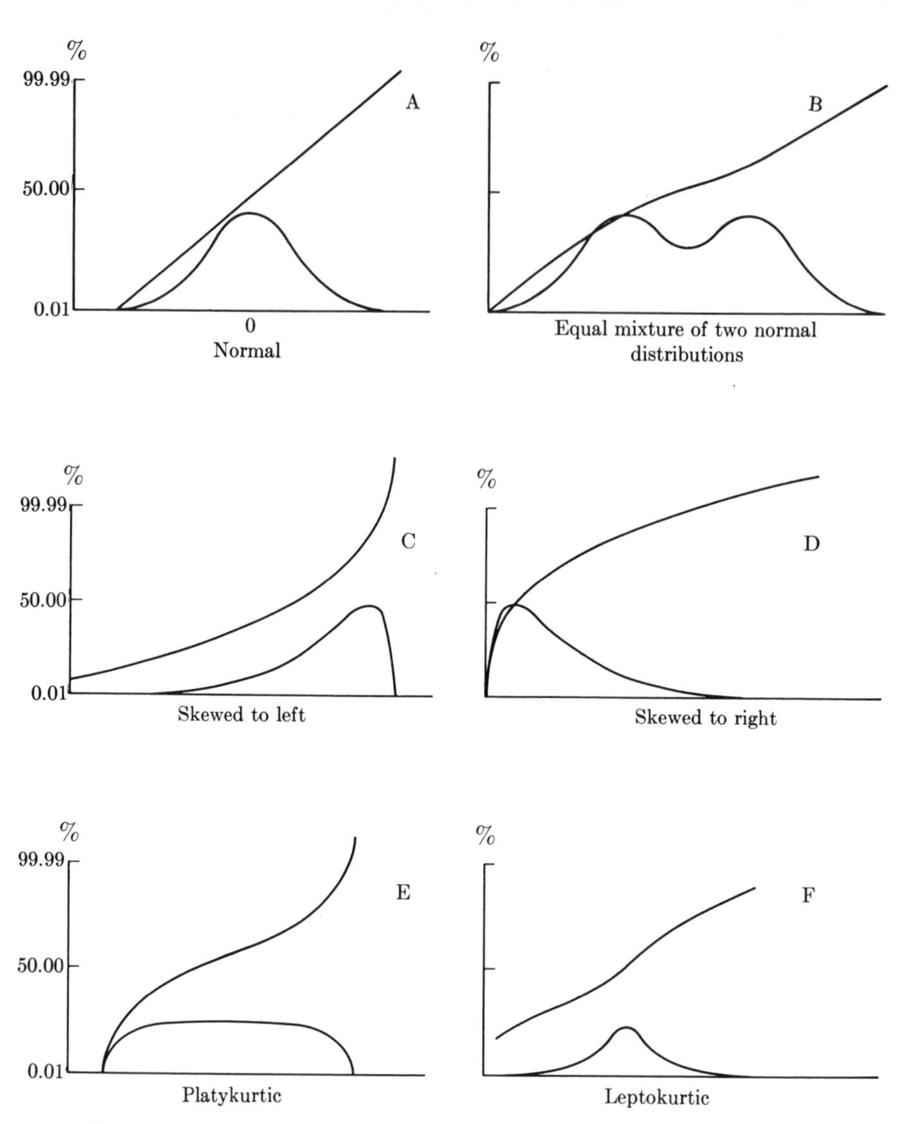

FIGURE 5.6

Examples of some frequency distributions with their cumulative distributions plotted with the ordinate in normal probability scale. (See Box 5.1 for explanation.)

edge graduated in probability scale, while the short edge is in linear scale. Note that there are no 0% or 100% points on the ordinate. This is not possible since the normal frequency distribution extends from negative to positive infinity, and however long we made our line we would never reach the limiting values of 0% and 100%. If we graph a cumulative normal distribution with the ordinate in normal probability scale, it will lie exactly on a straight line. Figure 5.6A shows such a graph drawn on probability paper. Box 5.1 shows

BOX 5.1

Graphic test for normality of a frequency distribution and estimate of mean and standard deviation. Use of arithmetic probability paper.

Birth weights of male Chinese in ounces, from Box 3.2.

(1) Class mark Y	*(2)* Upper class limit	*(3)* f	*(4)* Cumulative frequencies F	*(5)* Percent cumulative frequencies
59.5	63.5	2	2	0.02
67.5	71.5	6	8	0.08
75.5	79.5	39	47	0.50
83.5	87.5	385	432	4.6
91.5	95.5	888	1320	13.9
99.5	103.5	1729	3049	32.2
107.5	111.5	2240	5289	55.9
115.5	119.5	2007	7296	77.1
123.5	127.5	1233	8529	90.1
131.5	135.5	641	9170	96.9
139.5	143.5	201	9371	99.0
147.5	151.5	74	9445	99.79
155.5	159.5	14	9459	99.94
163.5	167.5	5	9464	99.99
171.5	175.5	1	9465	100.0
		9465		

Computational steps

1. Prepare a frequency distribution as shown in columns (1), (2), and (3).
2. Form a cumulative frequency distribution as shown in column (4). It is obtained by successive summation of the frequency values. In column (5) express the cumulative frequencies as percentages of total sample size n, which is 9465 in this example. Thus they are the values of column (4) divided by 9465.
3. Graph the upper class limit of each class along the abscissa (in linear scale) against percent cumulative frequency along the ordinate (in probability scale) on normal probability paper (see Figure 5.7). A straight line is fitted to the points by eye, preferably using a transparent plastic ruler, which permits all the points to be seen as the line is drawn. In drawing the line, most weight should be given to the points between cumulative frequencies of 25% to 75%. This is because a difference of a single item may make appreciable changes in the percentages at the tails. We notice that the upper frequencies deviate to the right of the straight line. This is typical of data that are skewed to the right (see Figure 5.6D).
4. Such a graph permits the rapid estimation of the mean and standard deviation of a sample. The mean is approximated by a graphic estimation of the median. The more normal the distribution is, the closer will the mean be to the median.

BOX 5.1 continued

The median is estimated by dropping a perpendicular from the intersection of the 50% point on the ordinate and the cumulative frequency curve to the abscissa (see Figure 5.7). The estimate of the mean of 110.7 oz. is quite close to the computed mean of 109.9 oz.

5. An estimate of the standard deviation is obtained by dropping similar perpendiculars from the intersections of the 15.9% and the 84.1% points with the cumulative curve, respectively. These points enclose the portion of a normal curve represented by $\mu \pm \sigma$. By measuring the difference between these perpendiculars and dividing this by 2, we obtain an estimate of one standard deviation. In this instance the estimate is $s = 13.6$, since the difference is 27.2 oz. divided by 2. This is a close approximation to the computed value of 13.59 oz.

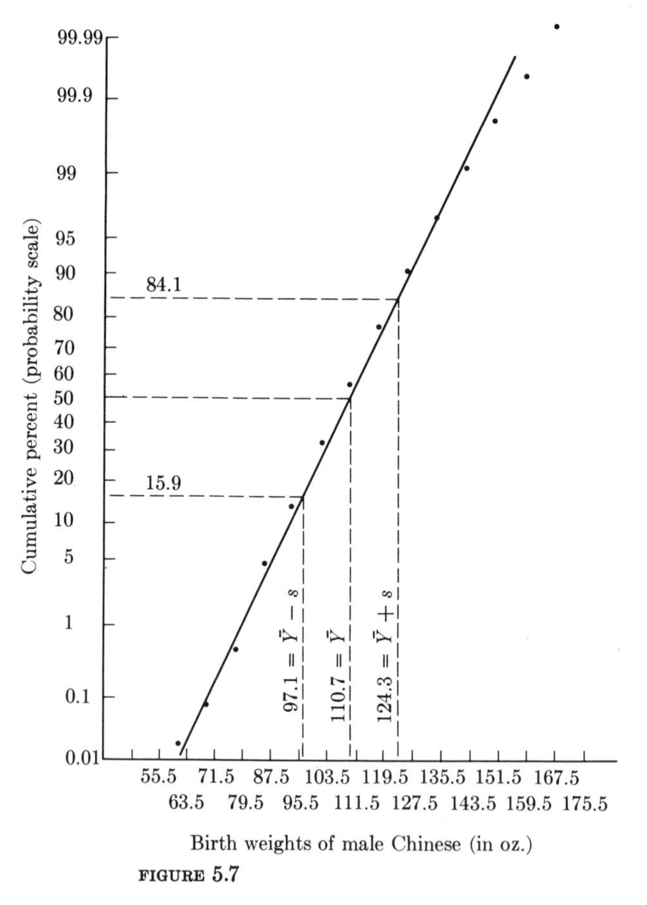

FIGURE 5.7

Graphic analysis of data from Box 5.1.

you how to use probability paper to examine a frequency distribution for normality and to obtain graphic estimates of its mean and standard deviation. The method works best for fairly large samples ($n > 50$). The method does not permit the plotting of the last cumulative frequency, 100%, since it corresponds to an infinite distance from the mean. If there is interest in plotting all observations, one can plot, instead of cumulative frequencies F, the quantity $F - \frac{1}{2}$ expressed as a percentage of n.

Figure 5.6 shows a series of frequency distributions variously departing from normality. They are graphed as ordinary frequency distributions with density on a linear scale (ordinate not shown) and as cumulative distributions on probability paper. They are useful as guide lines when examining the distributions of data on probability paper.

Exercises 5

5.1 Perform the following operations on the data of Exercise 2.4. (a) If you have not already done so, make a frequency distribution from the data and graph the results in the form of a histogram. (b) Compute the expected frequencies for each of the classes based on a normal distribution with $\mu = \bar{Y}$ and $\sigma = s$. (c) Graph the expected frequencies in the form of a histogram and compare them with the observed frequencies. (d) Comment on the degree of agreement between observed and expected frequencies.

5.2 Carry out the operations listed in Exercise 5.1 on the transformed data generated in Exercise 2.6.

5.3 Assume you know that the petal length of a population of plants of species X is normally distributed with a mean of $\mu = 3.2$ cm and a standard deviation of $\sigma = 1.8$. What proportion of the population would be expected to have a petal length (a) greater than 4.5 cm? (b) greater than 1.78 cm? (c) between 2.9 and 3.6 cm? ANS. (a) = 0.2353, (b) = 0.7845, and (c) = 0.154.

5.4 Perform a graphic analysis of the butterfat data given in Exercise 3.3, using probability paper. In addition, plot the data on probability paper with the abscissa in logarithmic units. Compare the results of the two analyses.

5.5 Assume that traits A and B are independent and normally distributed with parameters $\mu_A = 28.6$, $\sigma_A = 4.8$, $\mu_B = 16.2$, and $\sigma_B = 4.1$. You sample two individuals at random. What is the probability of obtaining samples where both individuals are less than 20 for the two traits? What is the probability that at least one of the individuals is greater than 30 for trait B?

5.6 Using the information given in Box 3.2, what is the probability of obtaining an individual with a negative birth weight? What is this probability if we assume that birth weights are normally distributed?

6

Estimation and Hypothesis Testing

In this chapter we provide answers to two fundamental statistical questions that every biologist must answer repeatedly in the course of his work: (1) how reliable are the results I have obtained? and (2) how probable is it that the differences between observed results and those expected on the basis of an hypothesis have been produced by chance alone? The first question, about reliability, is answered through the setting of confidence limits to sample statistics. The second question leads into hypothesis testing. Both subjects belong to the field of statistical inference. The subject matter in this chapter is fundamental to an understanding of any of the subsequent chapters.

In Section 6.1 we shall first of all consider the form of the distribution of means and their variance. In the next section (6.2) we examine the distributions and variances of statistics other than the mean. This brings us to the general subject of standard errors, which are statistics measuring the reliability of an estimate. Confidence limits provide bounds to our estimates of population parameters. We develop the idea of a confidence limit in Section 6.3 and show its application to samples where the true standard deviation is known. However, one usually deals with small, more or less normally distributed samples with unknown standard deviations, in which case the t-distribution must be used. We shall introduce the t-distribution in Section 6.4. The application of t to the computation of confidence limits for statistics of small samples with unknown population standard deviations is shown in Section 6.5. Another important distribution (chi-square) is explained in Section 6.6 and then applied to setting confidence limits for the variance (Section 6.7). The theory of hypothesis testing is introduced in Section 6.8 and applied to a variety of cases exhibiting the normal or t-distributions (Section 6.9). Finally, Section 6.10 illustrates hypothesis testing for variances by means of the chi-square distribution.

6.1 Distribution and variance of means

We commence our study of the distribution and variance of means with a sampling experiment.

Experiment 6.1. You were asked to retain from Experiment 5.1 the means of the seven samples of 5 housefly wing lengths and the seven similar means of milk yields. We can collect these means from every student in a class, possibly adding them to the sampling results of previous classes, and construct a frequency distribution of these means. For each variable we can also obtain the mean of the seven means, which is a mean of a sample of 35 items. Here again we shall make a frequency distribution of these means, although it takes a considerable number of samplers to accumulate a sufficient number of samples of 35 items for a meaningful frequency distribution.

In Table 6.1 we show a frequency distribution of 1400 means of samples of 5 housefly wing lengths. Consider only columns (1) and (3) for the time being. Actually these samples were not obtained by biostatistics classes but by a digital computer, enabling us to collect these values in a very short time. Their mean and standard deviation are given at the foot of the table. These values are plotted on probability paper in Figure 6.1. Note that the distribution appears quite normal, as does that of the means based on 200 samples of 35 wing lengths shown in Figure 6.2. This illustrates an important theorem. *The means of samples from a normally distributed population are themselves normally distributed regardless of sample size n.* Thus we note that the means of samples from the normally distributed housefly wing lengths are normally distributed whether they are based on 5 or 35 individual readings.

Similarly obtained distributions of means of the heavily skewed milk yields (Figures 6.3 and 6.4) appear to be close to normal distributions. However, the means based on five milk yields (Figure 6.3) do not agree with the

TABLE 6.1

Frequency distribution of means of 1400 random samples of 5 housefly wing lengths. (Data from Table 5.1.) Class marks chosen to give intervals of $\frac{1}{2}\sigma_{\bar{Y}}$ to each side of the parametric mean μ.

	(1) Class mark Y (in mm $\times 10^{-1}$)	(2) Class mark (in $\sigma_{\bar{Y}}$ units)	(3) f
	39.832	$-3\frac{1}{4}$	1
	40.704	$-2\frac{3}{4}$	11
	41.576	$-2\frac{1}{4}$	19
	42.448	$-1\frac{3}{4}$	64
	43.320	$-1\frac{1}{4}$	128
	44.192	$-\frac{3}{4}$	247
	45.064	$-\frac{1}{4}$	226
$\mu = 45.5 \rightarrow$			
	45.936	$\frac{1}{4}$	259
	46.808	$\frac{3}{4}$	231
	47.680	$1\frac{1}{4}$	121
	48.552	$1\frac{3}{4}$	61
	49.424	$2\frac{1}{4}$	23
	50.296	$2\frac{3}{4}$	6
	51.168	$3\frac{1}{4}$	3
			1400
	$\bar{Y} = 45.480$	$s = 1.778$	$\sigma_{\bar{Y}} = 1.744$

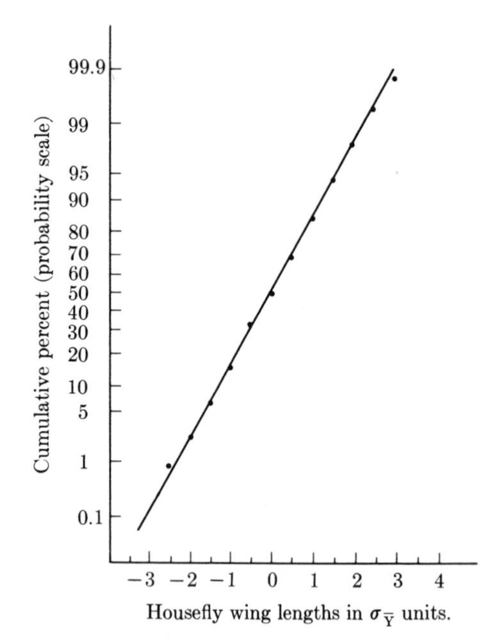

FIGURE 6.1

Graphic analysis of data from Table 6.1.

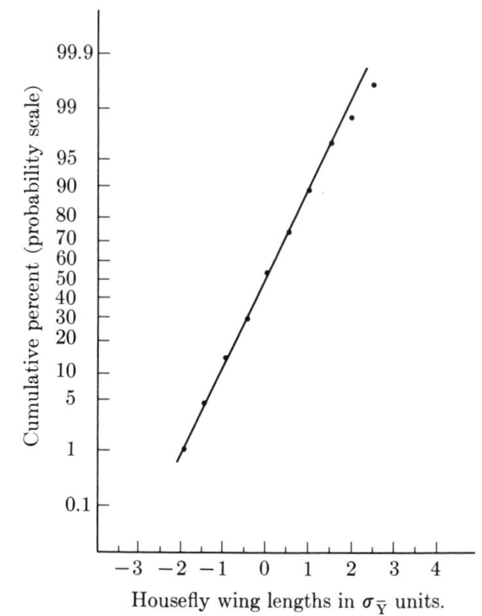

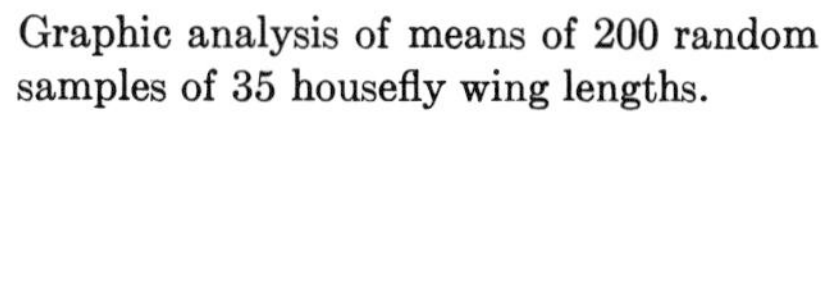

FIGURE 6.2

Graphic analysis of means of 200 random samples of 35 housefly wing lengths.

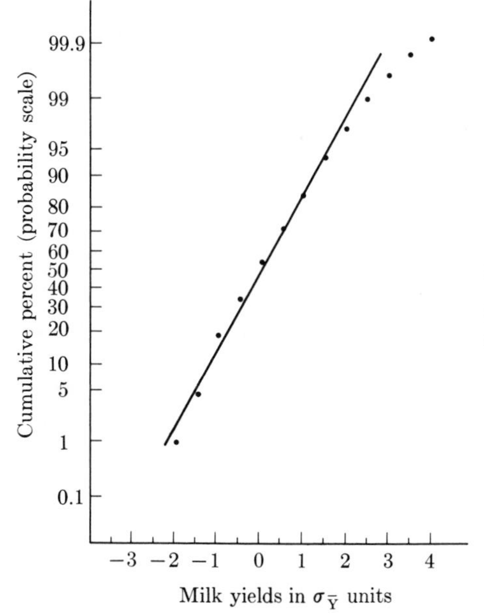

FIGURE 6.3

Graphic analysis of means of 1400 random samples of 5 milk yields.

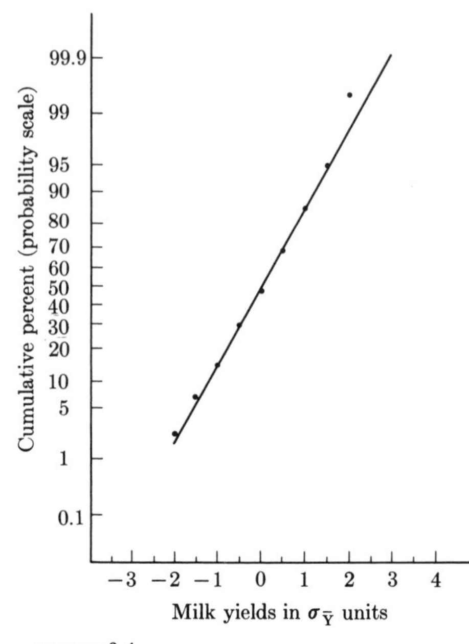

FIGURE 6.4

Graphic analysis of means of 200 random samples of 35 milk yields.

normal nearly as well as do the means of 35 items (Figure 6.4). This illustrates another theorem of fundamental importance in statistics. *As sample size increases, the means of samples drawn from a population of any distribution will approach the normal distribution.* This theorem, when rigorously stated (about sampling from populations with finite variances), is known as the *central limit theorem.* The importance of this theorem is that it permits us to use the normal distribution to make statistical inferences about means of populations in which the items are not at all normally distributed.

The next fact of importance is that the range of the means is considerably less than that of the original items. Thus the wing length means range from 39.4 to 51.6 in samples of 5 and from 43.9 to 47.4 in samples of 35, but the individual wing lengths range from 36 to 55. The milk yield means range from 54.2 to 89.0 in samples of 5 and from 61.9 to 71.3 in samples of 35, but the individual milk yields range from 51 to 98. Not only do means show less scatter than the items upon which they are based (an easily understood phenomenon if you give some thought to it), but the range of the distribution of the means diminishes as the sample size upon which the means are based increases.

The differences in ranges are reflected in differences in the standard deviation of these distributions. If we calculate the standard deviations of the means in the four distributions under consideration, we obtain the following values:

	Observed standard deviations of distributions of means	
	$n = 5$	$n = 35$
Wing lengths	1.778	0.584
Milk yields	5.040	1.799

Note that the standard deviations of the sample means based on 35 items are considerably less than those based on 5 items. This is also intuitively obvious. Means based on large samples should be close to the parametric mean and means based on large samples will not vary as much as will means based on small samples. The variance of means is therefore partly a function of the sample size on which the means are based. It is also a function of the variance of the items in the samples. Thus, in the text table above, the means of milk yields have a much greater standard deviation than means of wing lengths based on comparable sample size simply because the standard deviation of the individual milk yields (11.1597) is considerably greater than that of individual wing lengths (3.90).

Mathematical statisticians have worked out the expected value of the variance of sample means. By *expected value* we mean the average value to be obtained by infinitely repeated sampling. Thus if we were to take samples of a means of n items repeatedly and were to calculate the variance of these a means each time, the average of these variances would be the expected value. The expected value of the variance of means based on n items is

$$\sigma_{\bar{Y}}^2 = \frac{\sigma^2}{n} \tag{6.1}$$

Consequently the expected standard deviation of means is

$$\sigma_{\bar{Y}} = \frac{\sigma}{\sqrt{n}} \tag{6.2}$$

From this formula it is clear that the standard deviation of means is a function of the standard deviation of items as well as of sample size of means. The greater the sample size, the smaller will be the standard deviation of means. In fact, as sample size increases to a very large number, the standard deviation of means becomes vanishingly small. This makes good sense. Very large sample sizes, averaging many observations, should yield estimates of means less variable than those based on a few items.

When working with samples from a population we do not, of course, know its parametric standard deviation σ, but can only obtain a sample estimate, s, of the latter. Also, we would be unlikely to have numerous samples of size n from which to compute the standard deviation of means directly. Customarily, we therefore have to estimate the standard deviation of means from a single sample by using Expression (6.2), substituting s for σ:

$$s_{\bar{Y}} = \frac{s}{\sqrt{n}} \tag{6.3}$$

Thus we obtain, from the standard deviation of a single sample, an estimate of the standard deviation of means we would expect were we to obtain a collection of means based on equal-sized samples of n items from the same population. As we shall see, this estimate of the standard deviation of a mean is a very important and frequently used statistic.

Table 6.2 illustrates some estimates of the standard deviations of means that might be obtained from random samples of the two populations that we have been discussing. The means of 5 samples of wing lengths based on 5 individuals ranged from 43.6 to 46.8, their standard deviations from 1.095 to 4.827, and the estimate of standard deviation of the means from 0.490 to 2.159. Ranges for the other categories of samples in Table 6.2 similarly included the parametric values of these statistics. The estimates of the standard deviations of the means of the milk yields cluster around the expected value, since they are not dependent on normality of the variates. However, in a particular sample in which by chance the sample standard deviation is a poor estimate of the population standard deviation (as in the second sample of 5 milk yields), the estimate of the standard deviation of means is equally wide of the mark.

There is one point of difference between the standard deviation of items and the standard deviation of sample means. If we estimate a population standard deviation through the standard deviation of a sample, the magnitude of the estimate will not change as we increase our sample size. We may expect that the estimate will improve and will approach the true standard deviation of the population. However, its order of magnitude will be the same, whether the sample is based on 3, 30, or 3000 individuals. This can be clearly seen in Table 6.2. The values of s are closer to σ in the samples based on $n = 35$ than in samples of $n = 5$. Yet the general magnitude is the same in both instances. The standard deviation of means, however, decreases as sample size increases, as is obvious from Expression (6.3). Thus, means based on 3000 items will have a standard deviation only one-tenth that of means based on 30 items. This is obvious from

$$\frac{s}{\sqrt{3000}} = \frac{s}{\sqrt{30} \times \sqrt{100}} = \frac{s}{\sqrt{30} \times 10}$$

Since our sampled standard deviations show considerable variation (see Table 6.2), the estimates of the standard deviations of the means also vary accordingly. A poor estimate of σ will yield a poor estimate of $\sigma_{\bar{Y}}$.

TABLE 6.2

Means, standard deviations, and standard deviations of means (standard errors) of five random samples of 5 and 35 housefly wing lengths and Jersey cow milk yields, respectively. (Data from Table 5.1.) Parametric values for the statistics are given in the sixth line of each category.

	(1) $\bar{Y}$	(2) s	(3) $s_{\bar{Y}}$
	Wing lengths		
	45.8	1.095	0.490
	45.6	3.209	1.435
$n = 5$	43.6	4.827	2.159
	44.8	4.764	2.131
	46.8	1.095	0.490
	$\mu = 45.5$	$\sigma = 3.90$	$\sigma_{\bar{Y}} = 1.744$
	45.37	3.812	0.644
	45.00	3.850	0.651
$n = 35$	45.74	3.576	0.604
	45.29	4.198	0.710
	45.91	3.958	0.669
	$\mu = 45.5$	$\sigma = 3.90$	$\sigma_{\bar{Y}} = 0.659$
	Milk yields		
	66.0	6.205	2.775
	61.6	4.278	1.913
$n = 5$	67.6	16.072	7.188
	65.0	14.195	6.348
	62.2	5.215	2.332
	$\mu = 66.61$	$\sigma = 11.160$	$\sigma_{\bar{Y}} = 4.991$
	65.429	11.003	1.860
	64.971	11.221	1.897
$n = 35$	66.543	9.978	1.687
	64.400	9.001	1.521
	68.914	12.415	2.099
	$\mu = 66.61$	$\sigma = 11.160$	$\sigma_{\bar{Y}} = 1.886$

6.2 Distribution and variance of other statistics

Just as we obtained a mean and a standard deviation from each sample of the wing lengths and milk yields, so we could also have obtained other statistics from each sample, such as a variance, a median, or a coefficient of variation. After repeated sampling and computation we would have frequency distributions for these statistics and would be able to compute their standard deviations just as we did for the frequency distribution of means. In many cases the statistics are normally distributed, as was true for the means. In

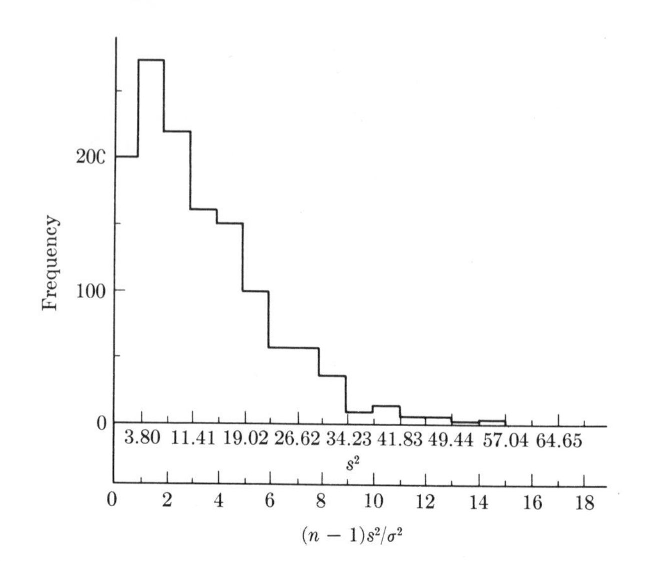

FIGURE 6.5

Histogram of variances based on 1400 samples of 5 housefly wing lengths from Table 5.1. Abscissa is given in terms of s^2 and $(n - 1)s^2/\sigma^2$.

others the statistics will be distributed normally only if they are based on samples from a normally distributed population, or if they are based on large samples, or if both these conditions hold. In some instances, as in variances, their distribution is never normal. This is illustrated by Figure 6.5, which shows a frequency distribution of the variances from the 1400 samples of housefly wing lengths. We notice that the distribution is strongly skewed to the right, which is characteristic of the distribution of variances.

Standard deviations of various statistics are generally known as *standard errors*. Beginners sometimes get confused by an imagined distinction between standard deviations and standard errors. The standard error of a statistic such as the mean (or CV), is the standard deviation of a distribution of means (or CV's) for samples of a given sample size n. Thus the terms standard error and standard deviation are used synonymously with the following exception: it is not customary to use standard error as a synonym of standard deviation of items in a sample or population. Standard error or standard deviation has to be qualified by referring to a given statistic, such as the standard deviation of CV, which is the same as the standard error of CV. Used without any qualification the term "standard error" conventionally implies the standard error of the mean. "Standard deviation" used without qualification generally means standard deviation of items in a sample or population. Thus when you read that means, standard deviations, standard errors, and coefficients of variation are shown in a table, this signifies that arithmetic means, standard deviations of items in samples, standard deviations of their means (= standard errors of means), as well as coefficients of variation, are displayed. The following summary of terms may be helpful:

Standard deviation $= s = \sqrt{\sum y^2/(n-1)}$
Standard deviation of a statistic St
$=$ standard error of a statistic $St = s_{St}$
Standard error $=$ standard error of a mean
$=$ standard deviation of a mean $= s_{\bar{Y}}$

Standard errors are usually not obtained from a frequency distribution by repeated sampling but are estimated from only a single sample and represent the expected standard deviation of the statistic in case a large number of such samples had been obtained. You will remember that we estimated the standard error of a distribution of means from a single sample in this manner in the previous section.

Box 6.1 lists the standard errors of four common statistics. Column (1) lists the statistic whose standard error is described; column (2) shows the formula for the estimated standard error; column (3) gives the degrees of freedom on which the standard error is based (their use is explained in Section 6.5); column (4) provides comments on the range of application of the standard error. The uses of these standard errors will be illustrated in subsequent sections.

BOX 6.1

Standard errors for common statistics.

	(1) *Statistic*	*(2)* *Estimate of standard error*	*(3)* *df*	*(4)* *Comments on applicability*
1	$\bar{Y}$	$s_{\bar{Y}} = \frac{s}{\sqrt{n}} = \frac{s_Y}{\sqrt{n}} = \sqrt{\frac{s^2}{n}}$	$n - 1$	True for any population with finite variance
2	Median	$s_{\text{med}} = (1.2533)s_{\bar{Y}}$	$n - 1$	Large samples from normal populations
3	s	$s_s = (0.7071068)\frac{s}{\sqrt{n}}$	$n - 1$	Samples from normal populations ($n > 15$)
4	CV	$s_{CV} = \frac{CV}{\sqrt{2n}}\sqrt{1 + 2\left(\frac{CV}{100}\right)^2}$	$n - 1$	Samples from normal populations
		$s_{CV} \approx \frac{CV}{\sqrt{2n}}$	$n - 1$	Used when $CV < 15$

6.3 Introduction to confidence limits

The various sample statistics we have been obtaining, such as means or standard deviations, are estimates of population parameters μ, or σ, respectively. So far we have not discussed the reliability of these estimates. We first of all wish to know whether the sample statistics are *unbiased estimators* of the population parameters, as discussed in Section 3.7. But knowing, for example, that $\bar{Y}$ is an unbiased estimate of μ is not enough. We would like to find out how reliable a measure of μ it is. The true values of the parameters will almost always remain unknown and we commonly estimate reliability of a sample statistic by setting confidence limits to it.

To begin our discussion of this topic, let us start with the unusual case of a population whose parametric mean and standard deviation are known to be μ and σ, respectively. The mean of a sample of n items is symbolized by $\bar{Y}$. The expected standard error of the mean is $\sigma/\sqrt{n}$. As we have seen, the sample means will be normally distributed. Therefore, from Section 5.3, the region from $1.96\sigma/\sqrt{n}$ below μ to $1.96\sigma/\sqrt{n}$ above μ includes 95% of the sample means of size n. Another way of stating this is to consider the ratio $(\bar{Y} - \mu)/(\sigma/\sqrt{n})$. This is the standard deviate of a sample mean from the parametric mean. Since they are normally distributed, 95% of such standard deviates will lie between -1.96 and $+1.96$. We can express this statement symbolically as follows:

$$P\left\{-1.96 \leq \frac{\bar{Y} - \mu}{\sigma/\sqrt{n}} \leq +1.96\right\} = 0.95$$

This means that the probability P that the sample means $\bar{Y}$ will differ by no more than 1.96 standard errors $\sigma/\sqrt{n}$ from the parametric mean μ equals 0.95. The expression between the brackets is an inequality, all terms of which can be multiplied by $\sigma/\sqrt{n}$ to yield

$$\{-1.96\sigma/\sqrt{n} \leq (\bar{Y} - \mu) \leq +1.96\sigma/\sqrt{n}\}$$

We can rewrite this expression as

$$\{-1.96\sigma/\sqrt{n} \leq (\mu - \bar{Y}) \leq +1.96\sigma/\sqrt{n}\}$$

because $-a \leq b \leq a$ implies $a \geq -b \geq -a$, which can be written as $-a \leq -b \leq a$. And finally we can transfer $-\bar{Y}$ across the inequality signs, just as in an equation it could be transferred across the equal sign. This yields the final desired expression:

$$P\left\{\bar{Y} - \frac{1.96\sigma}{\sqrt{n}} \leq \mu \leq \bar{Y} + \frac{1.96\sigma}{\sqrt{n}}\right\} = 0.95 \tag{6.4}$$

or

$$P\{\bar{Y} - 1.96\sigma_{\bar{Y}} \leq \mu \leq \bar{Y} + 1.96\sigma_{\bar{Y}}\} = 0.95 \tag{6.4a}$$

This means that the probability P that the term $\bar{Y} - 1.96\sigma_{\bar{Y}}$ is less than or equal to the parametric mean μ and that the term $\bar{Y} + 1.96\sigma_{\bar{Y}}$ is greater than or equal to μ is 0.95. We shall call the two terms, $\bar{Y} - 1.96\sigma_{\bar{Y}}$ and $\bar{Y} + 1.96\sigma_{\bar{Y}}$, L_1 and L_2, respectively, the lower and upper 95% *confidence limits* of the mean.

Another way of stating the relationship implied by Expression (6.4a) is that if we repeatedly obtained samples of size n from the population and constructed these limits for each, we could expect 95% of the intervals between these limits to contain the true mean, and only 5% of the intervals would miss μ. The interval from L_1 to L_2 is called a *confidence interval.*

If you were not satisfied to have the confidence interval contain the true mean only 95 times out of 100, you might employ 2.576 as a coefficient in place of 1.960. You may remember that 99% of the area of the normal curve lies between $\mu \pm 2.576\sigma$. Thus, to calculate 99% confidence limits, compute the two quantities $L_1 = \bar{Y} - 2.576\sigma/\sqrt{n}$ and $L_2 = \bar{Y} + 2.576\sigma/\sqrt{n}$ as lower and upper confidence limits, respectively. In this case 99 out of 100 confidence intervals obtained in repeated sampling would contain the true mean. The new confidence interval is wider than the 95% interval (since we have multiplied by a greater coefficient). If you were still not satisfied with the reliability of the confidence limit you could increase it, multiplying the standard error of the mean by 3.291 to obtain 99.9% confidence limits. This value (or 3.890 for 99.99% limits) could be found by inverse interpolation in a more extensive table of areas of the normal curve. The new coefficient would widen the interval further. Notice that you can construct confidence intervals that will be expected to contain μ an increasingly greater percentage of the time. First you would expect to be right 95 times out of 100, then 99 times out of 100, finally 9999 times out of 10,000. But as your confidence increases your statement becomes vaguer and vaguer, since the confidence interval lengthens. Let us examine this by way of an actual sample.

We obtain a sample of 35 housefly wing lengths from the population of Table 5.1 with known mean ($\mu = 45.5$) and standard deviation ($\sigma = 3.90$). Let us assume that the sample mean is 44.8. We can expect the standard deviation of means based on samples of 35 items to be $\sigma_{\bar{Y}} = \sigma/\sqrt{n} = 3.90/\sqrt{35} = 0.6592$. We compute confidence limits as follows:

$$\text{the lower limit is } L_1 = 44.8 - (1.960)(0.6592) = 43.51$$
$$\text{the upper limit is } L_2 = 44.8 + (1.960)(0.6592) = 46.09$$

Remember that this is an unusual case in which we happen to know the true mean of the population ($\mu = 45.5$) and hence we know that the confidence limits enclose the mean. We expect 95% of such confidence intervals obtained in repeated sampling to include the parametric mean. We could increase the reliability of these limits by going to 99% confidence intervals, replacing 1.960 in the above expression by 2.576 and obtaining $L_1 = 43.10$ and $L_2 = 46.50$. We could have greater confidence that our interval covers the mean, but we could be much less certain about the true value of the mean because of the wider limits. By increasing the degree of confidence still further, say to 99.99%, we could be virtually certain that our confidence limits ($L_1 = 42.24$, $L_2 = 47.36$) contain the population mean, but the bounds enclosing the mean are now so wide as to make our prediction far less useful than previously.

Experiment 6.2. For the seven samples of 5 housefly wing lengths and the seven similar samples of milk yields last worked with in Experiment 6.1 (Section 6.1), compute 95% confidence limits to the parametric mean for each sample and for the total sample based on 35 items. Base the standard errors of the means on the parametric standard deviations of these populations (housefly wing lengths $\sigma = 3.90$, milk yields $\sigma = 11.1597$). Record how many in each of the four classes of confidence limits (wing lengths and milk yields, $n = 5$ and $n = 35$) were correct—that is, contained the parametric mean of the population. Pool your results with those of other class members.

We tried the experiment on a computer for the 200 samples of 35 wing lengths each, computing confidence limits of the parametric mean employing the parametric standard error of the mean, $\sigma_{\bar{Y}} = 0.6592$. Of the 200 confidence intervals plotted parallel to the ordinate, 194 (97.0%) cross the parametric mean of the population.

To reduce the width of the confidence interval we have to reduce the standard error of the mean. Since $\sigma_{\bar{Y}} = \sigma/\sqrt{n}$, this can only be done by reducing the standard deviation of the items or by increasing the sample size. The first of these alternatives is frequently not available. If we are sampling from a population in nature we ordinarily have no way of reducing its standard deviation. However, in many experimental procedures we may be able to reduce the variance of the data. For example, if we are studying heart weight in rats and find that its variance is rather large, we might be able to reduce this variance by taking rats of only one age group, in which the variation of heart weight would be considerably less. Thus, by controlling one of the variables of the experiment, the variance of the response variable, heart weight, is reduced. Similarly, by keeping temperature or other environmental variables constant in a procedure, we can frequently reduce the variance of our response variable and hence obtain more precise estimates of population parameters.

A more common way to reduce the standard error is by increasing sample size. It is obvious from Expression (6.2) that as n increases the standard error decreases; hence as n approaches infinity, the standard error and the lengths of confidence intervals approach zero. This ties in with what we have learned before: in samples whose size approaches infinity, the sample mean would approach the parametric mean.

We must guard against a common mistake in expressing the meaning of the confidence limits of a statistic. When we have set lower and upper limits (L_1 and L_2, respectively) to a statistic, we imply that the probability of this interval covering the mean is 0.95 or, expressed in another way, that on the average 95 out of 100 confidence intervals similarly obtained would cover the mean. We *cannot state* that there is a probability of 0.95 that the true mean is contained within a given pair of confidence limits, although this may seem to be saying the same thing. The latter statement is incorrect because the true mean is a parameter; hence it is a fixed value and it is therefore either inside the interval or outside it. It cannot be inside the given interval

95% of the time. It is important, therefore, to learn the correct statement and meaning of confidence limits.

So far we have only considered means based on normally distributed samples with known parametric standard deviations. We can, however, extend the methods just learned to samples from populations with unknown standard deviations but where the population is known to be normally distributed and the samples are large, say $n \geq 100$. In such cases we use the sample standard deviation for computing the standard error of the mean.

However, when the samples are small ($n < 100$) and we lack knowledge of the parametric standard deviation, we must take into consideration the reliability of our sample standard deviation. To do so we must make use of the so-called t or Student's distribution. We shall learn how to set confidence limits employing the t-distribution in Section 6.5. Before that, however, we shall have to become familiar with this distribution in the next section.

6.4 Student's *t*-distribution

The deviations $\bar{Y} - \mu$ of sample means from the parametric mean of a normal distribution are themselves normally distributed. If these deviations are divided by the parametric standard deviation, $(\bar{Y} - \mu)/\sigma_{\bar{Y}}$, they are still normally distributed, with $\mu = 0$ and $\sigma = 1$. Subtracting the constant μ from every $\bar{Y}_i$ is simply an additive code (Section 3.8) and will not change the form of the distribution of sample means, which is normal (Section 6.1). Dividing each deviation by the constant $\sigma_{\bar{Y}}$ reduces the variance to unity, but proportionately so for the entire distribution, so that its shape is not altered and a previously normal distribution remains so.

If, on the other hand, we had calculated the variance s_i^2 of each of the samples and calculated the deviation for each mean $\bar{Y}_i$ as $(\bar{Y}_i - \mu)/s_{\bar{Y}_i}$, where $s_{\bar{Y}_i}$ stands for the estimate of the standard error of the mean of the ith sample, we would have found the distribution of the deviations to be wider and flatter than the normal distribution. This is illustrated in Figure 6.6, which shows the ratio $(\bar{Y}_i - \mu)/s_{\bar{Y}_i}$, for the 1400 samples of five housefly wing lengths of Table 6.1. The new distribution ranges wider than the corresponding normal distribution, because the denominator is the sample standard error rather than the parametric standard error and will sometimes be smaller, at other times greater than expected. This increased variation will be reflected in the greater variance of the ratio $(\bar{Y} - \mu)/s_{\bar{Y}}$. The expected distribution of this ratio is called the t-distribution, also known as "Student's" distribution, named after W. S. Gossett who first described it, publishing under the pseudonym "Student." The t-distribution is a function with a complicated mathematical formula that need not be presented here.

The t-distribution shares with the normal the properties of being symmetric and of extending from negative to positive infinity. However, it differs from the normal in that it assumes different shapes depending on the number

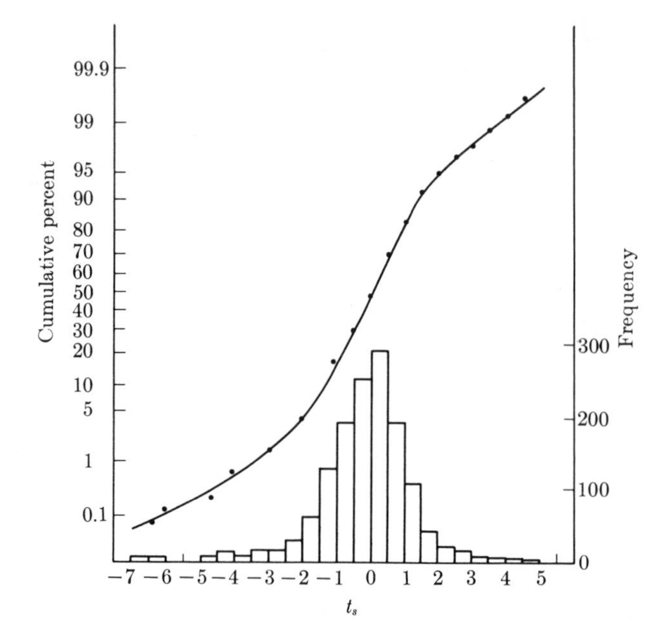

FIGURE 6.6

Distribution of quantity $t_s = (\bar{Y} - \mu)/s_{\bar{Y}}$ along abscissa computed for 1400 samples of 5 housefly wing lengths presented as a histogram and as a cumulative frequency distribution. Right-hand ordinate represents frequencies for the histogram, left-hand ordinate is cumulative frequency in probability scale.

of degrees of freedom. By degrees of freedom is meant the quantity $n - 1$, where n is the sample size upon which a variance has been based. It will be remembered that this quantity $n - 1$ is the divisor in obtaining an unbiased estimate of the variance from a sum of squares. The number of degrees of freedom pertinent to a given Student's distribution is the same as the number of degrees of freedom of the standard deviation in the ratio $(\bar{Y} - \mu)/s_{\bar{Y}}$. Degrees of freedom (abbreviated *df* or sometimes μ) can range from 1 to infinity. A t-distribution for $df = 1$ deviates most markedly from the normal. As the number of degrees of freedom increases, Student's distribution approaches the shape of the standard normal distribution ($\mu = 0$, $\sigma = 1$) ever more closely, and in a graph the size of this page a t-distribution of $df = 30$ is essentially indistinguishable from a normal distribution. At $df = \infty$ the t-distribution *is* the normal distribution. Thus we can think of the t-distribution as the general case, considering the normal as a special case of Student's distribution with $df = \infty$. Figure 6.7 shows t-distributions for 1 and 2 degrees of freedom compared with a normal frequency distribution.

We were able to employ a single table for the areas of the normal curve by coding the argument in standard deviation units. However, since the t-distributions differ in shape for differing degrees of freedom, it will be neces-

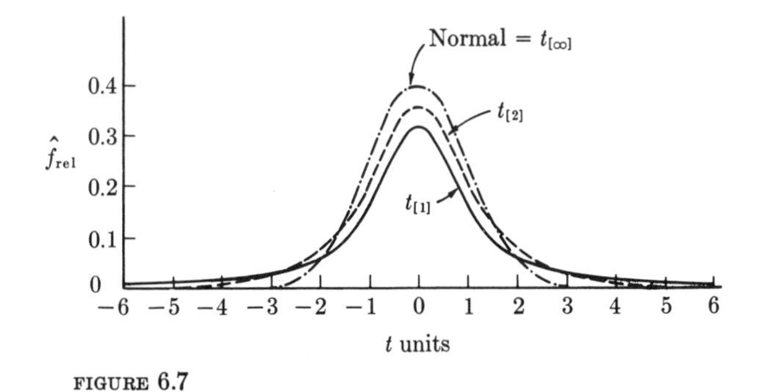

FIGURE 6.7

Frequency curves of t-distributions for 1 and 2 degrees of freedom compared with the normal distribution.

sary to have a separate t-table, corresponding in structure to the table of the areas of the normal curve, for each value of *df*. This would make for very cumbersome and elaborate sets of tables. Conventional t-tables are therefore differently arranged. Table **III** shows degrees of freedom and probability as arguments and the corresponding values of t as functions. The probabilities indicate the percent of the area in both tails of the curve (to the right and left of the mean) beyond the indicated value of t. Thus, looking up the *critical value* of t at probability $P = 0.05$ and $df = 5$, we find $t = 2.571$ in Table **III**. Since this is a two-tailed table, the probability of 0.05 means that there will be 0.025 of the area in each tail beyond a t-value of ± 2.571. You will recall that the corresponding value for infinite degrees of freedom (for the normal curve) is 1.960. Only those probabilities generally used are shown in Table **III**. You should become very familiar with looking up t-values in this table. This is one of the most important tables to be consulted. A fairly conventional symbolism is $t_{\alpha[\nu]}$, meaning the tabled t-value for ν degrees of freedom and proportion α in both tails ($\alpha/2$ in each tail), which is equivalent to the t-value for the cumulative probability of $1 - (\alpha/2)$. Try looking up some of these values to become familiar with the table. For example, convince yourself that $t_{.05[7]}$, $t_{.01[3]}$, $t_{.02[10]}$, and $t_{.05[\infty]}$ correspond to 2.365, 5.841, 2.764, and 1.960, respectively.

We shall now employ the t-distribution for the setting of confidence limits to means of small samples.

6.5 Confidence limits based on sample statistics

Armed with a knowledge of the t-distribution we are now able to set confidence limits to the means of samples from a normal frequency distribution whose parametric standard deviation is unknown. The limits are computed as $L_1 = \bar{Y} - t_{\alpha[n-1]}s_{\bar{Y}}$ and $L_2 = \bar{Y} + t_{\alpha[n-1]}s_{\bar{Y}}$ for confidence limits of probability $P = 1 - \alpha$. Thus for 95% confidence limits we use values of $t_{.05[n-1]}$.

BOX 6.2

Confidence limits for μ.

Aphid stem mother femur lengths from Boxes 2.1 and 3.1: $\bar{Y} = 4.004$; $s = 0.366$; $n = 25$.

Values for $t_{\alpha[n-1]}$ from a two-tailed t-table (Table **III**), where $1 - \alpha$ is the proportion expressing confidence and $n - 1$ are the degrees of freedom:

$$t_{.05[24]} = 2.064 \qquad t_{.01[24]} = 2.797$$

The 95% confidence limits for the population mean, μ, are given by the equations

$$L_1 \text{ (lower limit)} = \bar{Y} - t_{.05[n-1]} \frac{s}{\sqrt{n}}$$

$$= 4.004 - \left(2.064 \frac{0.366}{\sqrt{25}}\right) = 4.004 - 0.151$$

$$= 3.853$$

$$L_2 \text{ (upper limit)} = \bar{Y} + t_{.05[n-1]} \frac{s}{\sqrt{n}}$$

$$= 4.004 + 0.151$$

$$= 4.155$$

The 99% confidence limits are

$$L_1 = \bar{Y} - t_{.01[24]} \frac{s}{\sqrt{n}}$$

$$= 4.004 - \left(2.797 \frac{0.366}{\sqrt{25}}\right) = 4.004 - 0.205$$

$$= 3.799$$

$$L_2 = \bar{Y} + t_{.01[24]} \frac{s}{\sqrt{n}}$$

$$= 4.004 + 0.205$$

$$= 4.209$$

We can rewrite Expression (6.4a) as

$$P\{L_1 \le \mu \le L_2\} = P\{\bar{Y} - t_{\alpha[n-1]}s_{\bar{Y}} \le \mu \le \bar{Y} + t_{\alpha[n-1]}s_{\bar{Y}}\} = 1 - \alpha \qquad (6.5)$$

An example of the application of this expression is shown in Box 6.2. We can convince ourselves of the appropriateness of the t-distribution for setting confidence limits to means of samples from a normally distributed population with unknown σ through a sampling experiment.

Experiment 6.3. Repeat the computations and procedures of Experiment 6.2 (Section 6.3), but base standard errors of the means on the standard deviations computed for each sample and use the appropriate t-value in place of a standard normal deviate.

Figure 6.8 shows 95% confidence limits of 200 sampled means of 35 housefly wing lengths, computed with t and $s_{\bar{Y}}$ rather than with the normal curve and $\sigma_{\bar{Y}}$. We note that 191 (95.5%) of the 200 confidence intervals cross the parametric mean.

We can use the same technique for setting confidence limits to any given statistic as long as it follows the normal distribution. This will apply in an approximate way to all the statistics of Box 6.1. Thus, for example, we may set confidence limits to the coefficient of variation of the aphid femur lengths of Boxes 2.1 and 3.1. These are computed as

$$P\{CV - t_{\alpha[n-1]}s_{CV} \leq CV_P \leq CV + t_{\alpha[n-1]}s_{CV}\} = 1 - \alpha$$

where CV_P stands for the parametric value of the coefficient of variation. Since the standard error of the coefficient of variation approximately equals $s_{CV} = CV/\sqrt{2n}$, we proceed as follows:

$$CV = \frac{100s}{\bar{Y}} = \frac{100(0.3656)}{4.004} = 9.13$$

$$s_{CV} = \frac{9.13}{\sqrt{2 \times 25}} = \frac{9.13}{7.0711} = 1.29$$

$$\begin{aligned} L_1 &= CV - t_{.05[24]}s_{CV} \\ &= 9.13 - (2.064)(1.29) \\ &= 9.13 - 2.66 \\ &= 6.47 \\ L_2 &= CV + t_{.05[24]}s_{CV} \\ &= 9.13 + 2.66 \\ &= 11.79 \end{aligned}$$

When sample size is very large or when σ is known, the distribution is effectively normal. However, rather than turn to the table of areas of the normal curve, we usually simply use $t_{\alpha[\infty]}$, the t-distribution with infinite degrees of freedom.

Although confidence limits are a useful measure of the reliability of a sample statistic, they are not commonly given in scientific publications, the statistic ± its standard error being cited in their place. Thus, you will frequently see column headings such as "Mean ± S.E." This indicates that the reader is free to use the standard error to set confidence limits if he is interested. It should be obvious to you from your study of the t-distribution that you could not set confidence limits to a statistic without knowing the sample size on which it is based, n being necessary to compute the correct degrees of freedom. Thus, the occasional citation of means and standard errors without also stating sample size n is to be strongly deplored.

It is important to state a statistic and its standard error to a sufficient number of decimal places. The following rule of thumb helps. Divide the

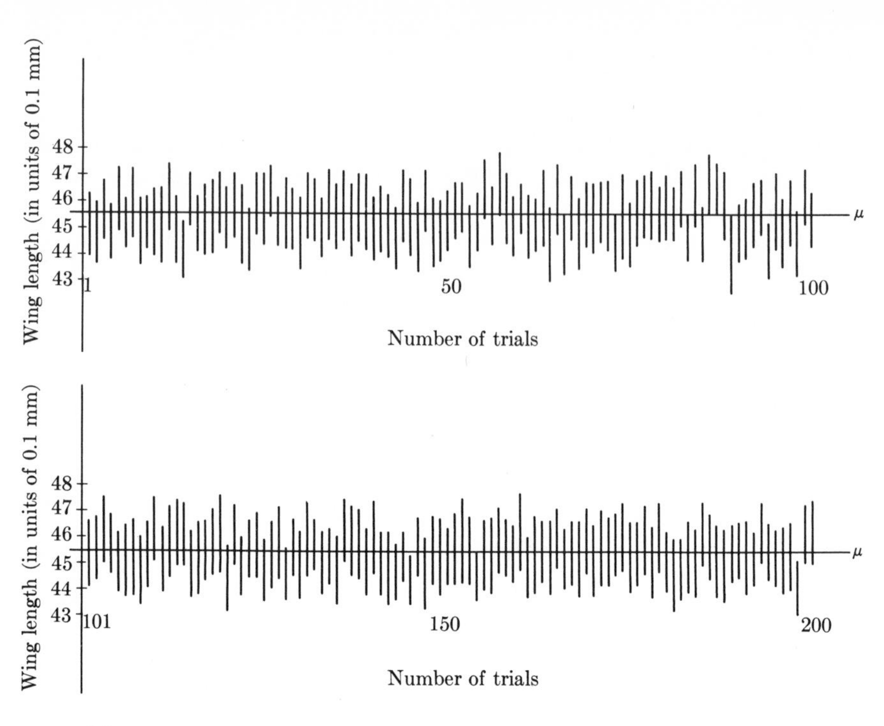

FIGURE 6.8

Ninety-five percent confidence intervals of means of 200 samples of 35 housefly wing lengths, based on sample standard errors, $s_{\bar{Y}}$. The heavy horizontal line is the parametric mean μ. The ordinate represents the variable.

standard error by three, then note the decimal place of the first nonzero digit of the quotient; give the statistic significant to that decimal place and provide one further decimal for the standard error. This rule is quite simple, as an example will illustrate. If the mean and standard error of a sample are computed as 2.354 ± 0.363, we divide 0.363 by 3 which yields 0.121. Therefore the mean should be reported to one decimal place, and the standard error should be reported to two decimal places. Thus we report this result as 2.4 ± 0.36. If, on the other hand, the same mean had a standard error of 0.243, dividing this standard error by 3 would have yielded 0.081 and the first nonzero digit would have been in the second decimal place. Thus the mean should have been reported as 2.35 ± 0.243.

6.6 The chi-square distribution

Another continuous distribution of great importance in statistics is the distribution of χ^2 (read *chi-square*). We need to learn it now in connection with the distribution and confidence limits of variances.

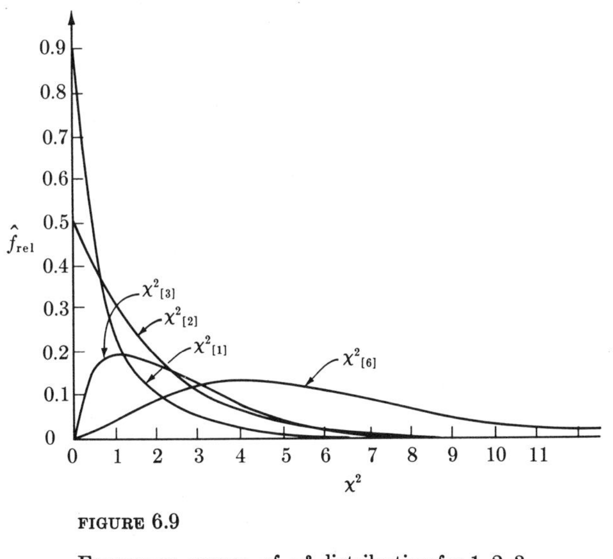

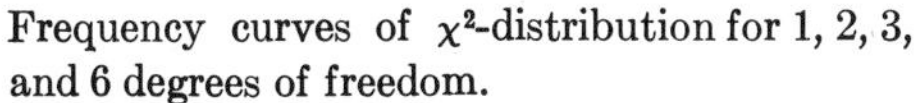

FIGURE 6.9

Frequency curves of χ^2-distribution for 1, 2, 3, and 6 degrees of freedom.

The chi-square distribution is a probability density function whose values range from zero to positive infinity. Thus, unlike the normal distribution or t, the function approaches the horizontal axis asymptotically only at the right-hand tail of the curve, not at both tails. The function describing the χ^2-distribution is complicated and will not be given here. As in t, there is not merely one χ^2-distribution, but there is one distribution for each number of degrees of freedom. Therefore, χ^2 is a function of ν, the number of degrees of freedom. Figure 6.9 shows probability density functions for the χ^2-distributions for 1, 2, 3, and 6 degrees of freedom. Notice that the curves are strongly skewed to the right, L-shaped at first, but more or less approaching symmetry for higher degrees of freedom.

We can generate a χ^2-distribution from a population of standard normal deviates. You will recall that we standardize a variable Y_i by subjecting it to the operation $(Y_i - \mu)/\sigma$. Let us symbolize a standardized variable as $Y'_i = (Y_i - \mu)/\sigma$. Now imagine repeated samples of n variates Y_i from a normal population with mean μ and standard deviation σ. For each sample we transform every variate Y_i to Y'_i, as defined above. The quantities $\sum^n Y'^2_i$ computed for each sample will be distributed as a χ^2-distribution with n degrees of freedom. Using the definition of Y'_i, we can rewrite $\sum^n Y'^2_i$ as

$$\sum^n \frac{(Y_i - \mu)^2}{\sigma^2} = \frac{1}{\sigma^2} \sum^n (Y_i - \mu)^2 \tag{6.6}$$

When we change the parametric mean μ to a sample mean in this expression, it becomes

$$\frac{1}{\sigma^2}\sum^{n}(Y_i - \bar{Y})^2 \tag{6.7}$$

which is simply the sum of squares of the variable divided by a constant, the parametric variance. Another common way of stating this expression is

$$\frac{(n-1)s^2}{\sigma^2} \tag{6.8}$$

which simply replaced the numerator of Expression (6.7) with $n - 1$ times the sample variance, which is of course the sum of squares.

If we were to sample repeatedly n items from a normally distributed population, Expression (6.8) computed for each sample would yield a χ^2-distribution with $n - 1$ degrees of freedom. Notice that, although we have samples of n items, we have lost a degree of freedom because we are now employing a sample mean rather than the parametric mean. Figure 6.5, a sample distribution of variances, has a second scale along the abscissa, which is the first scale multiplied by the constant $(n - 1)/\sigma^2$. This scale converts the sample variances s^2 of the first scale into Expression (6.8). Since the second scale is proportional to s^2, the distribution of the sample variance will serve to illustrate a sample distribution approximating χ^2. The distribution is strongly skewed to the right, as would be expected in a χ^2-distribution.

Conventional χ^2-tables as shown in Table **IV** give the probability levels customarily required and degrees of freedom as arguments and list the χ^2 corresponding to the probability and the *df* as the functions. Each chi-square in Table **IV** is the value of χ^2 beyond which the area under the χ^2-distribution for ν degrees of freedom represents the indicated probability. Just as we used subscripts to indicate the cumulative proportion of the area as well as the degrees of freedom represented by a given value of t, we shall subscript χ^2 as follows: $\chi^2_{\alpha[\nu]}$ indicates the χ^2-value to the right of which is found proportion α of the area under a χ^2-distribution for ν degrees of freedom.

Let us learn how to use Table **IV.** Looking at the distribution of $\chi^2_{[2]}$ we note that 90% of all values of $\chi^2_{[2]}$ would be to the right of 0.211, but only 5% of all values of $\chi^2_{[2]}$ would be greater than 5.991. Mathematical statisticians have shown that the expected value of $\chi^2_{[\nu]}$ (the mean of a χ^2-distribution) equals its degrees of freedom ν. Thus the expected value of a $\chi^2_{[5]}$-distribution is 5. When we examine 50% values (the medians) in the χ^2-table we notice that they are generally lower than the expected value (the means). Thus for $\chi^2_{[5]}$ the 50% point is 4.351. This illustrates the asymmetry of the χ^2-distribution, the mean being to the right of the median. Our first application of the χ^2-distribution will be in the next section. However, its most extensive use will be in connection with Chapter 13.

6.7 Confidence limits for variances

We saw in the last section that the ratio $(n-1)s^2/\sigma^2$ is distributed as χ^2 with $n-1$ degrees of freedom. We take advantage of this fact in setting confidence limits to variances.

First of all we can make the following statement about the ratio $(n-1)s^2/\sigma^2$:

$$P\left\{\chi^2_{(1-(\alpha/2))[n-1]} \leq \frac{(n-1)s^2}{\sigma^2} \leq \chi^2_{(\alpha/2)[n-1]}\right\} = 1-\alpha$$

This expression is similar to those encountered in Section 6.3 and implies that the probability P that this ratio will be within the indicated boundary values of $\chi^2_{[n-1]}$ is 0.95. Simple algebraic manipulation of the quantities in the inequality within brackets yields

$$P\{(n-1)s^2/\chi^2_{(\alpha/2)[n-1]} \leq \sigma^2 \leq (n-1)s^2/\chi^2_{(1-(\alpha/2))[n-1]}\} = 1-\alpha \quad (6.9)$$

Since $(n-1)s^2 = \sum y^2$, we can simplify Expression (6.9) to

$$P\{\textstyle\sum y^2/\chi^2_{(\alpha/2)[n-1]} \leq \sigma^2 \leq \sum y^2/\chi^2_{(1-(\alpha/2))[n-1]}\} = 1-\alpha \quad (6.10)$$

This still looks like a formidable expression but simply means that if we divide the sum of squares $\sum y^2$ by the two values of $\chi^2_{[n-1]}$ bounding $1-\alpha$ of the area of the $\chi^2_{[n-1]}$-distribution, the two quotients will enclose the true value of the variance σ^2 with a probability of $P = 1-\alpha$.

An actual numerical example will make this clear. Suppose we have a sample of 5 housefly wing lengths with a sample variance of $s^2 = 13.52$. If we wish to set 95% confidence limits to the parametric variance we evaluate Expression (6.10) for the sample variance s^2. We first calculate the sum of squares for this sample: $4 \times 13.52 = 54.08$. Then we look up the values for $\chi^2_{.025[4]}$ and $\chi^2_{.975[4]}$. Since 95% confidence limits are required, α in this case is equal to 0.05. These χ^2-values span between them 95% of the area under the χ^2-curve. They correspond to 11.143 and 0.484, respectively, and the limits in Expression (6.10) then become

$$L_1 = 54.08/11.143 \qquad \text{and} \qquad L_2 = 54.08/0.484$$

or

$$L_1 = 4.85 \qquad \text{and} \qquad L_2 = 111.74$$

This confidence interval is very wide but we must not forget that the sample variance is, after all, based on only 5 individuals. Note also that the interval is asymmetrical around 13.52, the sample variance. This is in contrast to the confidence intervals encountered earlier, which were symmetrical around the sample statistic.

BOX 6.3

Confidence limits for σ^2. Method of shortest unbiased confidence intervals.

Aphid stem mother femur lengths from Boxes 2.1 and 3.1: $n = 25$; $s^2 = 0.1337$.

The factors from Table **VII** for $\nu = n - 1 = 24$ *df* and confidence coefficient $(1 - \alpha) = 0.95$ are

$f_1 = 0.5943 \qquad f_2 = 1.8763$

and for a confidence coefficient of 0.99 they are

$f_1 = 0.5139 \qquad f_2 = 2.3513$

The 95% confidence limits for the population variance, σ^2, are given by the equations

$$L_1 = (\text{lower limit}) = f_1 s^2 = 0.5943(0.1337) = 0.07946$$

$$L_2 = (\text{upper limit}) = f_2 s^2 = 1.8763(0.1337) = 0.2509$$

The 99% confidence limits are

$$L_1 = f_1 s^2 = 0.5139(0.1337) = 0.06871$$

$$L_2 = f_2 s^2 = 2.3513(0.1337) = 0.3144$$

The method described above is called the equal tails method because an equal amount of probability is placed in each tail (for example, $2\frac{1}{2}\%$). It can be shown that in view of the skewness of the distribution of variances this method does not yield the shortest possible confidence intervals. One may wish the confidence interval to be "shortest" in the sense that the ratio L_2/L_1 be as small as possible. Box 6.3 shows how to obtain these shortest unbiased confidence intervals for σ^2 using Table **VII**, based on the method of Tate and Klett (1959). This table gives $(n - 1)/\chi^2_{p[n-1]}$, where p is an adjusted value of $\alpha/2$ or $1 - (\alpha/2)$ designed to yield the shortest unbiased confidence intervals. The computation is very simple.

6.8 Introduction to hypothesis testing

The most frequent application of statistics in biological research is to test some scientific hypothesis. Statistical methods are important in biology because results of experiments are usually not clearcut and therefore need statistical tests to support decisions between alternative hypotheses. A statistical test examines a set of sample data and, on the basis of an expected distribution of the data, leads to a decision on whether to accept the hypothesis underlying the expected distribution or to reject that hypothesis and accept an alternative one. The nature of the tests varies with the data and the hypothesis, but the same general philosophy of hypothesis testing is common to all tests and will be discussed in this section. Study the material below very carefully because it is fundamental to an understanding of every subsequent chapter in this book!

We would like to refresh your memory on the sample of 17 animals of species A, 14 of which were females and 3 of which were males. These data were examined for their fit to the binomial frequency distributon presented in Section 4.2 and their analysis is shown in Table 4.3. We concluded from Table 4.3 that if the sex ratio in the population was 1:1 ($p_{♀} = q_{♂} = 0.5$), the probability of obtaining a sample with 14 males and 3 females is 0.005,188, making it very unlikely that such a result could be obtained by chance alone. We learned that it is conventional to include all "worse" outcomes—that is, all those that deviate even more from the outcome expected on the hypothesis $p_{♀} = q_{♂} = 0.5$. Including all worse outcomes the probability is 0.006,363, still a very small value. The above computation is based on the idea of a one-tailed test, in which we are only interested in departures from the 1:1 sex ratio that show a preponderance of females. If we have no preconception about the direction of the departures from expectation, we must calculate the probability of obtaining a sample as deviant as 14 females and 3 males *in either direction* from expectation. This requires the probability either of obtaining a sample of 3 females and 14 males (and all worse samples) or of obtaining 14 females and 3 males (and all worse samples). Such a test is two-tailed and, since the distribution is symmetrical, we double the previously discussed probability to yield 0.012,726.

What does this probability mean? It is our hypothesis that $p_{♀} = q_{♂} = 0.5$. Let us call this hypothesis H_0, the *null hypothesis*, which is the hypothesis under test. It is called the null hypothesis because it assumes that there is no real difference between the true value of p in the population from which we sampled and the hypothesized value of $\hat{p} = 0.5$; for example, in the present example we believe that the only reason our sample does not exhibit a 1:1 sex ratio is because of sampling error. If the null hypothesis $p_{♀} = q_{♂} = 0.5$ is true, then approximately 13 samples out of 1000 will be as deviant or more deviant than this one in either direction *by chance alone*. Thus, it is quite *possible* to have arrived at a sample of 14 females and 3 males by chance, but it is not very *probable*, since so deviant an event would occur only about 13 out of 1000 times or 1.3% of the time. If we actually obtain such a sample, we may make one of two decisions. We may decide that the null hypothesis is in fact true (that is, the sex ratio is 1:1) and that the sample obtained by us just happened to be one of those in the tail of the distribution, or we may decide that so deviant a sample is too improbable an event to justify acceptance of the null hypothesis. We may therefore decide that the hypothesis about the sex ratio being 1:1 is not true. Either of these decisions may be correct, depending upon the truth of the matter. If in fact the 1:1 hypothesis is correct, then the first decision (to accept the null hypothesis) will be correct. If we decide to reject the hypothesis under these circumstances, we commit an error. *The rejection of a true null hypothesis is called a type I error*. On the other hand, if in fact the true sex ratio of the population is other than 1:1, the first decision (to accept the 1:1 hypothesis) is an error, a so-called *type II error, which is the acceptance of a false null hypothesis.* Finally, if the 1:1 hypothesis is not true and we do decide to reject it, then we again make the correct decision. Thus, there are two kinds of correct decisions—accepting a true null hypothesis and rejecting a false null hypothesis—and two kinds of errors—type I, rejecting a true null hypothesis, and type II, accepting a false null hypothesis.

Before we carry out a test, we have to decide what magnitude of type I error (rejection of true hypothesis) we are going to allow. Even when we sample from a population of known parameters, there will always be some samples that by chance are very deviant. The most deviant of these are likely to mislead us into believing our hypothesis H_0 to be untrue. If we permit 5% of samples to lead us into a type I error, then we shall reject 5 out of 100 samples from the population, deciding that these are not samples from the given population. In the distribution under study, this means that we would reject all samples of 17 animals containing 13 of one sex plus 4 of the other sex. This can be seen by referring to column (3) of Table 6.3, where the expected frequencies of the various outcomes on the hypothesis $p_♀ = q_♂ = 0.5$ are shown. This table is an extension of the earlier Table 4.3, which showed only a tail of this distribution. Actually, you would obtain a type I error slightly less than 5% if we sum relative expected frequencies for both tails starting with the class of 13 of one sex and 4 of the other. From Table 6.3 it can be seen that the relative expected frequency in the two tails will be $2 \times 0.0245209 = 0.0490418$. In a discrete frequency distribution, such as the binomial, we cannot calculate errors of exactly 5% as we can in a continuous frequency distribution, where we can measure off exactly 5% of the area. If we decide on an approximate 1% error, we would reject the hypothesis $p_♀ = q_♂$ for all samples of 17 animals having 14 or more of one sex (from Table 6.3 we find the $\hat{f}_{rel}$ in the tails equals $2 \times 0.0063629 = 0.0127258$). Thus, the smaller the type I error we are prepared to accept, the more deviant a sample has to be for us to reject the null hypothesis H_0. Your natural inclination might well be to have as little error as possible. You may decide to work with an extremely small type I error, such as 0.1% or even 0.01%, accepting the null hypothesis unless the sample is extremly deviant. The difficulty with such an approach is that although guarding against an error of the first kind, you might be falling into an error of the second kind (type II) of accepting the null hypothesis when in fact it is not true and an alternative hypothesis H_1 is true. Presently we shall show how this comes about.

First let us learn some more terminology. Type I error is most frequently expressed as a probability and is symbolized by α. When expressed as a percentage it is also known as the *significance level*. Thus a type I error of $\alpha =$

TABLE 6.3

Relative expected frequencies for samples of 17 animals under two hypotheses. Binomial distribution.

(1)	(2)	(3)	(4)
♀♀	♂♂	H_0: $p_♀ = q_♂ = \frac{1}{2}$ $\hat{f}_{rel}$	H_1: $p_♀ = 2q_♂ = \frac{2}{3}$ $\hat{f}_{rel}$
17	0	0.0000076	0.0010150
16	1	0.0001297	0.0086272
15	2	0.0010376	0.0345086
14	3	0.0051880	0.0862715
13	4	0.0181580	0.1509752
12	5	0.0472107	0.1962677
11	6	0.0944214	0.1962677
10	7	0.1483765	0.1542104
9	8	0.1854706	0.0963815
8	9	0.1854706	0.0481907
7	10	0.1483765	0.0192763
6	11	0.0944214	0.0061334
5	12	0.0472107	0.0015333
4	13	0.0181580	0.0002949
3	14	0.0051880	0.0000421
2	15	0.0010376	0.0000042
1	16	0.0001297	0.0000002
0	17	0.0000076	0.0000000
	Total	1.0000002	0.9999999

0.05 corresponds to a significance level of 5% for a given test. When we cut off on a frequency distribution areas proportional to α, the type I error, the portion of the abscissa under the area that has been cut off is called the *rejection region* or *critical region* of a test, and the portion of the abscissa that would lead to acceptance of the null hypotheses is called the *acceptance region.* Figure 6.10A is a bar diagram showing the expected distribution of outcomes in the sex ratio example, given H_0. The dashed lines separate approximate 1% rejection regions from the 99% acceptance region.

Now let us take a closer look at the type II error. This is the probability of accepting the null hypothesis when in fact it is false. If you try to evaluate the probability of type II error, you immediately run into a problem. If the null hypothesis H_0 is false, some other hypothesis H_1 must be true. But unless you can specify H_1 you are not in a position to calculate type II error. An example will make this clear immediately. Suppose in our sex ratio case we have only two reasonable possibilities—(1) our old hypothesis $H_0: p_♀ = q_♂$, or (2) an alternative hypothesis $H_1: p_♀ = 2q_♂$, which states that the sex ratio is 2:1 in favor of females so that $p_♀ = \frac{2}{3}$ and $q_♂ = \frac{1}{3}$. We now have to

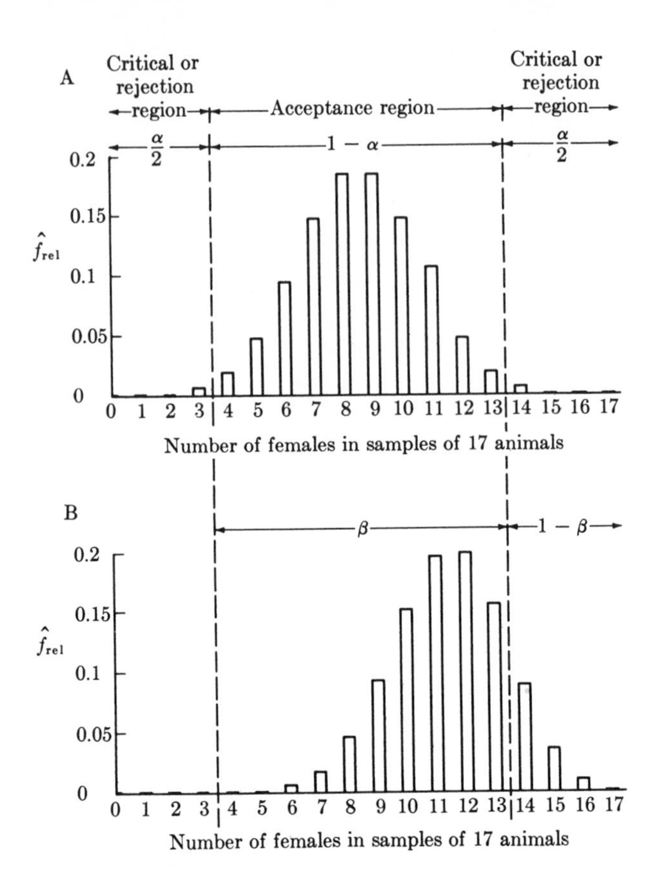

FIGURE 6.10

Expected distributions of outcomes when sampling 17 animals from two hypothetical populations. A. H_0: $p_♀ = q_♂ = \frac{1}{2}$. B. H_1: $p_♀ = 2q_♂ = \frac{2}{3}$. Dashed lines separate critical regions from acceptance region of the distribution of Figure A. Type I error α equals approximately 0.01.

calculate expected frequencies for the binomial distribution $(p_♀ + q_♂)^k = (\frac{2}{3} + \frac{1}{3})^{17}$ to find the probabilities of the various outcomes under this hypothesis. These are shown graphically in Figure 6.10B and are tabulated and compared with expected frequencies of the earlier distribution in Table 6.3.

Suppose we had decided on a type I error of $\alpha \approx 0.01$ ($\approx$ means "approximately equal to") as shown in Figure 6.10A. At this significance level we would accept the H_0 for all samples of 17 having 13 or fewer animals of one sex. Approximately 99% of all samples will fall into this category. However, what if H_0 is not true and H_1 is true? Clearly, from the population represented by hypothesis H_1 we could also obtain outcomes in which one sex was represented 13 or fewer times in samples of 17. We have to calculate what proportion of the curve representing hypothesis H_1 will overlap the acceptance region of the distribution representing hypothesis H_0. In this case we find that 0.8695 of the distribution representing H_1 overlaps the acceptance region of H_0 (see Figure 6.10B). Thus, if H_1 is really true (and H_0 correspondingly false), we would erroneously accept the null hypothesis 86.95% of the time. This percentage corresponds to the proportion of samples from H_1 that fall within the limits of the acceptance regions of H_0. This proportion is called β, the type II error expressed as a proportion. In this example β is quite large. Clearly a sample of 17 animals is unsatisfactory to discriminate between the two hypotheses. Though 99% of the samples under H_0 would fall in the acceptance region, fully 87% would do so under H_1. A single sample that falls in the acceptance region would not enable us to reach a decision between the hypotheses with a high degree of reliability. If the sample had 14 or more females, we would conclude that H_1 was correct. If it had 3 or less females we might conclude that neigher H_0 nor H_1 was true. As H_1 approaches H_0 (as in $H_1{:}p_{♀} = 0.55$, for example), the two distributions would overlap more and more and the magnitude of β would increase, making discrimination between the hypotheses even less likely. Conversely, if H_1 represented $p_{♀} = 0.9$, the distributions would be much farther apart and type II error β would be reduced. Clearly, then, the magnitude of β depends, among other things, on the parameters of the alternative hypothesis H_1 and cannot be specified without knowledge of the latter.

When the alternative hypothesis is fixed as in the previous example ($H_1{:}p_{♀} = 2q_{♂}$), the magnitude of the type I error α we are prepared to tolerate will determine the magnitude of the type II error β. The smaller the rejection region α in the distribution under H_0, the greater will be the acceptance region $1 - \alpha$ in this distribution. The greater $1 - \alpha$, however, the greater will be its overlap with the distribution representing H_1, and hence the greater will be β. Convince yourself of this in Figure 6.10. By moving the dotted lines outward we are reducing the critical regions representing type I error α in diagram A. But as the dashed lines move outward, more of the distribution of H_1 in diagram B will lie in the acceptance region of the null hypothesis. Thus, by decreasing α we are increasing β and in a sense defeating our own purposes. In most applications, scientists would wish to keep both of these errors small, since they do not wish to reject a null hypothesis when it is true, nor do they wish to accept it when another hypothesis is correct. We shall see below what steps can be taken to decrease β while holding α constant at a preset level.

Although significance levels α can be varied at will, investigators are frequently limited because, for many tests, cumulative probabilities of the appropriate distributions have not been tabulated and they must therefore make use of published probability levels. These are commonly 0.05, 0.01, and 0.001, although several others are occasionally encountered. When a null hypothesis has been rejected at a specified level of α, we say that the sample is *significantly different* from the parametric or hypothetical population at probability $P \leq \alpha$. Generally, values of α greater than 0.05 are not considered to be *statistically significant*. A significance level of 5% ($P = 0.05$) corresponds to one type I error in 20 trials, a level of 1% ($P = 0.01$) to one error in 100 trials. Significance levels less than 1% ($P \leq 0.01$) are nearly always adjudged significant; those between 5% and 1% may be considered significant at the discretion of the investigator. Since statistical significance has a special technical meaning (H_0 rejected at $P \leq \alpha$), we shall use the adjective significant only in this sense; its use in scientific papers and reports, unless such a technical meaning is clearly implied, should be discouraged. For general descriptive purposes synonyms such as important, meaningful, marked, noticeable, and others can serve to underscore differences and effects.

A brief remark on null hypotheses represented by asymmetrical probability distributions is in order here. Suppose our null hypothesis in the sex ratio case had been $H_0: p_{♀} = \frac{2}{3}$, as discussed above. The distribution of samples of 17 offspring from such a population is shown in Figure 6.10B. It is clearly assymmetrical and for this reason the critical regions have to be defined independently. For a given two-tailed test we can either double the probability P of a deviation in the direction of the closer tail and compare $2P$ with α, the conventional level of significance; or, we can compare P with $\alpha/2$, half the conventional level of significance. In this latter case 0.025 is the maximal value of P conventionally considered significant.

We shall review what we have learned by means of a second example, this time involving a continuous frequency distribution—the normally distributed housefly wing lengths—of parametric mean $\mu = 45.5$ and variance $\sigma^2 = 15.21$. Means based on 5 items sampled from these will also be normally distributed as was demonstrated in Table 6.1 and Figure 6.1. Let us assume that someone presents you with a single sample of 5 housefly wing lengths and you wish to test whether they could belong to the specified population. Your null hypothesis will be $H_0: \mu = 45.5$ or $H_0: \mu = \mu_0$, where μ is the true mean of the population from which you sampled and μ_0 stands for the hypothetical parametric mean of 45.5. We shall assume for the moment that we have no evidence that the variance of our sample is very much greater or smaller than the parametric variance of the housefly wing lengths. If it were, it would be unreasonable to assume that our sample comes from the specified population. There is a critical test of the assumption about the sample variance, which we shall take up later. The curve at the center of Figure 6.11 represents the expected distribution of means of samples of 5 housefly wing lengths from the specified population. Acceptance and rejection regions for a type I error $\alpha = 0.05$ are delimited along the abscissa. The boundaries of the critical regions are computed as follows (remember that $t_{[\infty]}$ is equivalent to the normal distribution):

$$L_1 = \mu_0 - t_{.05[\infty]}\sigma_{\bar{Y}} = 45.5 - (1.96)(1.744) = 42.08$$

and

$$L_2 = \mu_0 + t_{.05[\infty]}\sigma_{\bar{Y}} = 45.5 + (1.96)(1.744) = 48.92$$

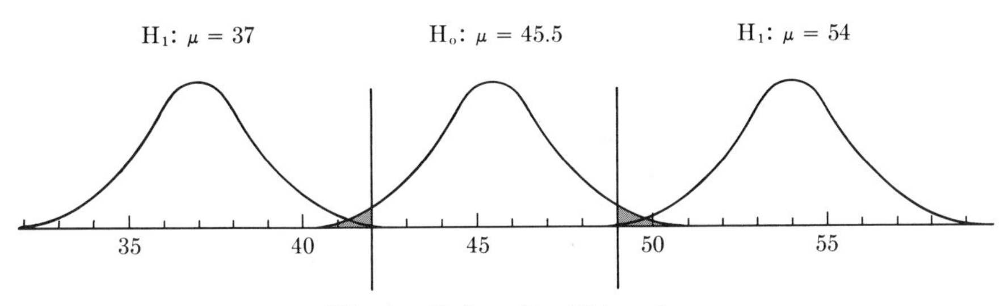

FIGURE 6.11

Expected distribution of means of samples of 5 housefly wing lengths from normal populations specified by μ as shown above curves and $\sigma_{\bar{Y}} = 1.744$. Center curve represents null hypothesis, $H_0{:}\mu = 45.5$, curves at sides represent alternative hypotheses, $\mu = 37$ or $\mu = 54$. Vertical lines delimit 5% rejection regions for the null hypothesis ($2\frac{1}{2}$% in each tail, shaded).

Thus we would consider it improbable for means less than 42.08 or greater than 48.92 to have been sampled from this population. For such sample means we would therefore reject the null hypothesis. The test we are proposing is two-tailed because we have no a priori assumption about the possible alternatives to our null hypothesis. If we could assume that the true mean of the population from which the sample was taken could only be equal to or greater than 45.5, the test would be one-tailed.

Now let us examine various alternative hypotheses. One alternative hypothesis might be that the true mean of the population from which our sample stems is 54.0, but that the variance is the same as before. We can express this assumption as $H_1{:}\mu = 54.0$ or $H_1{:}\mu = \mu_1$, where μ_1 stands for the alternative parametric mean 54.0. From the table of the areas of the normal curve and our knowledge of the variance of the means, we can calculate the proportion of the distribution implied by H_1 that would overlap the acceptance region implied by H_0. We find that 54.0 is 5.08 measurement units from 48.92, the upper boundary of the acceptance region of H_0. This corresponds to $5.08/1.744 = 2.91\sigma_{\bar{Y}}$ units. From the table of areas of the normal curve (Table **II**) we find that 0.0018 of the area will lie beyond 2.91σ at one tail of the curve. Thus under this alternative hypothesis 0.0018 of the distribution of H_1 will overlap the acceptance region of H_0. This is β, the type II error under this alternative hypothesis. Actually this is not entirely correct. Since the left tail of the H_1 distribution goes all the way to negative infinity, it will leave the acceptance region and cross over into the left-hand rejection region of H_0. However, this represents only an infinitesimal amount of the area of H_1 (the lower critical boundary of H_0, 42.08, is $6.83\sigma_{\bar{Y}}$ units from $\mu_1 = 54.0$) and can be ignored.

Our alternative hypothesis H_1 specified that μ_1 is 8.5 units greater than μ_0. However, as said before, we may have no a priori reason to believe that the true mean of our sample is either greater or less than μ. Therefore we may simply assume that it is 8.5 measurement units away from 45.5. In such a case we must similarly calculate β for the alternative hypothesis that $\mu_1 = \mu_0 - 8.5$. Thus the alternative hypothesis becomes $H_1: \mu = 54.0$ or 37.0, or $H_1: \mu = \mu_1$, where μ_1 represents either 54.0 or 37.0, the alternative parametric means. Since the distributions are symmetrical, β is the same for both alternative hypotheses. Type II error for hypothesis H_1 is therefore 0.0018, regardless of which of the two alternative hypotheses is correct. If H_1 is really true, 18 out of 10,000 samples would lead to an incorrect acceptance of H_0, a very low proportion of error. These relations are shown in Figure 6.11.

You may rightly ask what reason we have to believe that the alternative parametric value for the mean is 8.5 measurement units to either side of $\mu_0 = 45.5$. It would be quite unusual if we had any justification for such a belief. As a matter of fact the true mean may just as well be 7.5 or 6.0 or any number of units to either side of μ_0. If we draw curves for $H_1: \mu = \mu_0 \pm 7.5$, we find that β has increased considerably, the curves for H_0 and H_1 now being closer together. Thus the magnitude of β will depend on how far the alternative parametric mean is from the parametric mean of the null hypothesis. As the alternative mean approaches the parametric mean, β increases up to a maximum value of $1 - \alpha$, which is the area of the acceptance region under the null hypothesis. At this maximum the two distributions would be superimposed upon each other. Figure 6.12 illustrates the increase in β as μ_1 approaches μ, starting with the test illustrated in Figure 6.11. To simplify the graph the alternative distributions are shown for one tail only. Thus we clearly see that β is not a fixed value but varies with the nature of the alternative hypothesis.

An important concept in connection with hypothesis testing is the *power* of a test. It is $1 - \beta$, the complement of β, and is the probability of rejecting the null hypothesis when in fact it is false and the alternative hypothesis is correct. Obviously, for any given test we would like to have the quantity $1 - \beta$ be as large as possible and the quantity β as small as possible. Since we generally cannot specify a given alternative hypothesis, we have to describe β or $1 - \beta$ for a continuum of alternative values. When $1 - \beta$ is graphed in this manner the result is called a *power curve* for the test under consideration. Figure 6.13 shows the power curve for the housefly wing length example just discussed. This figure can be compared with Figure 6.12, from which it is directly derived. Figure 6.12 emphasizes the type II error β, and Figure 6.13 graphs the complement of this value, $1 - \beta$. We note that the power of the test falls off sharply as the alternative hypothesis approaches the null hypothesis. Common sense confirms these conclusions: we can make clear and firm decisions about whether our sample comes from a population of mean 45.5 or 60.0. The power is essentially 1. But if the alternative hypothesis is that $\mu_1 = 45.6$, differing only by 0.1 from the value assumed under the null hypothesis, it will be difficult to decide which of these hypotheses is true and the power will be very low.

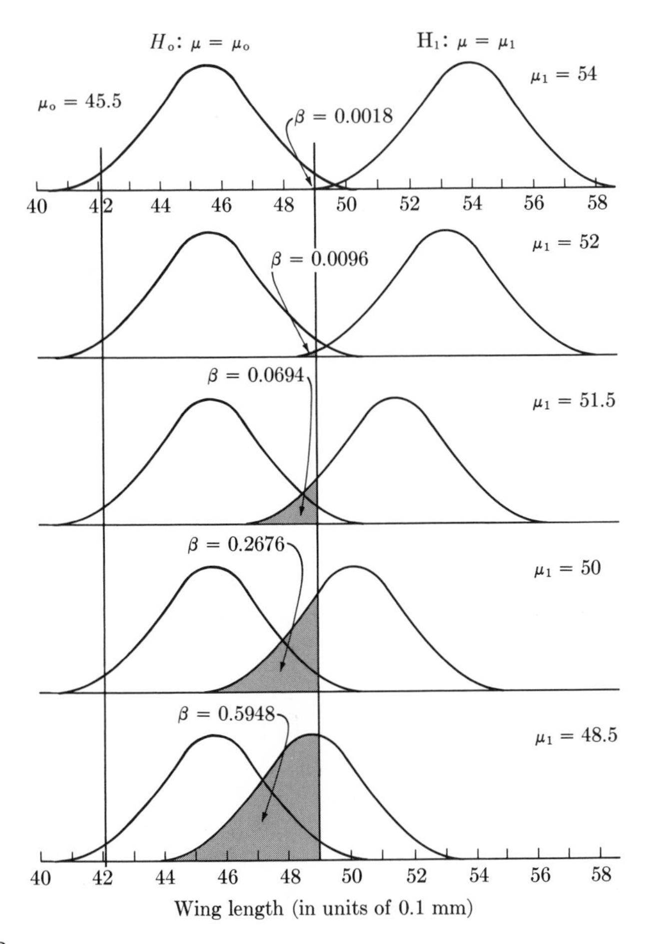

FIGURE 6.12

Diagram to illustrate increases in type II error, β, as alternative hypothesis, H_1, approaches null hypothesis H_0—that is, μ_1 approaches μ. Shading represents β. Vertical lines mark off 5% critical regions ($2\frac{1}{2}$% in each tail) for the null hypothesis. To simplify the graph the alternative distributions are shown for one tail only. Data identical to those in Figure 6.11.

To improve the power of a given test (or decrease β) while keeping α constant for a stated null hypothesis, we must increase sample size. If instead of sampling 5 wing lengths we had sampled 35, the distribution of means would be much narrower. Thus rejection regions for the identical type I error would now commence at 44.21 and 46.79. Although the acceptance and rejection regions have remained the same proportionately, the acceptance region has become much narrower in absolute value. Previously we could not, with confidence, reject the null hypothesis for a sample mean of 48.0. Now, when based on 35 individuals, a mean as deviant as 48.0 would occur only 15 times out of 100,000 and the hypothesis would, therefore, be

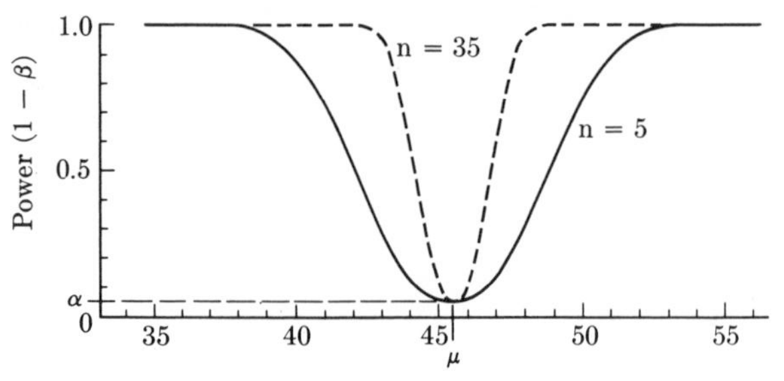

FIGURE 6.13

Power curves for testing $H_0: \mu = 45.5$, $H_1: \mu \neq 45.5$ for $n = 5$ (as in Figures 6.11 and 6.12) and for $n = 35$. (For explanation see text.)

rejected. What has happened to type II error? Since the distribution curves are not as wide as before, there is less overlap between them; if the alternative hypothesis $H_1: \mu = 54.0$ or 37.0 is true, the probability that the null hypothesis could be accepted by mistake (type II error) is infinitesimally small. If we let μ_1 approach μ_0, β will increase, of course, but it will always be smaller than the corresponding value for sample size $n = 5$. This comparison is shown in Figure 6.13, where the power for the test with $n = 35$ is much higher than that for $n = 5$. If we were to increase our sample size to 100 or 1000, the power would be still further increased. Thus we reach an important conclusion: if a given test is not sensitive enough we can increase its sensitivity (= power) by increasing sample size.

There is yet another way of increasing the power of a test. If we cannot increase sample size, the power may be raised by changing the nature of the test. Different statistical techniques testing roughly the same hypothesis may differ substantially in both the actual magnitude as well as in the slopes of their power curves. Tests that maintain higher power levels over substantial ranges of alternative hypotheses are clearly to be preferred. We have already mentioned the various nonparametric tests that in recent years have become increasingly popular and have begun to replace time-honored statistical tests such as the t-test and others. They are popular not only because they are simple to execute, but also because in many cases it has been shown that their power curves are less affected by failure of assumptions than are those of the parametric methods. However, it is also true that nonparametric tests have lower overall power than parametric ones, when all the assumptions of the parametric test are met.

Let us briefly look at a one-tailed test. The null hypothesis is $H_0: \mu_0 = 45.5$ as before. However, the alternative hypothesis assumes that we have

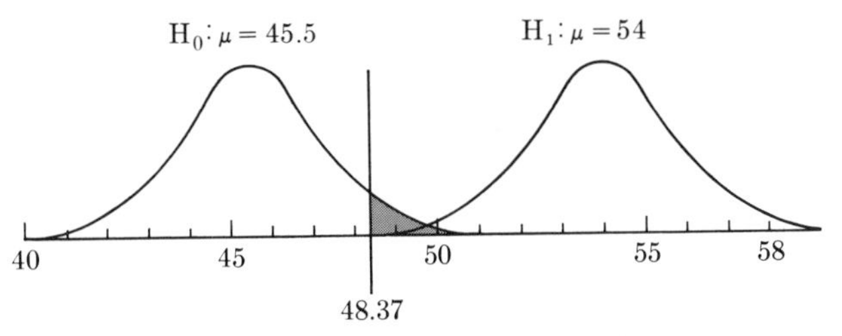

FIGURE 6.14

One-tailed significance test for the distribution of Figure 6.11. Vertical line now cuts off 5% rejection region from one tail of the distribution (corresponding area of curve has been shaded).

reason to believe that the parametric mean of the population from which our sample has been taken could not possibly be less than $\mu_0 = 45.5$. If it is different from that value it could only be greater than 45.5. We might have two grounds for such an hypothesis. First, we might have some biological reason for such a belief. Our parametric flies might be a dwarf population and any other population from which our sample could have come must be bigger. A second reason might be that we are interested in only one deviation of difference. For example, we may test the effect of a chemical in the larval food intended to increase the size of the sample of flies. Therefore, we would expect that $\mu_1 \geq \mu_0$, and we are not interested in testing for any μ_1 that is less than μ_0 because such an effect is the exact opposite of what we expect. Similarly, if we are investigating the effect of a certain drug as a cure for cancer, we might wish to compare the untreated population that has a mean fatality rate θ (from cancer) with the treated population, whose rate is θ_1. Our alternative hypotheses will be $H_1: \theta_1 < \theta$. That is, we are not interested in any θ_1 that is greater than θ, because if our drug will increase mortality from cancer it certainly is not much of a prospect for a cure.

When such a one-tailed test is performed, the rejection region along the abscissa is only under one tail of the curve representing the null hypothesis. Thus, for our housefly data (distribution of means of sample size $n = 5$) the rejection region will be in one tail of the curve only and for a 5% type I error will appear as shown in Figure 6.14. We compute the critical boundary as $45.5 + (1.645)(1.744) = 48.37$. The 1.645 is $t_{.10[\infty]}$, which corresponds to the 5% value for a one-tailed test. Compare this rejection region, which rejects the null hypothesis for all means greater than 48.37, with the two rejection regions in Figure 6.12, which reject the null hypothesis for means lower than 42.08 and greater than 48.92. The alternative hypothesis is considered for one tail of the distribution only, and the power curve of the test is not symmetrical but is drawn out with respect to one side of the distribution only.

6.9 Tests of simple hypotheses employing the *t*-distribution

We shall proceed to apply our newly won knowledge of hypothesis testing to a simple example involving the *t*-distribution.

Government regulations prescribe that the standard dosage in a certain biological preparation should be 600 activity units per cubic centimeter. We prepare 10 samples of this preparation and test each for potency. We find that the mean number of activity units per sample is 592.5 units per *cc* and the standard deviation of the samples is 11.2. Does our sample conform to the government standard? Stated more precisely, our null hypothesis is $H_0: \mu = \mu_0$. The alternative hypothesis is that the dosage is not equal to 600, or $H_1: \mu \neq \mu_0$. We proceed to calculate the significance of the deviation $\bar{Y} - \mu_0$ expressed in standard deviation units. The appropriate standard deviation is that of means (the standard error of the mean), *not* the standard deviation of items because the deviation is that of a sample mean around a parametric mean. We therefore calculate $s_{\bar{Y}} = s/\sqrt{n} = 11.2/\sqrt{10} = 3.542$. We next test the deviation $(\bar{Y} - \mu_0)/s_{\bar{Y}}$. We have seen earlier in Section 6.4 that a deviation divided by an estimated standard deviation will be distributed according to the *t*-distribution with $n - 1$ degrees of freedom. We therefore write

$$t_s = \frac{\bar{Y} - \mu_0}{s_{\bar{Y}}} \tag{6.11}$$

This indicates that we would expect this deviation to be distributed as a *t*-variate. Note that in Expression (6.11) we wrote t_s. In most textbooks you will find this ratio simply identified as t, but in fact the *t*-distribution is a parametric and theoretical distribution that generally is only approached, but never equaled, by observed, sampled data. This may seem a minor distinction, but readers should be quite clear that in any hypothesis testing of samples we are only *assuming* that the distributions of the tested variables follow certain theoretical probability distributions. To conform with general statistical practice, the *t*-distribution should really have a Greek letter (such as τ), with t serving as the sample statistic. Since this would violate long-standing practice, we prefer to use the subscript s to indicate the sample value.

The actual test is very simple. We calculate Expression (6.11),

$$t_s = \frac{592.5 - 600}{3.542} = \frac{-7.5}{3.542} = -2.12, \qquad df = n - 1 = 9,$$

and compare it with the expected values for t at 9 degrees of freedom. Since the *t*-distribution is symmetrical, we shall ignore the sign of t_s and always enter it in Table **III** under its positive value. The two values on either side of t_s are $t_{.05[9]} = 2.26$ and $t_{.10[9]} = 1.83$. These are *t*-values for two-tailed tests, appropriate in this instance because the alternative hypothesis is that $\mu \neq 600$; that is, it can be smaller or greater. It appears that the significance level of our value of t_s is between 5% and 10%; if the null hypothesis is actually true, the probability of obtaining a deviation as great or greater than 7.5 is somewhere between 0.05 and 0.10. By customary levels of significance this is insufficient for declaring the sample mean significantly different from the standard. We consequently accept the null hypothesis. In conventional language we would report the results of the statistical analysis as follows. "The sample mean is not significantly different from the accepted standard." Such a statement in a scientific report should always be backed up by a probability value, and the proper way of presenting this is to write $0.10 > P > 0.05$, which means that the probability of such a deviation is between 0.05 and 0.10. Another way of saying this is that the value of t_s is *not significant* (frequently abbreviated as *ns*).

A convention often encountered is the use of asterisks after the computed value of the significance test, as in $t_s = 2.86$**. The symbols generally represent the following probability ranges:

$$* = 0.05 > P > 0.01, \qquad ** = 0.01 > P > 0.001, \qquad *** = P < 0.001$$

However, since some authors occasionally imply other ranges by these asterisks, the meaning of the symbols has to be specified in each scientific report.

It might be argued that in a biological preparation the concern of the tester should not be whether the sample differs significantly from a standard, but whether it is significantly *below* the standard. This may be one of those biological preparations in which an excess of the active component is of no harm but a shortage would make the preparation ineffective at the conventional dosage. Then the test becomes one-tailed, performed in exactly the same manner except that the critical values of t for a one-tailed test are at half the probabilities of the two-tailed test. Thus $t_{.025[9]} = 2.26$, the former 0.05-value, and $t_{.05[9]} = 1.83$, the former 0.10-value, making our observed t_s-value of 2.12 "significant at the 5% level" or, more precisely stated, significant at $0.05 > P > 0.025$. If we are prepared to accept a 5% significance level, we would consider the preparation significantly below the standard.

You may be surprised that the same example, employing the same data and significance tests, should lead to two different conclusions, and you may begin to wonder whether some of the things you hear about statistics and statisticians are not, after all, correct. The explanation lies in the fact that the two results are answers to different questions. If we test whether our sample is significantly different from the standard in either direction, we must conclude that it is not different enough for us to reject the null hypothesis. If, on the other hand, we exclude from consideration the fact that the true sample mean μ could be greater than the established standard μ_0, the difference as found by us is clearly significant. It is obvious from this example that in any statistical test one must clearly state whether a one-tailed or a two-tailed test has been performed if the nature of the example is such that there could be any doubt about the matter. We should also point out that such a difference in the outcome of the results is not necessarily typical.

BOX 6.4

Testing the significance of a statistic—that is, the significance of a deviation from a parameter. For normally distributed statistics.

Computational steps

1. Compute t_s as the following ratio,

$$t_s = \frac{St - St_p}{s_{St}}$$

where St is a sample statistic, St_p is the parametric value against which the sample statistic is to be tested, and s_{St} is its estimated standard error, obtained from Box 6.1, or elsewhere in this book.

2. The pertinent hypotheses are

$$H_0: St = St_p \qquad H_1: St \neq St_p$$

for a two-tailed test.

$$H_0: St = St_p \qquad H_1: St > St_p$$

or

$$H_0: St = St_p \qquad H_1: St < St_p$$

for a one-tailed test.

3. In the two-tailed test look up the critical value of $t_{\alpha[\nu]}$, where α is the type I error agreed upon and ν is the degrees of freedom pertinent to the standard error employed (see Box 6.1). In the one-tailed test look up the critical value of $t_{2\alpha[\nu]}$ for a significance level of α.

4. Accept or reject the appropriate hypothesis in **2** on the basis of the t_s value in **1** compared with critical values of t in **3**.

It is only because the outcome in this case is in a borderline area between clear significance and nonsignificance. Had the difference between sample and standard been 10.5 activity units, the sample would have been unquestionably significantly different from the standard by the one-tailed or the two-tailed test.

The promulgation of a standard mean is generally insufficient for the establishment of a rigid standard for a product. If the variance among the samples is sufficiently large, it will never be possible to establish a significant difference between the standard and the sample mean. This is an important point that should be quite clear to you. Remember that the standard error can be increased in two ways—by lowering sample size or by increasing the standard deviation of the replicates. Both of these are undesirable aspects of any experimental setup.

The test described above for the biological preparation leads us to a general test for the significance of any statistic—that is, for the significance of a deviation of any statistic from a parameter, which is outlined in Box 6.4. Such a test applies whenever the statistics are expected to be normally distributed. When the standard error is estimated from the sample, the t-distribution is used. However, since the normal distribution is just a special case $t_{[\infty]}$, of the t-distribution, most statisticians uniformly apply the t-distribution with the appropriate degrees of freedom from 1 to infinity. An example of such a test is the t-test for the significance of a regression coefficient shown in step **2** of Box 11.4.

6.10 Testing the hypothesis $H_0: \sigma^2 = \sigma_0^2$

The method of Box 6.4 can be used only if the statistic is normally distributed. In the variance this is not so. As we have seen in Section 6.6, sums of squares divided by σ^2 follow the χ^2-distribution. Therefore, for testing the hypothesis that a sample variance is different from a parametric variance we must employ the χ^2-distribution.

Let us use the biological preparation of the last section as an example. We were told that the standard deviation was 11.2 based on 10 samples. Therefore the variance must have been 125.44. Suppose the government postulates that the variance of samples from the preparation should be no greater than 100.0. Is our sample variance significantly above 100.0? Remembering from Expression (6.8) that $(n - 1)s^2/\sigma^2$ is distributed as $\chi^2_{[n-1]}$, we proceed as follows. We first calculate

$$\begin{aligned} X^2 &= (n - 1)s^2/\sigma^2 \\ &= (9)125.44/100 \\ &= 11.290 \end{aligned}$$

Note that we compute the quantity X^2 rather than χ^2, again to emphasize that we are obtaining a sample statistic that we shall compare to the parametric distribution. The use of X^2 to denote the sample statistic approximating a χ^2-distribution has become quite widely established. Following the general outline of Box 6.4 we next establish our null and alternative hypotheses, which are $H_0: \sigma^2 \leq \sigma_0^2$ and $H_1: \sigma^2 > \sigma_0^2$; that is, we are to perform a one-tailed test. The critical value of χ^2 is found next as $\chi^2_{\alpha[\nu]}$, where α indicates the type I error and ν the pertinent degrees of freedom. Quantity α represents the proportion of the χ^2-distribution to the right of the given value as described in Section 6.6, and you see now why we used the symbol α for that portion of the area; it corresponds to the type I error. For 9 degrees of freedom we find in Table **IV** that

$$\chi^2_{.05[9]} = 16.919, \qquad \chi^2_{.10[9]} = 14.684, \qquad \chi^2_{.50[9]} = 8.343$$

We notice that the probability of getting a χ^2 as large as 11.290 is therefore higher than 0.10 but less than 0.50, assuming that the null hypothesis is true. Thus X^2 is not significant at the 5% level; we have no basis for rejecting the null hypothesis, and we must conclude that the variance of the 10 samples of the biological preparation may be no greater than the standard permitted by the government. If we had decided to test whether the variance is different from the standard, permitting it to deviate in either direction, the hypotheses for this two-tailed test would have been $H_0: \sigma^2 = \sigma_0^2$ and $H_1: \sigma^2 \neq \sigma_0^2$, and a 5% type I error would have yielded the following critical values for the two-tailed test:

$$\chi^2_{.975[9]} = 2.700, \qquad \chi^2_{.025[9]} = 19.023$$

The values represent chi-squares at points cutting off $2\frac{1}{2}\%$ rejection regions at each tail of the χ^2-distribution. A value of $X^2 < 2.700$ or > 19.023 would have been evidence that the sample variance did not belong to this population. Our value of $X^2 = 11.290$ would again have led to an acceptance of the null hypothesis.

In the next chapter we shall see that there is another significance test available to test the hypotheses about variances of the present section. This is the mathematically equivalent F-test, which is, however, a more general test, allowing us to test the hypothesis that two sample variances come from populations with equal variances.

Exercises 6

6.1 Differentiate between type I and type II errors. What do we mean by the power of a statistical test?

6.2 Since it is possible to test a statistical hypothesis with any size sample, why are larger sample sizes preferred?

6.3 The 95% confidence limits for μ as obtained in a given sample were 4.91 and 5.67 g. Is it correct to say that 95 times out of 100 the population mean, μ, falls inside the interval from 4.91 to 5.67 g? If not, what would the correct statement be?

6.4 Set 99% confidence limits to the mean, median, coefficient of variation, and variance for the birth weight data given in Box 3.2. ANS. The lower limits are 109.540, 109.060, 12.136, and 178.698 respectively.

6.5 Set 95% confidence limits to the means listed in Table 6.2. Are these limits all correct? (That is, do they contain μ?)

6.6 In a study of mating calls in the tree toad *Hyla ewingi*, Littlejohn (1965) found the note duration of the call in a sample of 39 observations from Tasmania to have a mean of 189 msec and a standard deviation of 32 msec. Set 95% confidence intervals to the mean and to the variance.

6.7 In Section 4.3 the coefficient of dispersion was given as an index of whether or not data agreed with a Poisson distribution. Since in a true Poisson distribution, the mean, μ, equals the parametric variance, σ^2, the coefficient of dispersion is analogous to Expression (6.8). Using the mite data from Table 4.5, test the hypothesis that the true variance is equal to the sample mean—in other words, that we have sampled from a Poisson distribution (in which the coefficient of dispersion should equal unity). Note that in these examples the chi-square table is not adequate, so that approximate critical values must be computed using the method given with Table **IV**. In Section 7.3 an alternative significance test that avoids this problem will be presented. ANS. $(n - 1) \times$ C.D. = 1308.30, $X^2_{.05[588]} \approx 645.708$.

6.8 Using the method described in Exercise 6.7, test the agreement of the observed distribution to a Poisson distribution by testing the hypothesis that the true coefficient of dispersion equals unity for the data of Table 4.6.

6.9 In direct klinokinetic behavior relating to temperature, animals turn more often in the warm end of a gradient and less often in the colder end, the direction of turning being at random, however. In a computer simulation of such behavior, the following results were found. The mean position along a temperature gradient was found to be -1.352. The standard deviation was 12.267 and n equaled 500 individuals. The gradient was marked off in units: zero corresponded to the middle of the gradient, the initial starting point of the animals; minus corresponded to the cold end; and plus corresponded to the warmer end. Test the hypothesis that direct klinokinetic behavior did not result in a tendency toward aggregation in either the warmer or colder end; that is, test the hypothesis that μ, the mean position along the gradient, was zero.

6.10 In a study of bill measurements of the dusky flycatcher, Johnson (1966) found that the bill length for the males had a mean of 8.14 ± 0.021 and a coefficient of variation of 4.67%. On the basis of this information, infer how many specimens must have been used? ANS. $n = 328$.

7

Introduction to Analysis of Variance

We now proceed to a study of the analysis of variance. This method, developed by R. A. Fisher, is fundamental to much of the application of statistics in biology and especially to experimental design. One way to approach the analysis of variance is to consider it a test of whether two or more sample means could have been obtained from populations with the same parametric mean with respect to a given variable. Alternatively, we could conclude that these means differ from each other to such an extent that we must assume they were sampled from different populations. Where only two samples are involved, the t-distribution has been traditionally used to test significant differences between means. However, the analysis of variance is a more general test, which permits testing two samples as well as many, and we are therefore introducing it at this early stage in order to equip the reader with this most powerful weapon for his statistical arsenal. We shall discuss the t-test for two samples as a special case in Section 8.4.

In Section 7.1 we shall approach the subject on familiar ground, the sampling experiment of the housefly wing lengths. From these samples we shall obtain two independent estimates of the population variance. We digress in Section 7.2 to introduce yet another continuous distribution, the F-distribution, needed for the significance test in analysis of variance. Section 7.3 is another digression in which we show how the newly learned F-distribution can be used to test whether two samples have the same variance. We are now ready for Section 7.4, in which we examine the effects of subjecting the samples to different treatments. The next section (7.5) describes the partitioning of sums of squares and of degrees of freedom, the actual *analysis* of variance. The last two sections (7.6 and 7.7) take up in a more formal way the two scientific models for which the analysis of variance is appropriate, the so-called fixed treatment effects model (Model I) and the variance component model (Model II).

Except for Section 7.3, the entire chapter is largely theoretical. We shall consider the practical details of computation in Chapter 8. However, a thorough understanding of the material in Chapter 7 is necessary for working out actual examples of analysis of variance in Chapter 8.

One final comment. We shall use J. W. Tukey's abbreviation "anova" interchangeably with "analysis of variance" throughout the text.

7.1 The variances of samples and their means

We shall approach analysis of variance through the familiar sampling experiment of housefly wing lengths (Experiment 5.1 and Table 5.1), in which we combined seven samples of 5 wing lengths to form samples of 35. We have reproduced one such sample in Table 7.1. The seven samples of 5, here called groups, are listed vertically in the upper half of the table. Before we proceed to explain Table 7.1 further, we must become familiar with added terminology and symbolism for dealing with this kind of problem. We call our samples *groups;* they are sometimes called *classes* and by yet other terms we shall learn later. In any analysis of variance problem we shall have two or more such samples or groups and we shall use the symbol a for the number of groups. Thus in the present example $a = 7$. Each group or sample is based on n items as before; in Table 7.1, $n = 5$. The total number of items in the table is a times n, which in this case equals 7×5 or 35.

The sums of the items in each group are shown in the row underneath the horizontal dividing line. In an anova, summation signs can no longer be as simple as heretofore. We can sum either the items of one group only or the items of the entire table. We therefore have to use superscripts with the summation symbol. In line with our policy of using the simplest possible notation, whenever this is not likely to lead to misunderstanding, we shall use $\sum^{n} Y$ to indicate the sum of the items of a group and $\sum^{an} Y$ to indicate the sum

TABLE 7.1

Seven samples (groups) of 5 wing lengths of houseflies randomly selected. (Data from Experiment 5.1 and Table 5.2.) Parametric mean, $\mu = 45.5$; variance, $\sigma^2 = 15.21$.

	a groups ($a = 7$)							*Computation of sum of squares of means*	*Computation of total sum of squares*
	1	*2*	*3*	*4*	*5*	*6*	*7*		
	41	48	40	40	49	40	41		
	44	49	50	39	41	48	46		
n individuals per group ($n = 5$)	48	49	44	46	50	51	54		
	43	49	48	46	39	47	44		
	42	45	50	41	42	51	42		
$\sum^{n} Y$	218	240	232	212	221	237	227	$\sum^{a} \bar{Y} = 317.4$	$\sum^{an} Y = 1587$
$\bar{Y}$	43.6	48.0	46.4	42.4	44.2	47.4	45.4	$\bar{\bar{Y}} = 45.34$	$\bar{Y} = 45.34$
$\sum^{n} Y^2$	9534	11,532	10,840	9034	9867	11,315	10,413	$\sum^{a} \bar{Y}^2 = 14{,}417.24$	$\sum^{an} Y^2 = 72535$
$\sum^{n} y^2$	29.2	12.0	75.2	45.2	98.8	81.2	107.2	$\sum^{a} (\bar{Y} - \bar{\bar{Y}})^2 = 25.417$	$\sum^{an} y^2 = 575.886$

of all the items in the table. The sum of the items of each group is shown in the first row under the horizontal line. The mean of each group, symbolized by $\bar{Y}$, is in the next row and is computed simply as $\sum^{n} Y/n$. The remaining two rows in that portion of Table 7.1 list $\sum^{n} Y^2$ and $\sum^{n} y^2$, separately for each group. These are the familiar quantities, the sum of the squared Y's and the sum of squares of Y.

From the sum of squares for each group we can obtain an estimate of the population variance of housefly wing length. Thus in the first group $\sum^{n} y^2 = 29.2$. Therefore our estimate of the population variance is

$$s^2 = \frac{\sum^{n} y^2}{(n-1)} = 29.2/4 = 7.3$$

a rather low estimate. Since we have a sum of squares for each group, we could obtain an estimate of the population variance from each of these. However, it stands to reason that we would get a better estimate if we averaged these separate variance estimates in some way. This is done by computing a *weighted average*. Actually, in this instance a simple average would suffice, since each estimate of the variance is based on samples of the same size. However, we prefer to give the general formula, which works equally well for this case as well as for instances of unequal sample sizes, where the weighted average is necessary. A general formula for calculating a weighted average of any statistic St is as follows:

$$\overline{St} = \frac{\sum_{i=1}^{i=m} w_i St_i}{\sum_{i=1}^{i=m} w_i} \tag{7.1}$$

where m statistics, St_i, each weighted by factor w_i, are being averaged. In this case each sample variance s_i^2 is weighted by its degrees of freedom, $w_i = n_i - 1$, resulting in a sum of squares $(\sum y^2)_i$, since $(n_i - 1)s_i^2 = \sum y_i^2$. Thus the numerator of Expression (7.1) is the sum of the sums of squares. The denominator is $\sum^{a}(n_i - 1) = 7 \times 4$, the sum of the degrees of freedom of each group. The average variance therefore is

$$s^2 = \frac{29.2 + 12.0 + 75.2 + 45.2 + 98.8 + 81.2 + 107.2}{28} = \frac{448.8}{28} = 16.029$$

This quantity is an estimate of 15.21, the parametric variance of housefly wing lengths. This estimate, based on 7 independent estimates of variances of groups, is called the *average variance within groups* or simply *variance within groups*. Note that we use the expression *within* groups, although in previous chapters we used the term variance *of* groups. The reason we do this is that the variance estimates used for computing the average variance have so far all come from sums of squares measuring the variation within one column. As we shall see below, one can also compute variances among groups, cutting across group boundaries.

To obtain a second estimate of the population variance we treat the seven group means, $\bar{Y}$, as though they were a sample of seven observations. The resulting statistics are shown in the lower right part of Table 7.1, headed Computation of sum of squares of means. There are seven means in this example; in the general case there will be a means. We first compute $\sum^{a} \bar{Y}$, the sum of the means. Note that this is rather sloppy symbolism. To be entirely proper, we should identify this quantity as $\sum_{i=1}^{i=a} \bar{Y}_i$, summing the means of group 1 through group a. The next quantity computed is $\bar{\bar{Y}}$, the grand mean of the group means, computed as $\bar{\bar{Y}} = \sum^{a} \bar{Y}/a$. The sum of the seven means is $\sum^{a} \bar{Y} = 317.4$ and the grand mean is $\bar{\bar{Y}} = 45.34$, a fairly close approximation to the parametric mean $\mu = 45.5$. The sum of squares represents the deviations of the group means from the grand mean, $\sum^{a} (\bar{Y} - \bar{\bar{Y}})^2$. For this we need, first of all, the quantity $\sum^{a} \bar{Y}^2$, which equals 14,417.24. The customary computational formula for sum of squares applied to these means is $\sum^{a} \bar{Y}^2 - [(\sum^{a} \bar{Y})^2/a] = 25.417$. From the sum of squares of the means we obtain a *variance among the means* in the conventional way as follows: $\sum^{a} (\bar{Y} - \bar{\bar{Y}})^2/(a - 1)$. We divide by $a - 1$ rather than $n - 1$ because the sum of squares was based on a items (means). Thus variance of the means $s_{\bar{Y}}^2 = 25.417/6 = 4.2362$. We learned in Chapter 6, Expression (6.1), that when randomly sampling from a single population

$$\sigma_{\bar{Y}}^2 = \frac{\sigma^2}{n}$$

and hence

$$\sigma^2 = n\sigma_{\bar{Y}}^2$$

Thus, we can estimate a variance of items by multiplying the variance of means by the sample size on which the means are based (assuming we have sampled from a single population). When we do this for our present example we obtain $s^2 = 5 \times 4.2362 = 21.181$. This is a second estimate of the parametric variance 15.21. It is not as close to the true value as the previous estimate based on the average variance within groups, but this is to be expected since it is only based on 7 "observations." We need a name describing this variance to distinguish it from the variance of means from which it was computed, as well as from the variance within groups with which it will be compared. We shall call it the *variance among groups;* it is n times the variance of means and is an independent estimate of the parametric variance σ^2 of the

TABLE 7.2

Data arranged for simple analysis of variance, single classification, completely randomized.

		a groups						
		1	*2*	*3*	...	*i*	...	*a*
n items	1	Y_{11}	Y_{21}	Y_{31}	...	Y_{i1}	...	Y_{a1}
	2	Y_{12}	Y_{22}	Y_{32}	...	Y_{i2}	...	Y_{a2}
	3	Y_{13}	Y_{23}	Y_{33}	...	Y_{i3}	...	Y_{a3}
	⋮	⋮	⋮	⋮	.	⋮	.	⋮
	j	Y_{1j}	Y_{2j}	Y_{3j}	...	Y_{ij}	...	Y_{aj}
	⋮	⋮	⋮	⋮	.	⋮	.	⋮
	n	Y_{1n}	Y_{2n}	Y_{3n}	...	Y_{in}	...	Y_{an}
Sums	$\sum^{n} Y$	$\sum^{n} Y_1$	$\sum^{n} Y_2$	$\sum^{n} Y_3$	...	$\sum^{n} Y_i$	...	$\sum^{n} Y_a$
Means	$\bar{Y}$	$\bar{Y}_1$	$\bar{Y}_2$	$\bar{Y}_3$	...	$\bar{Y}_i$	...	$\bar{Y}_a$

housefly wing lengths. It may not be clear at this stage why the two estimates of σ^2 that we have obtained, the variance within groups and the variance among groups, are independent. We ask you to take on faith that they are. Although this is in no way a proof of independence, note that in this example the two estimates are indeed different, being 16.029 for the variance within groups and 21.181 for the variance among groups.

Let us review what we have done so far by expressing it in a more formal way. Table 7.2 represents a generalized table for data such as the samples of housefly wing lengths. Each individual wing length is represented by Y, subscripted to indicate its position in the data table. The wing length of the jth fly from the ith sample or group is given by Y_{ij}. Thus you will notice that the first subscript changes with each column representing a group in the table, and the second subscript changes with each row representing an individual item. Using this notation we can compute the variance of sample 1 as

$$\frac{1}{n-1}\sum_{j=1}^{j=n}(Y_{1j}-\bar{Y}_1)^2$$

The variance within groups, which is the average variance of the samples, is computed as

$$\frac{1}{a(n-1)}\sum_{i=1}^{i=a}\sum_{j=1}^{j=n}(Y_{ij}-\bar{Y}_i)^2$$

Note the double summation. It means that we first set $i = 1$ (i being the index of the outer Σ), when we start with the first group. We sum the squared deviations of all items from the mean of the first group, changing index j

of the inner Σ from 1 to n in the process. We then return to the outer summation, set $i = 2$, and sum the squared deviations for group 2 from $j = 1$ to $j = n$. This process is continued until i, the index of the outer Σ, is set to a. In other words, we sum all the squared deviations within one group first and add this sum to similar sums from all the other groups. The variance among groups is computed as

$$\frac{n}{a-1} \sum_{i=1}^{i=a} (\bar{Y}_i - \bar{\bar{Y}})^2$$

Now that we have two independent estimates of the population variance, what shall we do with them? We might wish to find out whether they do in fact estimate the same parameter. To test this hypothesis we need a statistical test that will evaluate the probability that the two sample variances are from the same population. Such a test employs the F-distribution, which is taken up next.

7.2 The *F*-distribution

Let us devise yet another sampling experiment. This is quite a tedious one, so we will not ask you to carry it out. Assume that you are sampling at random from a normally distributed population, such as the housefly wing lengths with mean μ and variance σ^2. The sampling procedure consists of first sampling n_1 items and calculating their variance s_1^2, followed by sampling n_2 items and calculating their variance s_2^2. Sample sizes n_1 and n_2 may or may not be equal to each other but are fixed for any one sampling experiment. Thus, for example, we might always sample 8 wing lengths for the first sample (n_1) and 6 wing lengths for the second sample (n_2). After each pair of values (s_1^2 and s_2^2) has been obtained, we calculate

$$F_s = \frac{s_1^2}{s_2^2}$$

This will be a ratio near 1, because these variances are estimates of the same quantity. Its actual value will depend on the relative magnitudes of variances s_1^2 and s_2^2. If we repeatedly take samples of sizes n_1 and n_2, calculating the ratios F_s of their variances, the average of these ratios will in fact approach $(n_2 - 1)/(n_2 - 3)$, which is close to 1.0 when n_2 is large. The expected distribution of this statistic is called the *F-distribution* in honor of R. A. Fisher. This is another distribution described by a complicated mathematical function that need not concern us here. Unlike the t- and χ^2-distributions, the shape of the F-distribution is determined by *two* values for degrees of freedom, ν_1 and ν_2. Thus for every possible combination of values ν_1, ν_2, each ν ranging from 1 to infinity, there exists a separate F-distribution. Remember that the F-distribution is a theoretical probability distribution, like the t-

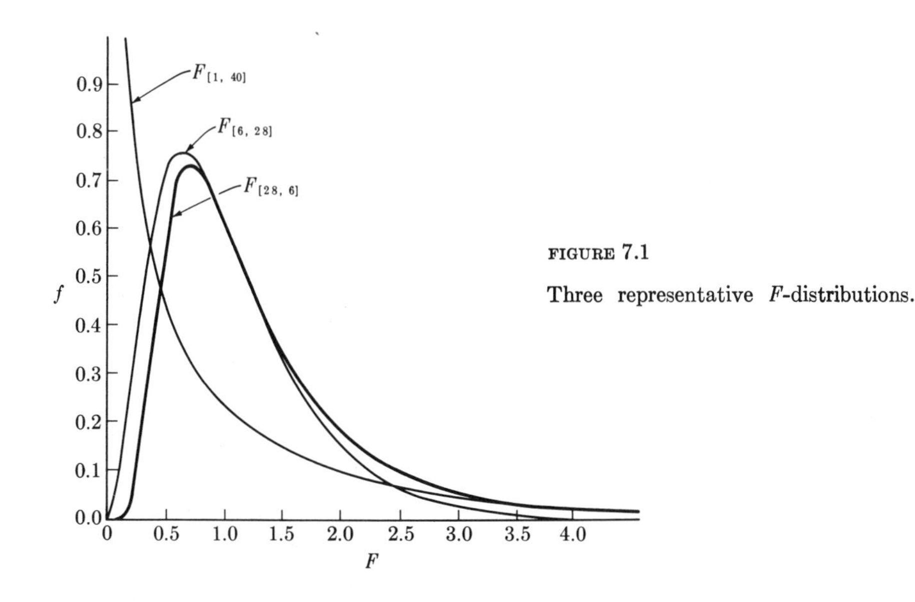

FIGURE 7.1

Three representative F-distributions.

distribution and the χ^2-distribution. Variance ratios based on sample variances, s_1^2/s_2^2, are sample statistics that may or may not follow the F-distribution. We have, therefore, distinguished the sample variance ratio by calling it F_s, conforming to our convention of separate symbols for sample statistics as distinct from probability distributions (such as t_s and X^2 for t and χ^2).

We have discussed the generation of an F-distribution by repeatedly taking two samples from the same normal distribution. We could also have generated it by sampling from two separate normal distributions differing in their mean but identical in their parametric variances—that is, with $\mu_1 \neq \mu_2$ but $\sigma_1^2 = \sigma_2^2$. Thus we obtain an F-distribution whether the samples come from the same normal population or from different ones, so long as their variances are identical.

Figure 7.1 shows several representative F-distributions. For very low degrees of freedom the distribution is L-shaped, but it becomes humped and strongly skewed to the right as both degrees of freedom increase. Table **V** shows the cumulative probability distribution of F for three selected probability values. The values in the table represent $F_{\alpha[\nu_1, \nu_2]}$, where α is the proportion of the F-distribution to the right of the given F-value (in one tail) and ν_1, ν_2 are the degrees of freedom pertaining to the numerator and the denominator of the variance ratio, respectively. The table is arranged so that across the top one reads ν_1, the degrees of freedom pertaining to the upper (numerator) variance, and along the left margin one reads ν_2, the degrees of freedom pertaining to the lower (denominator) variance. At each intersection of degree of freedom values we list three values of F decreasing in magnitude of α. For example, an F-distribution with $\nu_1 = 6$, $\nu_2 = 24$ is 2.51 at $\alpha = 0.05$. By

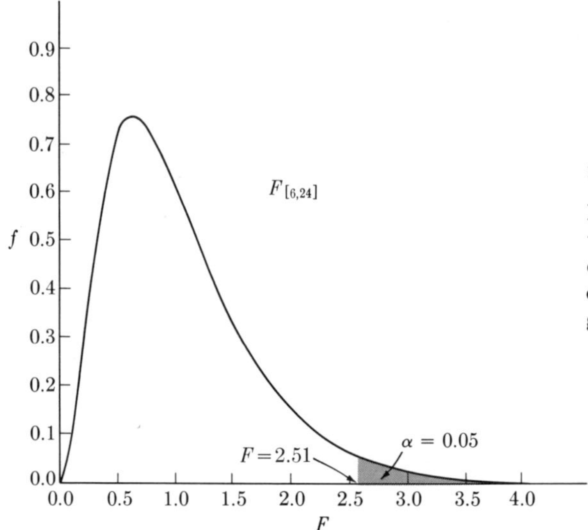

FIGURE 7.2

Frequency curve of the F-distribution for 6 and 24 degrees of freedom, respectively. A one-tailed 5% rejection region is marked off at $F = 2.51$.

that we mean that 0.05 of the area under the curve lies to the right of $F = 2.51$. Figure 7.2 illustrates this. Only 0.01 of the area under the curve lies to the right of $F = 3.67$. Thus if we have a null hypothesis $H_0: \sigma_1^2 = \sigma_2^2$, with the alternative hypothesis $H_1: \sigma_1^2 > \sigma_2^2$, we would use a one-tailed F-test, as illustrated by Figure 7.2.

We can now test the two variances obtained in the sampling experiment of Section 7.1 and Table 7.1. The variance among groups based on 7 means was 21.180 and the variance within 7 groups of 5 individuals was 16.029. Our null hypothesis is that the two variances estimate the same parametric variance; the alternative hypothesis in an anova is always that the parametric variance estimated by the variance among groups is greater than that estimated by the variance within groups. The reason for this restrictive alternative hypothesis, which leads to a one-tailed test, will be explained in Section 7.4. We calculate the variance ratio $F_s = s_1^2/s_2^2 = 21.181/16.029 = 1.32$. Before we can inspect the F-table we have to know the appropriate degrees of freedom for this variance ratio. We shall learn simple formulas for degrees of freedom in an anova later, but at the moment let us reason it out for ourselves. The upper variance (among groups) was based on the variance of 7 means; hence it should have 6 degrees of freedom. The lower variance was based on an average of 7 variances, each of them based on 5 individuals yielding 4 degrees of freedom per variance; $7 \times 4 = 28$ degrees of freedom. Thus the upper variance has 6, the lower variance 28 degrees of freedom. If we check Table **V** for $\nu_1 = 6$, $\nu_2 = 24$, the closest arguments in the table, we find that $F_{.05[6,24]} = 2.51$. For $F = 1.32$, corresponding to the F_s-value actually obtained, α is clearly >0.05. Thus, we may expect more than

5% of all variance ratios of samples based on 6 and 28 degrees of freedom, respectively, to have F_s-values greater than 1.32. We have no evidence to reject the null hypothesis and conclude that the two sample variances estimate the same parametric variance. This corresponds, of course, to what we knew anyway from our sampling experiment. Since the seven samples were taken from the same population, the estimate using the variance of their means is expected to yield another estimate of the parametric variance of housefly wing length.

Whenever the alternative hypothesis is that the two parametric variances are unequal (rather than the restrictive hypothesis $H_1: \sigma_1^2 > \sigma_2^2$), the sample variance s_1^2 could be smaller as well as greater than s_2^2. This leads to a two-tailed test and in such cases a 5% type I error means that rejection regions of $2\frac{1}{2}\%$ will occur at each tail of the curve.

It is sometimes necessary to obtain F-values for $\alpha > 0.5$ (that is, in the left half of the F-distribution). Since, as we have just learned, these values are rarely tabulated, they can be obtained by a simple rule:

$$F_{\alpha[\nu_1,\nu_2]} = \frac{1}{F_{(1-\alpha)[\nu_2,\nu_1]}} \tag{7.2}$$

For example, $F_{.05[5,24]} = 2.62$. If we wish to obtain $F_{.95[5,24]}$ (the F-value for 5 and 24 degrees of freedom, respectively, to the right of which lies 95% of the area of the F-distribution), we first have to find $F_{.05[24,5]} = 4.53$. Then $F_{.95[5,24]}$ is the reciprocal of 4.53, which equals 0.221. Thus 95% of an F-distribution for 5 and 24 degrees of freedom lies to the right of 0.221.

There is an important relationship between the F-distribution and the χ^2-distribution. You may remember that the ratio $X^2 = \sum y^2/\sigma^2$ was distributed as a χ^2 with $n - 1$ degrees of freedom. If you divide the numerator of this expression by $n - 1$, you obtain the ratio $F_s = s^2/\sigma^2$, which is a variance ratio with an expected distribution of $F_{[n-1,\infty]}$. The upper degrees of freedom are $n - 1$, the degrees of freedom of the sum of squares or sample variance; the lower degrees of freedom are considered infinity because only on the basis of an infinite number of items can we obtain the true, parametric variance of a population. Therefore, by dividing a value of X^2 by $n - 1$ degrees of freedom, we obtain an F_s-value with $n - 1$ and ∞ df, respectively. In general, $\chi^2_{[\nu]}/\nu = F_{[\nu,\infty]}$. We can convince ourselves of this by inspecting the F- and χ^2-tables. From the χ^2-table (Table **IV**) we find that $\chi^2_{.05[10]} = 18.307$. Dividing this value by 10 df we obtain 1.8307. From the F-table (Table **V**) we find, for $\nu_1 = 10$, $\nu_2 = \infty$, that $F_{.05[10,\infty]} = 1.83$. Thus the two statistics of significance are closely related and, lacking a χ^2-table, we could make do with an F-table alone, using the values of $\nu F_{[\nu,\infty]}$ in place of $\chi^2_{[\nu]}$.

Before we return to analysis of variance we shall first apply our newly won knowledge of the F-distribution to testing a hypothesis about two sample variances.

BOX 7.1

Testing the significance of differences between two variances.

Survival in days of the cockroach *Blattella vaga* when kept without food or water.

Females	$n_1 = 10$	$\bar{Y}_1 = 8.5$ days	$s_1^2 = 3.6$
Males	$n_2 = 10$	$\bar{Y}_2 = 4.8$ days	$s_2^2 = 0.9$

$H_0: \sigma_1^2 = \sigma_2^2 \qquad H_1: \sigma_1^2 \neq \sigma_2^2$

Source: Data modified from Willis and Lewis (1957).

The alternative hypothesis is that the two variances are unequal. We have no reason to suppose that one sex should be more variable than the other. In view of the alternative hypothesis this is a two-tailed test. Since only the right tail of the F-distribution is tabled extensively in Table **V** and in most other tables, we calculate F_s as the ratio of the greater variance over the lesser one:

$$F_s = \frac{s_1^2}{s_2^2} = \frac{3.6}{0.9} = 4.00$$

Because the test is two-tailed, we look up the critical value $F_{\alpha/2[\nu_1,\nu_2]}$, where α is the type I error accepted, $\nu_1 = n_1 - 1$ and $\nu_2 = n_2 - 1$, the degrees of freedom for the upper and lower variance, respectively. Whether we look up $F_{\alpha/2[\nu_1,\nu_2]}$ or $F_{\alpha/2[\nu_2,\nu_1]}$ depends on whether sample 1 or sample 2 has the greater variance and was placed in the numerator.

From Table **V** we find $F_{.025[9,9]} = 4.03$ and $F_{.05[9,9]} = 3.18$. Because this is a two-tailed test we double these probabilities. Thus, the F-value of 4.03 represents a probability of $\alpha = 0.05$, since the right-hand tail area of $\alpha = 0.025$ is matched by a similar left-hand area to the left of $F_{.975[9,9]} = 1/F_{.025[9,9]} = 0.248$. Therefore, assuming the null hypothesis is true, the probability of observing an F-value greater than 4.00 and smaller than $1/4.00 = 0.25$ is $0.10 > P > 0.05$. Strictly speaking, the two sample variances are not significantly different—the two sexes were equally variable in their duration of survival. However, the outcome is close enough to the 5% significance level to make us suspicious that possibly the variances were in fact different. It would be desirable to repeat this experiment in the hope that more decisive results would emerge.

7.3 The hypothesis $H_0: \sigma_1^2 = \sigma_2^2$

A test of the null hypothesis that two normal populations represented by two samples have the same variance is illustrated in Box 7.1. As will be seen later, the acceptance of this null hypothesis is a prerequisite for some tests leading to a decision about whether the means of two samples come from the same population. However, this test is of interest in its own right. Repeatedly it will be necessary to test whether two samples have the same variance. In genetics we may need to know whether an offspring generation is more variable for a character than the parent generation. In systematics we would like to find out whether two local populations are equally variable. In experimental biology we may wish to demonstrate under which of two experimental setups the readings will be more variable. In general, the less variable setup would be preferred; if both setups were equally variable the experimenter would pursue the one that is simpler or less costly to undertake.

7.4 Heterogeneity among sample means

We shall now modify the data of Table 7.1, discussed in Section 7.1. Suppose these seven groups of houseflies did not represent random samples from the same population but resulted from the following experiment. Each sample had been reared in a separate culture jar and the medium in each of the culture jars had been prepared in a different way. Some had more water added, others more sugar, yet others more solid matter. Let us assume that sample 7 represents the standard medium against which we propose to compare the other samples. The various changes in the medium affect the sizes of the flies that emerge from it; this in turn affects the wing lengths we have been measuring.

We shall assume the following effects resulting from treatment of the medium:

Medium 1—decreases average wing length of a sample by 5 units
2—decreases average wing length of a sample by 2 units
3—does not change average wing length of a sample
4—increases average wing length of a sample by 1 unit
5—increases average wing length of a sample by 1 unit
6—increases average wing length of a sample by 5 units
7—(control) does not change average wing length of a sample

We shall symbolize the effect of treatment i as α_i. (Please note that this use of α is not related to its use as a symbol for type I error. There are, regrettably, more symbols needed in statistics than there are letters in the Greek and Roman alphabets.) Thus α_i assumes the following values for the above treatment effects:

$$\alpha_1 = -5 \qquad \alpha_4 = 1$$
$$\alpha_2 = -2 \qquad \alpha_5 = 1$$
$$\alpha_3 = 0 \qquad \alpha_6 = 5$$
$$\alpha_7 = 0$$

Note that $\sum^{a}\alpha_i = 0$; that is, the sum of the effects cancels out. This is a convenient property that is generally postulated, but it is unnecessary for our argument. We can now modify Table 7.1 by adding the appropriate values of α_i to each sample. In sample 1 the value of α_1 is -5; therefore the

TABLE 7.3

Data of Table 7.1 with fixed-treatment effects α_i or random effects A_i added to each sample.

	a groups ($a = 7$)							Computation of sum of squares of means	Computation of total sum of squares
	1	*2*	*3*	*4*	*5*	*6*	*7*		
α_i or A_i	−5	−2	0	+1	+1	+5	0		
	36	46	40	41	50	45	41		
	39	47	50	40	42	53	46		
n individuals per group ($n = 5$)	43	47	44	47	51	56	54		
	38	47	48	47	40	52	44		
	37	43	50	42	43	56	42		
$\sum^{n} Y$	193	230	232	217	226	262	227	$\sum^{a} \bar{Y} = 317.4$	$\sum^{an} Y = 1587$
$\bar{Y}$	38.6	46.0	46.4	43.4	45.2	52.4	45.4	$\bar{\bar{Y}} = 45.34$	$\bar{Y} = 45.34$
	$\begin{pmatrix} 43.6 \\ -5 \end{pmatrix}$	$\begin{pmatrix} 48.0 \\ -2 \end{pmatrix}$	$\begin{pmatrix} 46.4 \\ +0 \end{pmatrix}$	$\begin{pmatrix} 42.4 \\ +1 \end{pmatrix}$	$\begin{pmatrix} 44.2 \\ +1 \end{pmatrix}$	$\begin{pmatrix} 47.4 \\ +5 \end{pmatrix}$	$\begin{pmatrix} 45.4 \\ +0 \end{pmatrix}$	$\left(45.34 + \frac{-5 - 2 + 1 + 1 + 5}{7}\right)$	
$\sum^{n} Y^2$	7479	10,592	10,840	9463	10,314	13,810	10,413	$\sum^{a} \bar{Y}^2 = 14{,}492.44$	$\sum^{an} Y^2 = 72{,}911$
$\sum^{n} y^2$	29.2	12.0	75.2	45.2	98.8	81.2	107.2	$\sum^{a} (\bar{Y} - \bar{\bar{Y}})^2 = 100.617$	$\sum^{an} y^2 = 951.886$

TABLE 7.4

Data of Table 7.3 arranged in the manner of Table 7.2.

		a groups				
		1	*2*	*3* ...	*i* ...	*a*
n items	1	$Y_{11} + \alpha_1$	$Y_{21} + \alpha_2$	$Y_{31} + \alpha_3$ ...	$Y_{i1} + \alpha_i$ ...	$Y_{a1} + \alpha_a$
	2	$Y_{12} + \alpha_1$	$Y_{22} + \alpha_2$	$Y_{32} + \alpha_3$ ...	$Y_{i2} + \alpha_i$ ...	$Y_{a2} + \alpha_a$
	3	$Y_{13} + \alpha_1$	$Y_{23} + \alpha_2$	$Y_{33} + \alpha_3$ ...	$Y_{i3} + \alpha_i$ ...	$Y_{a3} + \alpha_a$
	⋮	⋮	⋮	⋮	⋮	⋮
	j	$Y_{1j} + \alpha_1$	$Y_{2j} + \alpha_2$	$Y_{3j} + \alpha_3$ ...	$Y_{ij} + \alpha_i$ ...	$Y_{aj} + \alpha_a$
	⋮	⋮	⋮	⋮	⋮	⋮
	n	$Y_{1n} + \alpha_1$	$Y_{2n} + \alpha_2$	$Y_{3n} + \alpha_3$ ...	$Y_{in} + \alpha_i$ ...	$Y_{an} + \alpha_a$
Sums		$\sum^{n} Y_1 + n\alpha_1$	$\sum^{n} Y_2 + n\alpha_2$	$\sum^{n} Y_3 + n\alpha_3$ ...	$\sum^{n} Y_i + n\alpha_i$ ...	$\sum^{n} Y_a + n\alpha_a$
Means		$\bar{Y}_1 + \alpha_1$	$\bar{Y}_2 + \alpha_2$	$\bar{Y}_3 + \alpha_3$ ...	$\bar{Y}_i + \alpha_i$ ...	$\bar{Y}_a + \alpha_a$

first wing length, which was 41 (see Table 7.1), now becomes 36, the second wing length, formerly 44, becomes 39, and so on. For the second sample α_2 is -2, changing the first wing length from 48 to 46. Where α_i is 0, the wing lengths do not change; where α_i is positive, they are increased by the magnitude indicated. The changed values can be inspected in Table 7.3, which is arranged identically to Table 7.1.

We now repeat our previous computations. We first calculate the sum of squares of the first sample and find it to be 29.2. If you compare this value with the sum of squares of the first sample in Table 7.1, you find the two values to be identical. Similarly, all other values of $\sum^{n} y^2$, the sum of squares of each group, are identical to their previous values. Why is this so? The effect of adding α_i to each group is simply that of an additive code, since α_i is constant for any one group. We saw in Appendix A1.2 that additive codes do not affect sums of squares or variances. Therefore, not only is each separate sum of squares the same as before, but the average variance within groups is still 16.029. Now let us compute the variance of the means. It is $100.617/6 = 16.770$, which is a value much higher than the variance of means found before, 4.236. When we multiply times $n = 5$ to get an estimate of σ^2 we obtain the variance of groups, which now is 83.848 and is no longer even close to an estimate of σ^2. We repeat the F-test with the new variances and find that $F_s = 83.848/16.029 = 5.23$, which is much greater than the closest critical value of $F_{.05[6,24]} = 2.51$. In fact, the observed F_s is greater than $F_{.01[6,24]} = 3.67$. Clearly, the upper variance, representing the variance among groups, has become significantly larger. The two variances are most unlikely to represent the same parametric variance.

What has happened? We can easily explain it by means of Table 7.4, which represents Table 7.3 symbolically in the manner that Table 7.2 represented Table 7.1. We note that each group has a constant α_i added and that this constant changes the sums of the groups by $n\alpha_i$ and the means of these groups by α_i. In Section 7.1 we computed the variance within groups as

$$\frac{1}{a(n-1)}\sum_{i=1}^{i=a}\sum_{j=1}^{j=n}(Y_{ij}-\bar{Y}_i)^2$$

If we try to repeat this, our formula becomes more complicated because to each Y_{ij} and each $\bar{Y}_i$ there has now been added α_i. We therefore write

$$\frac{1}{a(n-1)}\sum_{i=1}^{i=a}\sum_{j=1}^{j=n}[(Y_{ij}+\alpha_i)-(\bar{Y}_i+\alpha_i)]^2$$

When we open the parentheses inside the square brackets, the second α_i changes sign and the α_i's cancel out, leaving the expression exactly as before, substantiating our earlier observation that the variance within groups does not change despite the treatment effects.

The variance of means was previously calculated by the formula

$$\frac{1}{a-1}\sum_{i=1}^{i=a}(\bar{Y}_i-\bar{\bar{Y}})^2$$

However, from Table 7.4 we see that the new grand mean equals

$$\frac{1}{a}\sum_{i=1}^{i=a}(\bar{Y}_i+\alpha_i)=\frac{1}{a}\sum_{i=1}^{i=a}\bar{Y}_i+\frac{1}{a}\sum_{i=1}^{i=a}\alpha_i=\bar{\bar{Y}}+\bar{\alpha}$$

When we substitute the new values for the group means and the grand mean, the formula appears as

$$\frac{1}{a-1}\sum_{i=1}^{i=a}[(\bar{Y}_i+\alpha_i)-(\bar{\bar{Y}}+\bar{\alpha})]^2$$

which in turn yields

$$\frac{1}{a-1}\sum_{i=1}^{i=a}[(\bar{Y}_i-\bar{\bar{Y}})+(\alpha_i-\bar{\alpha})]^2$$

Squaring the expression in the square brackets, we obtain the terms

$$\frac{1}{a-1}\sum_{i=1}^{i=a}(\bar{Y}_i-\bar{\bar{Y}})^2+\frac{1}{a-1}\sum_{i=1}^{i=a}(\alpha_i-\bar{\alpha})^2+\frac{2}{a-1}\sum_{i=1}^{i=a}(\bar{Y}_i-\bar{\bar{Y}})(\alpha_i-\bar{\alpha})$$

The first of these terms we immediately recognize as the previous variance of the means, $s_{\bar{Y}}^2$. The second is a new quantity but is familiar by general appearance. It clearly is a variance or at least a quantity akin to a variance. The third expression is a new type. It is a so-called covariance, which we have not yet encountered. We shall not be concerned with it at this stage except to say that in cases such as the present one, where the magnitude of treatment effects α_i is independent of the $\bar{Y}_i$ to which they are added, the expected value of this quantity is zero; hence it would not contribute to the new variance of means.

The independence of the treatments effects and the sample means is an important concept that we must clearly understand. If we had not applied different treatments to the medium jars, but simply treated all jars as controls, we would still have obtained differences among the wing length means. Those are the differences found in Table 7.1 with random sampling from the same population. By chance some of these means are greater, some are smaller. In our planning of the experiment we had no way of predicting which sample means would be small and which would be large. Therefore, in planning our treatments we had no way of matching up large treatment effect, such as that of medium 6, with the mean that by chance would be the greatest, as that for sample 2. Also, the smallest sample mean (sample 4) is not associated with the smallest treatment effect. Only if the magnitude of the treatment effects were deliberately correlated with the sample means (this would be difficult to do in the experiment designed here), would the third term in the expression, the covariance, have an expected value other than zero.

The second term in the above expression is clearly added as a result of the treatment effects. It is analogous to a variance but cannot be called such since it is not based on a random variable but rather on deliberately chosen treatments largely under our control. By changing the magnitude and nature of the treatments we can more or less alter the variancelike quantity at will. We shall, therefore, call it the *added component due to treatment effects*. Since the α_i's are arranged so that $\bar{\alpha} = 0$, we can rewrite the middle term as

$$\frac{1}{a-1}\sum_{i=1}^{i=a}(\alpha_i - \bar{\alpha})^2 = \frac{1}{a-1}\sum_{i=1}^{i=a}\alpha_i^2 = \frac{1}{a-1}\sum^{a}\alpha^2$$

In analysis of variance we multiply the variance of the means by n in order to estimate the parametric variance of the items. As you know, we call the quantity so obtained the variance of groups. When we do this for the case in which treatment effects are present we obtain

$$n\left(s_{\bar{Y}}^2 + \frac{1}{a-1}\sum^{a}\alpha^2\right) = s^2 + \frac{n}{a-1}\sum^{a}\alpha^2$$

Thus we see that the estimate of the parametric variance of the population is increased by the quantity

$$\frac{n}{a-1}\sum^{a}\alpha^2$$

which is n times the added component due to treatment effects. We found the variance ratio F_s to be significantly greater than could be reconciled with the null hypothesis. It is now obvious why this is so. We were testing the variance ratio expecting to find F approximately equal to $\sigma^2/\sigma^2 = 1$. In fact, however, we have

$$F \approx \frac{\sigma^2 + \dfrac{n}{a-1}\displaystyle\sum^{a}\alpha^2}{\sigma^2}$$

It is clear from this formula, deliberately displayed in this lopsided manner, that the F-test is sensitive to the presence of the added component due to treatment effects.

At this point you should have your first insight into the purpose of analysis of variance. It permits us to test whether there are added treatment effects—that is, whether a group of means can simply be considered random samples from the same population or whether treatments that have affected each group separately have resulted in shifting these means sufficiently so that they can no longer be considered samples from the same population. If that is so, an added component due to treatment effects will be present and may be detected by an F-test in the significance test of the analysis of variance. In such a study we are generally not interested in the magnitude of

$$\frac{n}{a-1} \sum^{a} \alpha^2$$

but we are interested in the magnitude of the separate values of α_i. In our example these are the effects of different formulations of the medium on wing length. If instead of housefly wing length we were measuring blood pressure in samples of rats, and the different groups had been subjected to different drugs or different doses of the same drug, the quantities α_i would represent the effects of drugs on the blood pressure, which is clearly the issue of interest to the investigator. We may also be interested in studying differences of the type $\alpha_1 - \alpha_2$, leading us to the question of the significance of the differences between the effects of any two types of medium or any two drugs. But we are a little ahead of our story.

When analysis of variance involves treatment effects of the type just studied, we call it a Model I anova. Later in this chapter (Section 7.6), Model I will be defined precisely. There is another model, called a Model II anova, in which the added effects for each group are not fixed treatments but are random effects. By this we mean that we have not deliberately planned or fixed the treatment for any one group but that the actual effects on each group are random and only partly under our control. Suppose that the seven samples of houseflies in Table 7.3 represented the offspring of seven randomly selected females from a population reared on a uniform medium. There would be genetic differences among these females and their seven broods would reflect this. The exact nature of these differences is unclear and unpredictable. Before actually measuring them, we have no way of knowing whether brood 1 would have longer wings than brood 2, nor have we any way of controlling this experiment so that brood 1 would in fact grow longer wings. So far as we can ascertain, the genetic factors for wing length are distributed in an unknown manner in the population of houseflies (we might hope that they are normally distributed), and our sample of seven is a random sample of these factors. Another example for a Model II might be that instead of making up our seven cultures from a single batch of medium, we have prepared seven batches separately, one right after the other, and are now analyzing the variation among the batches. We would not be interested in the exact differences from batch to batch. Even if these were measured we would not be in a position to interpret them. Not having deliberately varied batch 3, we have no idea why it should produce longer wings than batch 2, for example. We would, however, be interested in the magnitude of the variance of the added effects. Thus, if we used seven jars of medium derived from one batch, we could expect the variance of the jar means to be $\sigma^2/5$, since there were 5 flies per jar. But when based on different batches of medium the variance could be expected to be greater, because all the imponderable accidents of formulation and environmental differences during medium preparation that make one batch of medium different from another would come into play. Interest would focus on the added variance component arising from differences among batches. Similarly, in the other example we would be interested in the added variance component arising from genetic differences among the females.

We shall now take a rapid look at the algebraic formulation of the anova in the case of Model II. In Table 7.3 the second row at the head of the data columns shows not only α_i but also A_i, which is the symbol we shall use for a random group effect. We use a capital letter to indicate that the effect is a variable. The algebra of calculating the two estimates of the population variance is the same as in Model I, except that in place of α_i we imagine A_i substituted in Table 7.4. The estimate of the variance among means now represents the quantity

$$\frac{1}{a-1}\sum_{i=1}^{i=a}(\bar{Y}_i - \bar{\bar{Y}})^2 + \frac{1}{a-1}\sum_{i=1}^{i=a}(A_i - \bar{A})^2 + \frac{2}{a-1}\sum_{i=1}^{i=a}(\bar{Y}_i - \bar{\bar{Y}})(A_i - \bar{A})$$

The first term is the variance of means $s_{\bar{Y}}^2$ as before, and the last term is the covariance between the group means and the random effects A_i, the expected value of which is zero (as before) because the random effects are independent of the magnitude of the means. The middle term is a true variance since A_i is a random variable. We symbolize it by s_A^2 and call it the *added variance component among groups*. It would represent the added variance component among females or among medium batches, depending on which of the designs discussed above we think of. The existence of this added variance component is demonstrated by the F-test. If the groups are random samples we may expect F to approximate $\sigma^2/\sigma^2 = 1$, but with an added variance component the expected ratio is

$$F \approx \frac{\sigma^2 + n\sigma_A^2}{\sigma^2}$$

Note that σ_A^2, the parametric value of s_A^2, is multiplied by n, since we have to multiply the variance of means by n to obtain an independent estimate of the variance of the population. In a Model II we are not interested in the magnitude of any A_i or in differences such as $A_1 - A_2$, but we are interested in the magnitude of σ_A^2 and its relative magnitude with respect to σ^2, which is generally expressed as the percentage $100s_A^2/(s^2 + s_A^2)$. Since the variance among groups estimates $\sigma^2 + n\sigma_A^2$, we can calculate s_A^2 as

$$\frac{1}{n}\,(\text{variance among groups} - \text{variance within groups})$$

$$= \frac{1}{n}\,[(s^2 + ns_A^2) - s^2] = \frac{1}{n}\,(ns_A^2) = s_A^2$$

For the present example $s_A^2 = \frac{1}{5}(83.848 - 16.029) = 13.56$. This added variance component among groups is

$$\frac{100 \times 13.56}{16.029 + 13.56} = \frac{1356}{29.589} = 45.83\%$$

of the sum of the variances among and within groups. Model II will be formally discussed at the end of this chapter (Section 7.7); the methods of estimating variance components are treated in detail in the next chapter.

7.5 Partitioning the total sum of squares and degrees of freedom

So far we have ignored one other variance that can be computed from the data in Table 7.1. If we remove the classification into groups, we can consider the housefly data to be a single sample of $an = 35$ wing lengths and calculate the mean and variance of these items in the conventional manner. The various quantities necessary for this computation are shown in the last column at the right in Tables 7.1 and 7.3, headed Computation of total sum of squares. We obtain a mean of $\bar{Y} = 45.34$ for the sample in Table 7.1, which is, of course, the same as the quantity $\bar{\bar{Y}}$ computed previously from the seven group means. The sum of squares of the 35 items is 575.886, which gives a variance of 16.938 when divided by 34 degrees of freedom. Repeating these computations for the data in Table 7.3 we obtain $\bar{Y} = 45.34$ (the same as in Table 7.1 because $\sum^{a}\alpha_i = 0$) and $s^2 = 27.997$, which is considerably greater than the corresponding variance from Table 7.1. The total variance computed from all an items is another estimate of σ^2. It is a good estimate in the first case, but in the second sample (Table 7.3), where added components due to treatment effects or added variance components are present, it is a poor estimate of the population variance.

However, the purpose of calculating the total variance in an anova is not for using it as yet another estimate of σ^2 but for ease of computation. This is best seen when we arrange our results in a conventional *analysis of variance table* as shown in Table 7.5. Such a table is divided into four columns. The first identifies the source of variation as among groups, within groups,

TABLE 7.5

Anova table for data in Table 7.1.

	(1) Source of variation	*(2)* df	*(3)* Sums of squares SS	*(4)* Mean squares MS
$\bar{Y} - \bar{\bar{Y}}$	Among groups	6	127.086	21.181
$Y - \bar{Y}$	Within groups	28	448.800	16.029
$Y - \bar{\bar{Y}}$	Total	34	575.886	16.938

and total (groups amalgamated to form a single sample). The column headed *df* gives the degrees of freedom by which the sums of squares pertinent to that source of variation must be divided in order to yield the variances. The degrees of freedom for variation among groups is $a - 1$, that for variation within groups is $a(n - 1)$, and that for the total variation is $an - 1$. The next two columns show sums of squares and variances, respectively. Notice that the sums of squares entered in the anova table are the sum of squares among groups, the sum of squares within groups, and the sum of squares of the total sample of an items. You will note that variances are not called such in anova, but are generally called *mean squares*, since, in a Model I anova, they do not estimate a population variance. These quantities are not true *mean* squares because the sums of squares are divided by the degrees of freedom rather than sample size. The sums of squares and mean squares are frequently abbreviated SS and MS, respectively.

The sums of squares and mean squares in Table 7.5 are the same as those obtained previously, except for minute rounding errors. Note, however, an important property of the sums of squares. They were obtained independently of each other, but when we add the SS among groups to the SS within groups we obtain the total SS. The sums of squares are additive! Another way of saying this is that we can decompose the total sum of squares into a portion due to variation among groups and another portion due to variation within groups. Observe that the degrees of freedom are also additive and that the total of 34 *df* can be decompsed into 6 *df* among groups and 28 *df* within groups. Thus, if we know any two of the sums of squares (and their appropriate degrees of freedom), we can compute the third and complete our analysis of variance. Note that the mean squares are not additive. This is obvious since generally $(a + b)/(c + d) \neq a/c + b/d$.

We shall use the computational formula for sum of squares [Expression (3.7)] to demonstrate why these sums of squares are additive. Although it is an algebraic derivation, it is placed here rather than in the Appendix because these formulas will also lead us to the actual computational formulas for analysis of variance. As you may have suspected, the method we used to obtain the sums of squares is not the most rapid computational procedure; we shall employ a far simpler method for routine computation. The sum of squares of means in simplified notation is

$$SS_{\text{means}} = \sum^{a} (\bar{Y} - \bar{\bar{Y}})^2 = \sum^{a} \bar{Y}^2 - \frac{\left(\sum^{a} \bar{Y}\right)^2}{a}$$

$$= \sum^{a} \left(\frac{1}{n} \sum^{n} Y\right)^2 - \frac{1}{a}\left[\sum^{a} \left(\frac{1}{n} \sum^{n} Y\right)\right]^2$$

$$= \frac{1}{n^2} \sum^{a} \left(\sum^{n} Y\right)^2 - \frac{1}{an^2}\left(\sum^{a}\sum^{n} Y\right)^2$$

Note that the deviation of means from the grand mean is first rearranged to fit the computational formula [Expression (3.7)] and then each mean is written in terms of its constituent variates. Collection of denominators outside the summation signs yields the final desired form. To obtain the sum of squares of groups, we multiply SS_{means} by n, as before. This yields

$$SS_{\text{groups}} = n \times SS_{\text{means}} = \frac{1}{n} \sum^{a} \left(\sum^{n} Y\right)^2 - \frac{1}{an}\left(\sum^{a}\sum^{n} Y\right)^2$$

Next we evaluate the sum of squares within groups:

$$SS_{\text{within}} = \sum^{a}\sum^{n} (Y - \bar{Y})^2 = \sum^{a} \left[\sum^{n} Y^2 - \frac{1}{n}\left(\sum^{n} Y\right)^2\right]$$

$$= \sum^{a}\sum^{n} Y^2 - \frac{1}{n}\sum^{a}\left(\sum^{n} Y\right)^2$$

The total sum of squares represents

$$SS_{\text{total}} = \sum^{a}\sum^{n} (Y - \bar{\bar{Y}})^2$$

$$= \sum^{a}\sum^{n} Y^2 - \frac{1}{an}\left(\sum^{a}\sum^{n} Y\right)^2$$

We now copy the formulas for these sums of squares, slightly rearranged as follows:

$$\begin{aligned}
SS_{\text{groups}} &= \quad \frac{1}{n} \sum^{a} \left(\sum^{n} Y\right)^2 && - \frac{1}{an}\left(\sum^{a}\sum^{n} Y\right)^2 \\
SS_{\text{within}} &= -\frac{1}{n} \sum^{a} \left(\sum^{n} Y\right)^2 + \sum^{a}\sum^{n} Y^2 && \\
\hline
SS_{\text{total}} &= \qquad \sum^{a}\sum^{n} Y^2 && - \frac{1}{an}\left(\sum^{a}\sum^{n} Y\right)^2
\end{aligned}$$

Adding the expression for SS_{groups} to that for SS_{within} we obtain a quantity that is identical to the one we just developed as SS_{total}. This demonstration explains why the sums of squares are additive.

TABLE 7.6

Anova table for data in Table 7.3.

	(1) Source of variation	*(2)* df	*(3)* Sums of squares SS	*(4)* Mean squares MS
$\bar{Y} - \bar{\bar{Y}}$	Among groups	6	503.086	83.848
$Y - \bar{Y}$	Within groups	28	448.800	16.029
$Y - \bar{\bar{Y}}$	Total	34	951.886	27.997

We shall not go through any derivation but simply state that the degrees of freedom pertaining to the sums of squares are also additive. The total degrees of freedom are split up into the degrees of freedom corresponding to variation among groups and those of variation of items within groups.

Before we continue, let us review the meaning of the three mean squares in the anova. The total *MS* is a statistic of dispersion of the 35 (an) items around their mean, the grand mean 45.34. It describes the variance in the entire sample due to all and sundry causes and estimates σ^2 when there are no added treatment effects or variance components among groups. The within-group *MS*, also known as the *individual* or *intragroup* or *error mean square,* gives the average dispersion of the 5 (n) items in each group around the group means. If the a groups are random samples from a common homogeneous population, the within-group *MS* should estimate σ^2. The *MS* among groups is based on the variance of group means, which describes the dispersion of the 7 (a) group means around the grand mean. If the groups are random samples from a homogeneous population, the expected variance of their mean will be σ^2/n. Therefore, in order to have all three variances of the same order of magnitude, we multiply the variance of means by n to obtain the variance among groups. If there are no added treatment effects or variance components the *MS* among groups is an estimate of σ^2. Otherwise it is an estimate of

$$\sigma^2 + \frac{n}{a-1}\sum^{a}\alpha^2 \qquad \text{or} \qquad \sigma^2 + n\sigma_A^2$$

depending on whether the anova at hand is Model I or II.

The additivity relations we have just learned are independent of the presence of added treatment or random effects. We could show this algebraically, but it is simpler to inspect Table 7.6, which summarizes the anova of Table 7.3 in which α_i or A_i are added to each sample. The additivity relation still holds, although the values for group *SS* and total *SS* are different from those of Table 7.5.

Another way of looking at the partitioning of the variation is to study the deviation from means in a particular case. Referring to Table 7.1 we can look at the wing length of the first individual in the 7th group, which happens to be 41. Its deviation from its group mean is

$$Y_{71} - \bar{Y}_7 = 41 - 45.4 = -4.4$$

The deviation of the group mean from the grand mean is

$$\bar{Y}_7 - \bar{\bar{Y}} = 45.4 - 45.34 = 0.06$$

and the deviation of the individual wing length from the grand mean is

$$Y_{71} - \bar{\bar{Y}} = 41 - 45.34 = -4.34$$

Note that these deviations are additive. The deviation of the item from the group mean and that of the group mean from the grand mean add to the total deviation of the item from the grand mean. These deviations are stated algebraically as $(Y - \bar{Y}) + (\bar{Y} - \bar{\bar{Y}}) = (Y - \bar{\bar{Y}})$. Squaring and summing these deviations for an items will result in

$$\sum^{a} \sum^{n} (Y - \bar{Y})^2 + n \sum^{a} (\bar{Y} - \bar{\bar{Y}})^2 = \sum^{an} (Y - \bar{\bar{Y}})^2$$

Before squaring, the deviations were in the relationship $a + b = c$. After squaring we would expect them in the form $a^2 + b^2 + 2ab = c^2$. What happened to the cross-product term corresponding to $2ab$? This is

$$2 \sum^{an} (Y - \bar{Y})(\bar{Y} - \bar{\bar{Y}}) = 2 \sum^{a} \left[(\bar{Y} - \bar{\bar{Y}}) \sum^{n} (Y - \bar{Y})\right]$$

a covariance type term that is always zero, since $\sum^{n} (Y - \bar{Y}) = 0$ for each of the a groups (proof in Appendix A1.1).

We identify the deviations represented by each level of variation at the left margins of the tables giving the analysis of variance results (Tables 7.5 and 7.6). Note that the deviations add up correctly: the deviation among groups plus the deviation within groups equals to the total deviation of items in the analysis of variance, $(\bar{Y} - \bar{\bar{Y}}) + (Y - \bar{Y}) = (Y - \bar{\bar{Y}})$.

7.6 Model I anova

An important point to remember is that the basic setup of data, as well as the actual computation and significance test, in most cases is the same for both models. The purposes of analysis of variance differ for the two models, as do some of the supplementary tests and computations following the initial significance test.

Let us now try to resolve the variation found in an analysis of variance case. This will not only lead us to a more formal interpretation of anova but also give us a deeper understanding of the nature of variation itself. For purposes of discussion we shall refer back to the housefly wing lengths of Table 7.3. We ask the question, "What makes any given housefly wing length assume the value it does?" The third wing length of the first sample of flies is recorded as 43 units. How can we explain such a reading?

First of all, knowing nothing else about this individual housefly, our best guess of its wing length is the grand mean of the population, which we know to be $\mu = 45.5$. However, we have additional information about this fly. It is a member of group 1, which has undergone a treatment shifting the mean of the group downward by 5 units. Therefore, $\alpha_1 = -5$ and we would expect our individual Y_{13} (the third individual of group 1) to measure $45.5 - 5 = 40.5$ units. In fact, however, it is 43 units, which is 2.5 units above this latest expectation. To what can we ascribe this deviation? It is individual variation of the flies within a group because of the variance of individuals in the population ($\sigma^2 = 15.21$). All the genetic and environmental effects that make one housefly different from another housefly come into play to produce this variance. By means of carefully designed experiments we might learn something about the causation of this variance and attribute it to certain specific genetic or environmental factors. We might also be able to eliminate some of the variance. For instance, by using only full sibs (brothers and sisters) in any one culture jar we would decrease the genetic variation in individuals, and undoubtedly the variance within groups would be smaller. However, it is hopeless to try to eliminate all variance completely. Even if we could remove all genetic variance, there would still be environmental variance, and even in the most improbable case in which we could remove both there would remain measurement error, so that we would never obtain exactly the same reading even on the same individual fly. The within groups MS always remains as a residual, greater or smaller from experiment to experiment—part of the nature of things. This is why the within groups variance is also called the error variance or error mean square. It is not an error in the sense of someone having made a mistake, but in the sense of providing you with a measure of the variation you have to contend with when trying to estimate significant differences among the groups. The error variance is composed of individual deviations for each individual, symbolized by ϵ_{ij}, the random component of the jth individual variate in the ith group. In our case, $\epsilon_{13} = 2.5$, since the actual observed value is 2.5 units above its expectation of 40.5.

We shall now state this relationship more formally. In a Model I analysis of variance we assume that the differences among group means, if any, are due to the fixed treatment effects determined by the experimenter. The purpose of the analysis of variance is to estimate the true differences among the group means. Any single variate can be decomposed as follows:

$$Y_{ij} = \mu + \alpha_i + \epsilon_{ij} \tag{7.3}$$

where $i = 1, \ldots, a$, $j = 1, \ldots, n$, ϵ_{ij} represents an independent, normally distributed variable with mean $\bar{\epsilon}_{ij} = 0$, and variance $\sigma_\epsilon^2 = \sigma^2$. Therefore a given reading is composed of the grand mean μ of the population, a fixed deviation α_i of the mean of group i from the grand mean μ, and a random deviation ϵ_{ij} of the jth individual of group i from its expectation, which is $(\mu + \alpha_i)$. Remember that both α_i and ϵ_{ij} can be positive as well as negative. The expected value (mean) of the ϵ_{ij}'s is zero and their variance is the parametric variance of the population, σ^2. For all the assumptions of the analysis of variance to hold, the distribution of ϵ_{ij} must be normal.

In a Model I anova we test for differences of the type $\alpha_1 - \alpha_2$ among the group means by testing for the presence of an added component due to treatments. If we find that such a component is present, we reject the null hypothesis that the groups come from the same population and accept the alternative hypothesis that at least some of the group means are different from each other, which indicates that at least some of the α_i's are unequal in magnitude. Next we generally wish to test which α_i's are different from each other. This is done by significance tests, with alternative hypotheses such as $H_1: \alpha_1 > \alpha_2$ or $H_1: \frac{1}{2}(\alpha_1 + \alpha_2) > \alpha_3$. In words, these test whether the mean of group 1 is significantly greater than the mean of group 2, or whether the mean of group 3 is smaller than the average of the means of groups 1 and 2.

Some examples of Model I analyses of variance in various biological disciplines follow. An experiment in which we try the effects of different drugs on batches of animals results in a Model I anova. We are interested in the results of the treatments and the differences between them. The treatments are fixed and determined by the experimenter. This is true also when we test the effects of different doses of a given factor—a chemical or the amount of light to which a plant has been exposed or temperatures at which culture bottles of insects have been reared. The treatment does not have to be entirely understood and manipulated by the experimenter; so long as it is fixed and repeatable, Model I will apply. If we had wanted to compare the birth weights of the Chinese children in the hospital in Malaya with weights of Chinese children born in a hospital on the Chinese mainland, it would also have been a Model I anova. The treatment effects would then be "mainland versus Malaya," which sums up a whole series of different factors, genetic and environmental—some known to us but most of them not understood. However, this is a definite treatment we can describe and also repeat; namely, we can, if we wish, again sample birth weights of infants in Malaya as well as on the Chinese mainland.

Another example of Model I anova would be a study of body weights for several age groups of animals. The treatments would be the ages, which are fixed. If we find that there is a significant difference in weight among the ages, we might follow this up with the question of whether there is a difference from age 2 to age 3 or only from age 1 to age 2. To a very large extent Model I anovas are the result of an experiment and of deliberate manipulation of factors by the experimenter. However, the study of differences such as the comparison of birth weights from two countries, while not an experiment proper, also falls into this category.

7.7 Model II anova

The structure of variation in a Model II anova is quite similar to that in Model I:

$$Y_{ij} = \mu + A_i + \epsilon_{ij} \tag{7.4}$$

where $i = 1, \ldots, a, j = 1, \ldots, n, \epsilon_{ij}$ represents an independent, normally distributed variable with mean $\bar{\epsilon}_{ij} = 0$ and variance $\sigma_\epsilon^2 = \sigma^2$, and A_i represents a normally distributed variable, independent of all ϵ's, with mean $\bar{A}_i = 0$ and variance σ_A^2. The main distinction is that in place of fixed treatment effects α_i we now consider random effects A_i, differing from group to group. Since the effects are random, it is futile to estimate the magnitude of these random effects for any one group, or the differences from group to group, but we can estimate their variance, the added variance component among groups σ_A^2. We test for its presence and estimate its magnitude s_A^2, as well as its percentage contribution to the variation in a Model II analysis of variance.

Some examples will illustrate the applications of Model II anova. Suppose we wish to determine the *DNA* content of rat liver cells. We take five rats and make three preparations from each of the five livers obtained. The assay readings will be for $a = 5$ groups with $n = 3$ readings per group. The five rats presumably are sampled at random from the colony available to the experimenter. They must be different in various ways, genetically and environmentally, but we have no definite information about the nature of these differences. Thus, if we learn that rat 2 has slightly more *DNA* in its liver cells than rat 3, we can do little with this information, because we are unlikely to have any basis for following up this problem. We will, however, be interested in estimating the variance of the three replicates within any one liver and the variance among the five rats; that is, is there variance σ_A^2 among rats in addition to the variance σ^2 expected on the basis of the three replicates? The variance among the three preparations presumably arises only from differences in technique and possibly from differences in *DNA* content in different parts of the liver (unlikely in a homogenate). Added variance among rats, if it existed, might be due to differences in ploidy or related phenomena. The relative amounts of variation among rats and "within" rats (= among preparations) would guide us in designing further studies of this sort. If there was little variance among the preparations and relatively more variation among the rats, we would need fewer preparations and more rats. On the other hand, if the variance among rats is proportionately smaller, we would use fewer rats and more preparations per rat.

In a study of the amount of variation in skin pigment in human populations we might wish to study different families within a homogeneous racial group and brothers and sisters within each family. The variance within families would be the error mean square and we would test for an added variance component among families. We would expect an added variance component σ_A^2 because there are genetic differences among families that determine amount of skin pigmentation. We would be especially interested in the relative proportions of the two variances σ^2 and σ_A^2 because they would provide us with important genetic information. From our knowledge of genetic theory we would expect the variance among families to be greater than the variance among brothers and sisters within a family.

The above examples illustrate the two types of problems involving Model II analysis of variance that are most likely to arise in biological work. One is concerned with the general problem of the design of an experiment and the magnitude of the experimental error at different levels of replication, such as error among replicates within rat livers and among rats, error among batches, experiments. and so forth. The other relates to variation among and within families, among and within females, among and within populations, and so forth, being concerned with the general problem of the relation between genetic and phenotypic variation.

Exercises 7

7.1 In a study comparing the chemical composition of the urine of chimpanzees and gorillas (Gartler, Firschein, and Dobzhansky, 1956), the following results were obtained. For 37 chimpanzees the variance for the amount of glutamic acid in milligrams per milligram of creatinine was 0.01069. A similar study based on six gorillas yielded a variance of 0.12442. Is there a significant difference between the variability in chimpanzees and gorillas? ANS. $F_s = 11.639$, $F_{.025[5,36]} \approx 2.90$.

7.2 The following data are from an experiment by Sewall Wright. He crossed Polish and Flemish giant rabbits and obtained 27 F_1 rabbits. These were inbred and 112 F_2 rabbits were obtained. We have extracted the following data on femur length of these rabbits.

	n	$\bar{Y}$	s
F_1	27	83.39	1.65
F_2	112	80.5	3.81

Is there a significantly greater amount of variability in femur lengths among the F_2 than among the F_1 rabbits? What well-known genetic phenomenon is illustrated by these data?

7.3 Show that it is possible to represent the value of an individual variate as follows: $\bar{Y}_{ij} = (\bar{\bar{Y}}) + (Y_i - \bar{\bar{Y}}) + (Y_{ij} - \bar{Y}_i)$. What do each of the terms in parentheses estimate in a Model I anova and in a Model II anova?

7.4 For the data in Table 7.3, make tables to represent partitioning of the value of each variate into its three components, $\bar{\bar{Y}}$, $(\bar{Y}_i - \bar{\bar{Y}})$, $(Y_{ij} - \bar{Y}_i)$. The first table would then consist of 35 values, all equal to the grand mean. In the second table all entries in a given column would be equal to the difference between the mean of that column and the grand mean. And the last table will consist of the deviations of each individual variate from its column mean. These tables represent one's estimates of the individual components of Expression (7.3). Compute the mean and sums of squares for each table.

8

Single Classification Analysis of Variance

We are now ready to study actual cases of analysis of variance in a variety of applications and designs. The present chapter deals with the simplest kind of anova, *single classification analysis of variance*. By this is meant that the groups of samples are classified by only a single criterion. Both interpretations of the seven samples of housefly wing lengths studied in the last chapter, different medium formulations (Model I), and progenies of different females (Model II) would represent a single criterion for classification. Other examples would be different temperatures at which groups of animals were raised or different soils in which samples of plants have been grown.

We shall start out in Section 8.1 by stating the basic computational formulas for analysis of variance, based on the topics covered in the previous chapter. Section 8.2 gives an example of the common case with equal sample sizes. We shall illustrate this case by means of a Model I anova. Since the basic computations for the analysis of variance are the same in either model, it is not necessary to repeat the illustration with a Model II. The latter model is featured in Section 8.3, which shows the minor computational complications resulting from unequal sample sizes, since all groups in the anova need not necessarily have the same sample size. Some computations special to a Model II anova are also shown—estimating variance components. Formulas become especially simple for the two-sample case (Section 8.4). In Model I of this case the mathematically equivalent t-test can be applied as well.

When a Model I analysis of variance has been found to be significant, leading to the conclusion that the means are not from the same population, we wish to test the means in a variety of ways to discover which pairs of means are different from each other, and whether the means can be divided into groups that are significantly different from each other. These so-called multiple comparisons tests are covered in Sections 8.5 and 8.6. The earlier section deals with so-called planned comparisons designed before the test is run, the latter section with a posteriori tests that suggest themselves to the experimenter as a result of his analysis.

8.1 Computational formulas

We saw in Section 7.5 that the total sum of squares and degrees of freedom can be additively partitioned into those pertaining to variation among groups and those of variation within groups. It is simplest to calculate the total sum of squares and the sum of squares among groups, leaving the sum of squares within groups to be obtained by the subtraction $SS_{\text{total}} - SS_{\text{groups}}$. This rule does not apply to digital computers; there the tedium of computing sums of squares within groups is of no consequence but accuracy is. In Section 7.5 we arrived at the following computational formulas for these sums of squares:

$$SS_{\text{total}} = \sum^{a}\sum^{n} Y^2 - \frac{1}{an}\left(\sum^{a}\sum^{n} Y\right)^2$$

$$SS_{\text{groups}} = \frac{1}{n}\sum^{a}\left(\sum^{n} Y\right)^2 - \frac{1}{an}\left(\sum^{a}\sum^{n} Y\right)^2$$

These formulas assume equal sample size n for each group and will be modified in Section 8.3 for unequal sample sizes. However, they suffice in their present form to illustrate some general points about computational procedures in analysis of variance.

We note first of all that the second, subtracted term in each sum of squares is identical. This term represents the sum of all the variates in the anova (the grand total), quantity squared and divided by the total number of variates. It is comparable to the second term in the computational formula for the ordinary sum of squares [Expression (3.7)]. This term is often called the *correction term* (abbreviated CT).

The first term for the total sum of squares is simple. It is the sum of all squared variates in the anova table. Thus the total sum of squares, which describes the variation of a single unstructured sample of an items, is simply the familiar sum of squares formula of Expression (3.7).

The first term of the sum of squares among groups is obtained by squaring the sum of the items of each group, dividing each square by its sample size and summing the quotients from this operation for each group. Since the sample size of each group is equal in the above formulas, we can first sum all the squares of the group sums and then divide their sum by the constant n.

From the formula for the sum of squares among groups emerges an important computational rule of analysis of variance. *To find the sum of squares among any set of groups, square the sum of each group and divide it by its sample size; sum the quotients of these operations and subtract from this sum a correction term. To find this correction term, sum all the items in the set, square this sum, and divide it by the number of items on which this sum is based.*

8.2 Equal n

We shall illustrate a single classification anova with equal sample sizes by a Model I example. The computation up to and including the first test of significance is identical for both models. Thus the computation of Box 8.1 could also serve for a Model II anova with equal sample size.

The data are from an experiment in plant physiology. They record the length in coded units of pea sections grown in tissue culture with auxin present. The purpose of the experiment was to test the effects of the addition of various sugars on growth as measured by length. Four experimental groups, representing three different sugars and one mixture of sugars, were used, plus one control without sugar. Ten observations (replicates) were made for each treatment. The term "treatment" already implies a Model I anova. It is obvious that the five groups do not represent random samples from all possible experimental conditions but were deliberately designed to test the effects of certain sugars on the growth rate. We are interested in the effect of the sugars on length, and our null hypothesis will be that there is no added component due to treatment effects among the five groups; that is, the population means are all assumed to be equal.

The computation is illustrated in Box 8.1. After quantities **1** through **7** have been evaluated they are entered into an analysis of variance table, as shown in the box. General formulas for such a table are shown first, followed by a table filled in for the specific example. We note 4 degrees of freedom among groups, there being five treatments, and 45 df within groups, representing 5 times $(10 - 1)$ degrees of freedom. We find that the mean square among groups is considerably greater than the error mean square, giving rise to a suspicion that an added component due to treatment effects is present. If the MS_{groups} is equal to or less than the MS_{within}, we do not bother going on

BOX 8.1

Single classification anova with equal sample sizes.

The effect of the addition of different sugars on length, in ocular units ($\times 0.114 =$ mm), of pea sections grown in tissue culture with auxin present: $n = 10$ (replications per group). This is a Model I anova.

	Treatments ($a = 5$)				
Observations, i.e., replications	*Control*	*2% glucose added*	*2% fructose added*	*1% glucose + 1% fructose added*	*2% sucrose added*
1	75	57	58	58	62
2	67	58	61	59	66
3	70	60	56	58	65
4	75	59	58	61	63
5	65	62	57	57	64
6	71	60	56	56	62
7	67	60	61	58	65
8	67	57	60	57	65
9	76	59	57	57	62
10	68	61	58	59	67
$\sum^{n} Y$	701	593	582	580	641
$\bar{Y}$	70.1	59.3	58.2	58.0	64.1

Source: Data by W. Purves.

Preliminary computations

1. Grand total $= \sum^{a}\sum^{n} Y = 701 + 593 + \cdots + 641 = 3097$

2. Sum of the squared observations $= \sum^{a}\sum^{n} Y^2$

$$= 75^2 + 67^2 + \cdots + 68^2 + 57^2 + \cdots + 67^2 = 193{,}151$$

3. Sum of the squared group totals divided by $n = \frac{1}{n}\sum^{a}\left(\sum^{n} Y\right)^2$

$$= \tfrac{1}{10}[701^2 + 593^2 + \cdots + 641^2] = \tfrac{1}{10}[1{,}929{,}055] = 192{,}905.50$$

4. Grand total squared and divided by total sample size = correction term

$$CT = \frac{1}{an}\left(\sum^{a}\sum^{n} Y\right)^2 = \frac{(3097)^2}{5 \times 10} = \frac{9{,}591{,}409}{50} = 191{,}828.18$$

5. $SS_{\text{total}} = \sum^{a}\sum^{n} Y^2 - CT$

$$= \text{quantity } \mathbf{2} - \text{quantity } \mathbf{4} = 193{,}151 - 191{,}828.18 = 1322.82$$

BOX 8.1 continued

6. $SS_{\text{groups}} = \frac{1}{n}\sum^{a}\left(\sum^{n} Y\right)^2 - CT$

$= \text{quantity } \mathbf{3} - \text{quantity } \mathbf{4} = 192{,}905.50 - 191{,}828.18 = 1077.32$

7. $SS_{\text{within}} = SS_{\text{total}} - SS_{\text{groups}}$

$= \text{quantity } \mathbf{5} - \text{quantity } \mathbf{6} = 1322.82 - 1077.32 = 245.50$

The anova table is constructed as follows.

Source of variation	*df*	*SS*	*MS*	F_s	*Expected MS*
$\bar{Y} - \bar{\bar{Y}}$ Among groups	$a - 1$	6	$\frac{6}{(a-1)}$	$\frac{MS_{\text{groups}}}{MS_{\text{within}}}$	$\sigma^2 + \frac{n}{a-1}\sum^{a}\alpha^2$
$Y - \bar{Y}$ Within groups	$a(n-1)$	7	$\frac{7}{a(n-1)}$		σ^2
$Y - \bar{\bar{Y}}$ Total	$an - 1$	5			

Substituting the computed values into the above table we obtain the following.

Anova table

Source of variation	*df*	*SS*	*MS*	F_s
$\bar{Y} - \bar{\bar{Y}}$ among groups (among treatments)	4	1077.32	269.33	49.33
$Y - \bar{Y}$ within groups (error, replicates)	45	245.50	5.46	
$Y - \bar{\bar{Y}}$ Total	49	1322.82		

$F_{.05[4,45]} = 2.58$ $\quad F_{.01[4,45]} = 3.77$ $\quad F_{.001[4,45]} = 5.57$

Conclusions.—There is a highly significant ($P \ll 0.01$) added component due to treatment effects in the mean square among groups (treatments). The different sugar treatments clearly have a significant effect on growth of the pea sections.

See Sections 8.5 and 8.6 for the completion of a Model I analysis of variance: that is, the method for determining which means are significantly different from each other.

with the analysis, for we would not have evidence for the presence of an added variance component. You may wonder how it could be possible for the MS_{groups} to be less than the MS_{within}. You must remember that these two are independent estimates. If there is no added variance component among groups, the estimate of the variance among groups is as likely to be less as it is to be greater than the variance within groups.

Expressions for the expected values of the mean squares are also shown in the first anova table of Box 8.1. They are the expressions you learned in the previous chapter for a Model I anova.

It may seem that we are carrying an unnecessary number of digits in the computations in Box 8.1. This is often necessary to ensure that the error sum of squares, quantity **7**, has sufficient accuracy.

Since ν_2 is relatively large, the critical values of F have been computed by harmonic interpolation in Table **V** (see footnote for Table **III** on harmonic interpolation). The critical values have been given here only to present a complete record of the analysis. Ordinarily, when confronted with this example, you would not bother working out these values of F. Comparison of the observed variance ratio $F_s = 49.33$ with $F_{.01[4,40]} = 3.83$, the conservative critical value (the next tabled F with less degrees of freedom) would convince you to reject the null hypothesis. The probability that the five groups differ as much as they do by chance is almost infinitesimally small. Clearly the sugars produce an added treatment effect, apparently inhibiting growth and consequently reducing the length of the pea sections.

At this stage we are not in a position to say whether each treatment is different from every other treatment, or whether the sugars are different from the control but not different from each other. Such tests are necessary to complete a Model I analysis but we defer their discussion until Sections 8.5 and 8.6.

8.3 Unequal *n*

This time we shall use a Model II analysis of variance for an example. Remember that up to and including the F-test for significance the computations are exactly the same whether the anova is Model I or Model II. We shall point out the stage in the computations at which there would be a divergence of operations depending on the model.

The example is shown in Table 8.1. It concerns a series of morphological measurements of the width of the scutum (dorsal shield) of samples of tick larvae obtained from four different host individuals of the cottontail rabbit. These four hosts were obtained at random from one locality. We know nothing about their origins or their genetic constitution. They represent a random sample of the population of host individuals from the given locality. We would not be in a position to interpret differences between larvae from any given

TABLE 8.1

Data and anova table for a single classification anova with unequal sample sizes. Width of scutum (dorsal shield) of larvae of the tick *Haemaphysalis leporispalustris* in samples from 4 cottontail rabbits. Measurements in microns. This is a Model II anova.

	Hosts ($a = 4$)			
	1	*2*	*3*	*4*
	380	350	354	376
	376	356	360	344
	360	358	362	342
	368	376	352	372
	372	338	366	374
	366	342	372	360
	374	366	362	
	382	350	344	
		344	342	
		364	358	
			351	
			348	
			348	
$\sum^{n_i} Y$	2978	3544	4619	2168
n_i	8	10	13	6
$\sum^{n_i} Y^2$	1,108,940	1,257,272	1,642,121	784,536

Source: Data by P. A. Thomas.

Anova table

	Source of variation	*df*	*SS*	*MS*	F_s
$\bar{Y} - \bar{\bar{Y}}$	Among groups (among hosts)	3	1808	602.7	5.26**
$Y - \bar{Y}$	Within groups (error; among larvae on a host)	33	3778	114.5	
$Y - \bar{\bar{Y}}$	Total	36	5586		

$F_{.05[3,33]} = 2.89 \qquad F_{.01[3,33]} = 4.44$

** = $P < 0.01$

Conclusion.—There is a significant ($P < 0.01$) added variance component among hosts for width of scutum in larval ticks.

host, since we know nothing of the origins of the individual rabbits. Population biologists are nevertheless interested in such analyses because they provide an answer to the following question. Are the variances of means of larval characters among hosts greater than expected on the basis of variances of the characters within hosts? We can calculate the average variance of width of larval scutum on a host. This will be our "error" term in the analysis of variance. We then test the observed mean square among groups and see if it contains an added component of variance. What would such an added component of variance represent? The mean square within host individuals (that is, of larvae on any one host) represents genetic differences among larvae and differences in environmental experiences of these larvae. Added variance among hosts demonstrates significant differentiation among the larvae possibly due to differences among the hosts affecting the larvae. It also may be due to genetic differences among the larvae should each host carry a family of ticks, or at least a population whose individuals are more related to each other than they are to tick larvae on other host individuals. The emphasis in this example is on the magnitudes of the variances. In view of the random choice of hosts this is a clear case of a Model II anova.

The computation follows the outline furnished in Box 8.1, except that the symbol $\sum^{n}$ now need to be written $\sum^{n_i}$ since sample sizes differ for each group. Steps **1**, **2**, and **4** through **7** are carried out as before. Only step **3** needs to be modified appreciably. It is

3. Sum of the squared group totals, each divided by its sample size

$$= \sum^{a} \frac{\left(\sum^{n_i} Y\right)^2}{n_i} = \frac{(2978)^2}{8} + \frac{(3544)^2}{10} + \cdots + \frac{(2168)^2}{6} = 4{,}789{,}091$$

The critical 5% and 1% values of F are shown below the anova table in Table 8.1 (2.89 and 4.44, respectively). You should confirm them for yourself in Table **V**. Note that the argument $\nu_2 = 33$ is not given. You therefore have to interpolate between arguments representing 30 and 40 degrees of freedom, respectively. The values shown were computed using harmonic interpolation. However, again, it was not necessary to carry out such an interpolation. The conservative value of F, $F_{\alpha[3,30]}$, is 2.92 and 4.51 for $\alpha = 0.05$ and $\alpha = 0.01$, respectively. The observed value of F_s is 5.26, considerably above the interpolated as well as the conservative value of $F_{.01}$. We therefore reject the null hypothesis ($H_0: \sigma_A^2 = 0$) that there is no added variance component among groups and that the two mean squares estimate the same variance, allowing a type I error of less than 1%. We accept instead the alternative hypothesis of the existence of an added variance component σ_A^2.

What is the biological meaning of this conclusion? For some reason or other the ticks on different host individuals differ more from each other than do individual ticks on any one host. This may be due to some modifying influence of individual hosts on the ticks (biochemical differences in blood, differences in the skin, differences in the environment of the host individual; all of them rather unlikely in this case) or may be due to genetic differences among the ticks. Possibly the ticks on each host represent a sibship (the descendants of a single pair of parents), and the differences of the ticks among host individuals represent genetic differences among families; or selection has acted differently on the tick populations on each host; or the hosts have migrated to the collection locality from different geographic areas in which the ticks differ in width of scutum. Of these various possibilities, genetic differences among sibships seem most reasonable, in view of the biology of the organism.

The computations up to this point would have been identical in a Model I anova. If this had been Model I, the conclusion would have been that there is a significant treatment effect rather than an added variance component. Now, however, we must complete the computations appropriate to a Model II anova. These will include the estimation of the added variance component and the calculation of percentage variation at the two levels.

Since sample size n_i differs among groups in this example, we cannot write $\sigma^2 + n\sigma_A^2$ for the expected MS_{groups}. It is obvious that no single value of n would be appropriate in the formula. We therefore use an average n; this, however, is not simply $\bar{n}$, the arithmetic mean of the n_i's, but is

$$n_0 = \frac{1}{a-1}\left(\sum^a n_i - \frac{\sum^a n_i^2}{\sum^a n_i}\right) \tag{8.1}$$

which is an average usually close to but always less than $\bar{n}$, unless sample sizes are equal, when $n_0 = \bar{n}$. In this example

$$n_0 = \frac{1}{4-1}\left([8 + 10 + 13 + 6] - \frac{8^2 + 10^2 + 13^2 + 6^2}{8 + 10 + 13 + 6}\right) = 9.009$$

From the expressions for the expected mean squares shown in the anova table, it is obvious how the variance component among groups σ_A^2 and the error variance σ^2 are obtained. Of course the values that we obtain are simply estimates and therefore are written as s_A^2 and s^2. The added variance component s_A^2 is estimated as $(MS_{\text{groups}} - MS_{\text{within}})/n$. Whenever sample sizes are unequal the denominator becomes n_0. In this example $(602.7 - 114.5)/9.009 = 54.190$. We are frequently not so much interested in the absolute values of these variance components but in their relative magnitudes. For this purpose we sum them and express each as a percentage of this sum. Thus $s^2 + s_A^2 = 114.5 + 54.190 = 168.690$ and s^2 and s_A^2 are 67.9% and 32.1% of this sum, respectively. Relatively more variation occurs within groups (larvae on a host) than among groups (hosts).

8.4 Two groups

A frequent test in statistics is to establish the *significance of the difference between two means*. This can easily be done by means of an *analysis of variance for two groups*. Box 8.2 shows this procedure for a Model I anova, the common case.

The example in Box 8.2 concerns the onset of reproductive maturity in water fleas, *Daphnia longispina*. This is measured as the average age (in days) at beginning of reproduction. Each variate in the table is in fact an average and a possible flaw in the analysis might be that these averages are not based on equal sample sizes. However, we are not given this information and have to proceed on the assumption that each reading in the table is an equally reliable variate. The two series represent different genetic crosses and the seven replicates in each series are clones derived from the same genetic cross. This example is clearly a Model I, since the question to be answered is whether series I differs from series II in average age at the beginning of reproduction. Inspection of the data shows that the mean age at beginning of reproduction is very similar for the two series. It would surprise us therefore to find that they are significantly different. However, we shall carry out a test anyway. As you realize by now, one cannot tell from the absolute magnitude of a difference whether it is significant. This depends on the magnitude of the error mean square, representing the variance within series.

The computations for the analysis of variance are not shown. They would be the same as in Box 8.1. With equal sample sizes and only two groups there is one further computational shortcut. Quantity **6**, SS_{groups}, can be directly computed by the following simple formula:

$$SS_{\text{groups}} = \frac{\left(\sum^{n} Y_1 - \sum^{n} Y_2\right)^2}{2n} = \frac{(52.6 - 52.9)^2}{14} = 0.00643$$

There is only 1 degree of freedom between the two groups. The critical value of $F_{.05[1,12]}$ is given underneath the anova table but it is really not necessary to consult it. Inspection of the mean squares in the anova shows that MS_{groups} is much smaller than MS_{within}; therefore the value of F_s is far below unity and there cannot possibly be an added component due to treatment effects between the series. In cases where $MS_{\text{groups}} \leq MS_{\text{within}}$, we do not usually bother to calculate F_s because the analysis of variance could not possibly be significant.

There is another method of solving a Model I two-sample analysis of variance. This is a *t-test of the differences between two means*. This t-test is the traditional method of solving such a problem, and may already be familiar to you from previous acquaintance with statistical work. It has no real advantage in either ease of computation or understanding, and as you will see below it is mathematically equivalent to the anova in Box 8.2. It is presented

BOX 8.2

Testing the difference in means between two groups.

Average age (in days) at beginning of reproduction in *Daphnia longispina* (each variate is a mean based on approximately similar numbers of females). Two series derived from different genetic crosses and containing seven clones each are compared; $n = 7$ clones per series. This is a Model I anova.

	Series ($a = 2$)	
	I	*II*
	7.2	8.8
	7.1	7.5
	9.1	7.7
	7.2	7.6
	7.3	7.4
	7.2	6.7
	7.5	7.2
$\sum^n Y$	52.6	52.9
$\bar{Y}$	7.5143	7.5571
$\sum^n Y^2$	398.28	402.23

Source: Data by Ordway from Banta (1939).

Single classification anova with two groups with equal sample sizes

Anova table

Source of variation	*df*	*SS*	*MS*	F_s
$\bar{Y} - \bar{\bar{Y}}$ Between groups (series)	1	0.00643	0.00643	0.0141
$Y - \bar{Y}$ Within groups (error; clones within series)	12	5.48571	0.45714	
$Y - \bar{\bar{Y}}$ Total	13	5.49214		

$$F_{.05[1,12]} = 4.75$$

Conclusions.—Since $F_s \ll F_{.05[1,12]}$, the null hypothesis is accepted. The means of the two series are not significantly different; that is, the two series do not differ in average age at beginning of reproduction.

A t*-test of the hypothesis that two sample means come from a population with equal* μ*: also confidence limits of the difference between two means*

This test assumes that the variances in the populations from which the two samples were taken are identical. If in doubt about this hypothesis, test by method of Box 7.1, Section 7.3.

BOX 8.2 continued

The appropriate formula for t_s is one of the following.

Expression (8.2), when sample sizes are unequal and n_1 or n_2 or both sample sizes are small (< 30): $df = n_1 + n_2 - 2$

Expression (8.3), when sample sizes are identical (regardless of size): $df = 2(n - 1)$

Expression (8.4), when both n_1 and n_2 are unequal but large (> 30): $df = n_1 + n_2 - 2$

For the present data, since sample sizes are equal, we choose Expression (8.3):

$$t_s = \frac{(\bar{Y}_1 - \bar{Y}_2) - (\mu_1 - \mu_2)}{\sqrt{\frac{1}{n}(s_1^2 + s_2^2)}}$$

We do not have variances computed, but we can modify the formula to one based on sums of squares, which are obtained one computational step earlier from the quantities furnished.

$$t_s = \frac{\bar{Y}_1 - \bar{Y}_2 - (\mu_1 - \mu_2)}{\sqrt{\dfrac{\sum^n y_1^2 + \sum^n y_2^2}{n(n-1)}}}$$

We are testing the null hypothesis that $\mu_1 - \mu_2 = 0$. Therefore we replace this quantity by zero in this example and obtain $\sum y_1^2 = 3.029$ and $\sum y_2^2 = 2.457$. Then

$$t_s = \frac{7.5143 - 7.5571}{\sqrt{(3.029 + 2.457)/7 \times 6}} = \frac{-0.0428}{\sqrt{5.486/42}} = \frac{-0.0428}{0.3614} = -0.1184$$

The degrees of freedom for this example are $2(n - 1) = 2 \times 6 = 12$. The critical value of $t_{.05[12]} = 2.179$. Since the absolute value of our observed t_s is less than the critical t-value, the means are found to be not significantly different, which is the same result as was obtained by the anova.

Confidence limits of the difference between two means

$$L_1 = (\bar{Y}_1 - \bar{Y}_2) - t_{\alpha[\nu]} s_{\bar{Y}_1 - \bar{Y}_2}$$

$$L_2 = (\bar{Y}_1 - \bar{Y}_2) + t_{\alpha[\nu]} s_{\bar{Y}_1 - \bar{Y}_2}$$

In this case $\bar{Y}_1 - \bar{Y}_2 = -0.0428$, $t_{.05[12]} = 2.18$, and $s_{\bar{Y}_1 - \bar{Y}_2} = 0.3614$ as computed earlier as the denominator of the t-test. Therefore

$$L_1 = -0.0428 - (2.18)(0.3614) = -0.8307$$

$$L_2 = -0.0428 + (2.18)(0.3614) = 0.7451$$

The 95% confidence limits contain the zero point (no difference), as was to be expected, since the difference $\bar{Y}_1 - \bar{Y}_2$ was found to be not significant.

here mainly for the sake of completeness. It would seem too much of a break with tradition not to have the "t-test" in a biostatistics text.

In Section 6.4 we learned about the t-distribution and saw that a t-distribution of $n - 1$ df could be obtained from a distribution of the term $(\bar{Y}_i - \mu)/s_{\bar{Y}_i}$, where $s_{\bar{Y}_i}$ has $n - 1$ degrees of freedom and $\bar{Y}$ is normally distributed. The numerator of this term represents a deviation of a sample mean from a parametric mean and the denominator a standard error for such a deviation. We now learn that the expression

$$t_s = \frac{(\bar{Y}_1 - \bar{Y}_2) - (\mu_1 - \mu_2)}{\sqrt{\left[\frac{(n_1 - 1)s_1^2 + (n_2 - 1)s_2^2}{n_1 + n_2 - 2}\right]\left(\frac{n_1 + n_2}{n_1 n_2}\right)}} \tag{8.2}$$

is also distributed as t. Expression (8.2) looks complicated, but it really has the same structure as the simpler term for t. The numerator is a deviation, this time not between a single sample mean and the parametric mean, but between a single difference between two sample means, $\bar{Y}_1$ and $\bar{Y}_2$, and the true difference between the means of the populations represented by these means. In a test of this sort our null hypothesis is that the two samples come from the same population; that is, they must have the same parametric mean. Thus the difference $\mu_1 - \mu_2$ is expected to be zero. We therefore test the deviation of the difference $\bar{Y}_1 - \bar{Y}_2$ from zero. The denominator of Expression (8.2) is a standard error, the standard error of the difference between two means $s_{\bar{Y}_1 - \bar{Y}_2}$. The left portion of the expression, which is in square brackets, is a weighted average of the variances of the two samples, s_1^2 and s_2^2, computed in the manner of Section 7.1. The right term of the standard error is the computationally easier form of $(1/n_1) + (1/n_2)$, which is the factor by which the average variance within groups must be multiplied in order to convert it into a variance of the difference of means. The analogy with the multiplication of a sample variance s^2 by $1/n$ to transform it into a variance of a mean $s_{\bar{Y}}^2$ should be obvious.

The test as outlined here assumes equal variances in the two populations sampled. This is also an assumption of the analyses of variance carried out so far, although we have not stressed this. With only two variances, equality may be tested by the procedure in Box 7.1.

When sample sizes are equal in a two-sample test, Expression (8.2) simplifies to

$$t_s = \frac{(\bar{Y}_1 - \bar{Y}_2) - (\mu_1 - \mu_2)}{\sqrt{\frac{1}{n}(s_1^2 + s_2^2)}} \tag{8.3}$$

which is the formula applied in the present example in Box 8.2. When the sample sizes are unequal but rather large, so that the differences between n_i and $n_i - 1$ are relatively trivial, Expression (8.2) reduces to the simpler

$$t_s = \frac{(\bar{Y}_1 - \bar{Y}_2) - (\mu_1 - \mu_2)}{\sqrt{\frac{s_1^2}{n_2} + \frac{s_2^2}{n_1}}} \tag{8.4}$$

The simplification of Expression (8.2) to Expressions (8.3) and (8.4) is shown in Appendix A1.4. The pertinent degrees of freedom for Expressions (8.2) and (8.4) are $n_1 + n_2 - 2$ and for Expression (8.3) df is $2(n - 1)$.

The test of significance for differences between means using the t-test is shown in Box 8.2. This is a two-tailed test because our alternative hypothesis is $H_1\!: \mu_1 \neq \mu_2$. The results of this test are identical to those of the anova in the same box: the two means are not significantly different. We mentioned earlier that these two results are in fact mathematically identical. We can show this simplest by squaring the obtained value for t_s, which should be identical to the F_s-value of the corresponding analysis of variance. Since $t_s = -0.1184$ in Box 8.2, $t_s^2 = 0.0140$. Within rounding error this is equal to the F_s obtained in the anova ($F_s = 0.0141$). Why is this so? We shall show two reasons for this. We learned that $t_{[\nu]} = (\bar{Y} - \mu)/s_{\bar{Y}}$, where ν is the degrees of freedom of the variance of the mean $s_{\bar{Y}}^2$; therefore $t_{[\nu]}^2 = (\bar{Y} - \mu)^2/s_{\bar{Y}}^2$. However, this expression can be regarded as a variance ratio. The denominator is clearly a variance with ν degrees of freedom. The numerator is also a variance. It is a single deviation squared, which represents a sum of squares possessing 1 rather than zero degrees of freedom (since it is a deviation from the true mean μ rather than a sample mean). A sum of squares based on 1 degree of freedom is at the same time a variance. Thus t^2 is a variance ratio; specifically, $t_{[\nu]}^2 = F_{[1,\nu]}$, as we have seen above. In Appendix A1.5 we demonstrate algebraically that the t_s^2 and the F_s-value obtained in Box 8.2 are identical quantities.

We can also demonstrate the relationship between t and F from the statistical tables. In Table **V** we find that $F_{.05[1,10]} = 4.96$. This value is supposed to be equal to $t_{.05[10]}^2$. The square root of 4.96 is 2.227 and when we look up $t_{.05[10]}$ in Table **III** we find it to be 2.228, which is within acceptable rounding error of the previous value. Since t approaches the normal distribution as its degrees of freedom approach infinity, $F_{\alpha[1,\nu]}$ approaches the distribution of the square of the normal deviate as $\nu \longrightarrow \infty$.

The relation between t^2 and F leads to another relation. We have just learned that when $\nu_1 = 1$, $F_{[\nu_1,\nu_2]} = t_{[\nu_2]}^2$. We know from Section 7.2 that $\chi_{[\nu_1]}^2/\nu_1 = F_{[\nu_1,\infty]}$. Therefore, when $\nu_1 = 1$ and $\nu_2 = \infty$, $\chi_{[1]}^2 = F_{[1,\infty]} = t_{[\infty]}^2$. This can be demonstrated from Tables **IV**, **V**, and **III**, respectively.

$$\chi_{.05[1]}^2 = 3.841$$

$$F_{.05[1,\infty]} = 3.84$$

$$t_{.05[\infty]} = 1.960 \qquad t_{.05[\infty]}^2 = 3.8416$$

The t-test for differences between two means is useful when we wish to set confidence limits to such a difference. Box 8.2 shows how to calculate

95% confidence limits to the difference between the series means in the *Daphnia* example. The appropriate standard error and degrees of freedom depend on whether Expression (8.2), (8.3), or (8.4) is chosen for t_s. It does not surprise us to find that the confidence limits of the difference in this case enclose the value of zero, ranging from -0.8307 to $+0.7451$. This must be so when a difference is found to be not significantly different from zero. We can interpret this by saying that we cannot exclude zero as the true value of the difference between the means of the two series.

8.5 Comparisons among means: a priori tests

We have seen that after the initial significance test a Model II analysis of variance is completed by estimation of the added variance components. We usually complete a Model I anova of more than two groups by examining the data in greater detail, testing which means are different from which other ones or which groups of means are different from other such groups or from single means. Let us refer back to the Model I anovas treated so far in this chapter. We can dispose right away of the two-sample case in Box 8.2, the average age of water fleas at beginning of reproduction. As you will recall, there was no significant difference in age between the two genetic series. But even if there had been such a difference, no further tests are possible. However, the data on length of pea sections given in Box 8.1 show a significant difference among the five treatments (based on 4 degrees of freedom). Although we know that the means are not all equal, we do not know which ones differ from which other ones. This leads us to the subject of tests among pairs and groups of means. Thus, for example, we might test the control against the 4 experimental treatments representing added sugars. The question to be tested would be, Does the addition of sugars have an effect on length of pea sections? We might also test for differences among the sugar treatments. A reasonable test might be pure sugars (glucose, fructose, and sucrose) versus the mixed sugar treatment (1% glucose, 1% fructose).

An important point about such tests is that they are designed and chosen independently of the results of the experiment. They should be planned *before* the experiment has been actually carried out and the results obtained. Such comparisons are called *planned* or *a priori comparisons*. Such tests are applied regardless of the results of the preliminary overall anova. By contrast, after the experiment has been carried out we might wish to compare certain means that we notice to be markedly different. For instance, sucrose with a mean of 64.1 appears to have had less of a growth-inhibiting effect than fructose with a mean of 58.2. We might therefore wish to test whether there is in fact a significant difference between the effects of fructose and sucrose. Such comparisons, which suggest themselves as a result of the completed experiment, are called *unplanned* or *a posteriori comparisons*. These tests are performed only if the preliminary overall anova is significant. They include tests of the comparisons between all possible pairs of means. When there are a means, there can, of course, be $a(a - 1)/2$ possible comparisons between pairs of means. The reason we make this distinction between a priori and a posteriori comparisons is that the tests of significance appropriate for the two comparisons are different. A simple, analogous example will explain why this is so.

Let us assume we have sampled from an approximately normal population of heights of men. We have computed their mean and standard deviation. If we sample two men at a time from this population, we can predict the difference between them on the basis of ordinary statistical theory. Some men will be very similar, others relatively very different. Their differences will be distributed normally with a mean of 0 and an expected variance of $2\sigma^2$, for reasons that will be learned in a later chapter (Section 12.2). Thus, if we obtain a large difference between a randomly sampled pair of men, it will have to be a sufficient number of standard deviations greater than zero for us to reject our null hypothesis that the two men come from the specified population. If, on the other hand, we were to look at the heights of the people before sampling them and then take pairs that seem to be very different, it is obvious that we would repeatedly obtain differences between pairs of men that are several standard deviations apart. Such differences would be outliers in the expected frequency distribution of differences, and time and again we would reject our null hypothesis when in fact it is true. The men were sampled from the same population, but because they were not sampled at random but were inspected before being sampled, the probability distribution on which our hypothesis testing rested is no longer valid. It is obvious that the tails in a large sample from a normal distribution will be anywhere from 5 to 7 standard deviations apart, and that if we deliberately take individuals from each tail and compare them, they will appear to be highly significantly different from each other, even though they belong to the same population.

When we compare means differing greatly from each other as the result of some treatment in the analysis of variance, we are doing exactly the same thing as taking the tallest and the shortest men from the frequency distribution of heights. If we wish to know whether these are significantly different from each other, we cannot use the ordinary probability distribution on which the analysis of variance rested but we have to use special tests of significance. These a posteriori tests will be discussed in the next section. The present section concerns itself with the carrying out a priori tests, those comparisions planned before the execution of the experiment.

The general rule for making a planned comparison is extremely simple and relates back to the rule for obtaining the sum of squares for any set of groups (discussed at the end of Section 8.1). To compare k groups of any size n_i, take the sum of each group, square it, divide it by its sample size n_i, and sum the k quotients so obtained. From the sum of these quotients sub-

TABLE 8.2

Means, group sums, and sample sizes from the data in Box 8.1. Length of pea sections grown in tissue culture (in ocular units).

	Control	*2% glucose*	*2% fructose*	*1% glucose + 1% fructose*	*2% sucrose*	Σ
$\bar{Y}$	70.1	59.3	58.2	58.0	64.1	$(61.94 = \bar{\bar{Y}})$
$\sum^{n} Y$	701	593	582	580	641	3097
n	10	10	10	10	10	50

tract a correction term, which will be the grand sum of all the groups in this comparison, quantity squared and dividied by the number of items in this grand sum. If the comparison includes all the groups in the anova, the correction term will be the main CT of the study. If, however, the comparison includes only some of the groups of the anova, the CT will be different, being restricted only to these groups.

These rules can best be learned by means of an example. Table 8.2 lists the means, group sums, and sample sizes of the experiment with the pea sections from Box 8.1. You will recall that there were highly significant differences among the groups. We now wish to test whether the controls are different from the four treatments representing addition of sugar. There will thus be two groups, one the control group and the other the "sugars" group, with a sum of 2396 and a sample size of 40. We therefore compute

SS (control versus sugars)

$$= \frac{(701)^2}{10} + \frac{(593 + 582 + 580 + 641)^2}{40} - \frac{(701 + 593 + 582 + 580 + 641)^2}{50}$$

$$= \frac{(701)^2}{10} + \frac{(2396)^2}{40} - \frac{(3097)^2}{50} = 832.32$$

In this case the correction term is the same as for the anova because it involves all the groups of the study. The result is a sum of squares for the comparison between these two groups. Since a comparison between two groups has only 1 degree of freedom, the sum of squares is at the same time a mean square. This mean square is tested over the error mean square of the anova to give the following comparison:

$$F_s = \frac{MS\text{ (control versus sugars)}}{MS_{\text{within}}} = \frac{832.32}{5.46} = 152.44$$

$$F_{.05[1,45]} = 4.05, \qquad F_{.01[1,45]} = 7.23,$$

This comparison is highly significant, showing that the additions of sugars significantly retarded the growth of the pea sections.

Next we test whether the mixture of sugars is significantly different from the pure sugars. Using the same technique we calculate

SS (mixed sugars versus pure sugars)

$$= \frac{(580)^2}{10} + \frac{(593 + 582 + 641)^2}{30} - \frac{(593 + 582 + 580 + 641)^2}{40}$$

$$= \frac{(580)^2}{10} + \frac{(1816)^2}{30} - \frac{(2396)^2}{40} = 48.13$$

Here the CT is different, being based on the sum of the sugars only. The appropriate test statistic is

$$F_s = \frac{MS \text{ (mixed sugars versus pure sugars)}}{MS_{\text{within}}} = \frac{48.13}{5.46} = 8.82$$

This is significant in view of the critical values of $F_{\alpha[1,45]}$ given in the preceding paragraph.

A final test is among the three sugars. This mean square has 2 degrees of freedom, since it is based on three means. Thus we compute

$$SS \text{ (among pure sugars)} = \frac{(593)^2}{10} + \frac{(582)^2}{10} + \frac{(641)^2}{10} - \frac{(1816)^2}{30} = 196.87$$

$$MS \text{ (among pure sugars)} = \frac{SS \text{ (among pure sugars)}}{df} = \frac{196.87}{2} = 98.435$$

$$F_s = \frac{MS \text{ (among pure sugars)}}{MS_{\text{within}}} = \frac{98.435}{5.46} = 18.06$$

This F_s is highly significant, since even $F_{.01[2,40]} = 5.18$.

We conclude that the addition of the three sugars retards growth in the pea sections, that mixed sugars affect the sections differently from pure sugars, and that the pure sugars are significantly different among themselves, probably because of the sucrose, which has a far higher mean. We cannot test the sucrose against the other two, because that would be an a posteriori test, which suggested itself to us after we had looked at the results. To carry out such a test we need the methods of the next section.

However, our a priori tests might have been quite different, depending entirely on our initial hypotheses. Thus, we could have tested control versus sugars initially, followed by disaccharides (sucrose) versus monosaccharides (glucose, fructose, glucose + fructose), followed by mixed versus pure monosaccharides and finally by glucose versus fructose.

The pattern and number of a priori tests are determined by one's hypotheses about the data. However, there are certain restrictions. It would clearly not be proper to decide a priori that one wished to compare every mean against every other mean [$a(a - 1)/2$ comparisons]. For a groups the sum of the degrees of freedom of the separate a priori tests should not exceed

TABLE 8.3

Anova table from Box 8.1. with treatment sum of squares decomposed into planned comparisons.

Source of variation	*SS*	*df*	*MS*	F_s
Treatments	1077.32	4	269.33	49.33**
Control vs sugars	832.32	1	832.32	152.44**
Mixed vs pure sugars	48.13	1	48.13	8.82**
Among pure sugars	196.87	2	98.44	18.03**
Within	245.50	45	5.46	
Total	1322.82	49		

$a - 1$. In addition, it is desirable to structure the tests in such a way that each test tests an independent relationship among the means (as was done in the example above). For example, we would prefer not to test if means one, two, and three differed if we had already found that mean one differed from mean three, since significance of the latter implies significance of the former.

Since these tests are independent, the three sums of squares we have so far obtained, base on 1, 1, and 2 *df*, respectively, together add up to the sum of squares among treatments of the original analysis of variance based on 4 degrees of freedom. Thus:

		df
SS (control versus sugars)	= 832.32	1
SS (mixed versus pure sugars)	= 48.13	1
SS (among pure sugars)	= 196.87	2
SS (among treatments)	= 1077.32	4

This again illustrates the elegance of analysis of variance. The treatment sums of squares can be decomposed into separate parts that are sums of squares in their own right, with degrees of freedom pertaining to them. One sum of squares measures the difference between the controls and the sugars, the second that between the mixed sugars and the pure sugars, and the third the remaining variation among the three sugars. We can present all of these results as an anova table, as shown in Table 8.3.

8.6 Comparisons among means: a posteriori tests

If a single classification anova is significant it means, of course, that

$$\frac{MS_{\text{groups}}}{MS_{\text{within}}} \geq F_{\alpha[a-1,a(n-1)]} \tag{8.5}$$

Since $MS_{\text{groups}}/MS_{\text{within}} = SS_{\text{groups}}/[(a - 1)\, MS_{\text{within}}]$, we can rewrite Expression (8.5) as

$$SS_{\text{groups}} \geq (a - 1)\, MS_{\text{within}}\, F_{\alpha[a-1,a(n-1)]} \tag{8.6}$$

For example, in Box 8.1, where the anova is significant, $SS_{\text{groups}} = 1077.32$. Substituting into Expression (8.6) we obtain

$$1077.32 > (5 - 1)(5.46)(2.58) = 56.35 \text{ for } \alpha = 0.05$$

It is therefore possible to compute a critical SS value for a test of significance of an anova. Thus another way of calculating overall significance would be to see whether the SS_{groups} is greater than this critical SS. It is of interest to investigate why the SS_{groups} is as large as it is and to test for the significance of the various contributions made to this SS by differences among the sample means. This was discussed in the previous section, where separate sums of squares were computed based on comparisons among means planned before the data had been examined. A comparison was called significant if its F_s-ratio was $> F_{\alpha[k-1,a(n-1)]}$, where k is the number of means being compared. We can now also state this in terms of sums of squares. An SS is significant if it is greater than $(k - 1)\ MS_{\text{within}} F_{\alpha[k-1,a(n-1)]}$.

The above tests were a priori comparisons. One procedure for testing a posteriori comparisons would be to set $k = a$ in the last formula, no matter how many means we compare. Thus the critical value of the SS will be larger than in the previous method, making it more difficult to demonstrate the significance of a sample SS. This is done to allow for the fact that we choose for testing those differences between group means that appear to be contributing substantially to the significance of the overall anova.

For an example, let us return to the effects of sugars on growth in pea sections (Box 8.1). We write down the means in ascending order of magnitude: 58.0 (glucose + fructose), 58.2 (fructose), 59.3 (glucose), 64.1 (sucrose), 70.1 (control). We notice that the first three treatments have quite similar means and suspect that they do not differ significantly among themselves and hence do not contribute substantially to the significance of the SS_{groups}.

To test this we compute the SS among these three means by the usual formula:

$$SS = \frac{593^2 + 582^2 + 580^2}{10} - \frac{(593 + 582 + 580)^2}{3(10)}$$
$$= 102{,}677.3 - 102{,}667.5 = 9.8$$

The differences among these means are not significant because this SS is less than the critical SS (56.35) calculated above.

The sucrose mean looks suspiciously different from the means of the other sugars. To test this we compute

$$SS = \frac{641^2}{10} + \frac{(593 + 582 + 580)^2}{30} - \frac{(641 + 593 + 582 + 580)^2}{10 + 30}$$
$$= 41{,}088.1 + 102{,}667.5 - 143{,}520.4 = 235.2$$

which is greater than the critical SS. We conclude, therefore, that sucrose retards growth significantly less than the other sugars tested. We may continue in this fashion, testing all the differences that look suspicious or even testing all possible sets of means, considering them 2, 3, 4, and 5 at a time. This latter approach may require a computer if there are more than 5 means to be compared, since there are very many possible tests that could be made. This procedure was proposed by Gabriel (1964), who called it a *sum of squares simultaneous test procedure* (*SS-STP*).

In the *SS-STP* and in the original anova, the chance of making any type I error at all is α, the probability selected for the critical F value from Table V. By "making any type I error at all" we mean making such an error in the overall test of significance of the anova and in any of the subsidiary comparisons among means or sets of means needed to complete the analysis of the experiment. This probability α is called the "experimentwise" error rate. It should be noted that though the probability of any error at all is α, the probability of error for any particular test of some subset, such as a test of the difference among 3 or between two means, is necessarily less than α. Thus for the test of each subset one is really using a significance level α', which may be much less than the experimentwise α, and if there are many means in the anova this actual error rate α' may be a 1/10th, 1/100th, or even a 1/1000th of the experimentwise α (Gabriel, 1964). For this reason the a posteriori tests discussed so far and the overall anova are not very sensitive to differences of individual means or differences within small subsets. Obviously, not many differences are going to be considered significant if α' is minute. This is the price one pays for not planning one's comparisons before the data are examined. If a priori tests are made the error rate of each would still be α.

Numerous other techniques of a posteriori multiple comparisons tests have been described in the statistical literature, but the simultaneous test procedure given above should suffice as a conservative introduction to the subject.

Exercises 8

8.1 The following is an example with easy numbers to help you become familiar with the analysis of variance. A plant ecologist wishes to test the hypothesis that the height of plant species X depends on the type of soil it grows in. He measured the height of three plants in each of four plots representing different soil types, all four plots being contained in an area of two miles square. His results are tabulated below. (Height is given in centimeters.) Does your analysis support this hypothesis? ANS. $F_s = 6.951$, $F_{.05[3,8]} = 4.07$.

Observation number	*Localities*			
	1	*2*	*3*	*4*
1	15	25	17	10
2	9	21	23	13
3	14	19	20	16

8.2 The following are measurements (in coded micrometer units) of the thorax length of the aphid *Pemphigus populi-transversus*. The aphids were collected in 28 galls on the cottonwood, *Populas deltoides*. Four alate (winged) aphids were randomly selected from each gall and measured. The alate aphids of each gall are isogenic (identical twins), being descended parthenogenetically from one stem mother. Thus, any variance within galls can be due to environment only. Variance between different galls may be due to differences in genotype and also to environmental differences between galls. If this character, thorax length, is affected by genetic variation, significant intergall variance must be present. The converse is not necessarily true; significant variance between galls need not indicate genetic variation; it could as well be due to environmental differences between galls (data by Sokal, 1952). Analyze the variance of thorax length. Is there significant intergall variance present? Give estimates of the added component of intergall variance, if present. What percentage of the variance is controlled by intragall and what percentage by intergall factors? Discuss your results. (Remember to check your computations step by step; otherwise an error committed early in the calculation may ruin your entire effort. For computational purposes ignore the decimal point in the variates.)

Gall No.		*Gall No.*	
1.	6.1, 6.0, 5.7, 6.0	15.	6.3, 6.5, 6.1, 6.3
2.	6.2, 5.1, 6.1, 5.3	16.	5.9, 6.1, 6.1, 6.0
3.	6.2, 6.2, 5.3, 6.3	17.	5.8, 6.0, 5.9, 5.7
4.	5.1, 6.0, 5.8, 5.9	18.	6.5, 6.3, 6.5, 7.0
5.	4.4, 4.9, 4.7, 4.8	19.	5.9, 5.2, 5.7, 5.7
6.	5.7, 5.1, 5.8, 5.5	20.	5.2, 5.3, 5.4, 5.3
7.	6.3, 6.6, 6.4, 6.3	21.	5.4, 5.5, 5.2, 6.3
8.	4.5, 4.5, 4.0, 3.7	22.	4.3, 4.7, 4.5, 4.4
9.	6.3, 6.2, 5.9, 6.2	23.	6.0, 5.8, 5.7, 5.9
10.	5.4, 5.3, 5.0, 5.3	24.	5.5, 6.1, 5.5, 6.1
11.	5.9, 5.8, 6.3, 5.7	25.	4.0, 4.2, 4.3, 4.4
12.	5.9, 5.9, 5.5, 5.5	26.	5.8, 5.6, 5.6, 6.1
13.	5.8, 5.9, 5.4, 5.5	27.	4.3, 4.0, 4.4, 4.6
14.	5.6, 6.4, 6.4, 6.1	28.	6.1, 6.0, 5.6, 6.5

8.3 Millis and Seng (1954) published a study on the relation of birth order to the birth weights of infants. The data below on first-born and eighth-born infants are extracted from a table of birth weights of male infants of Chinese third-class patients at the Kandang Kerbau Maternity Hospital in Singapore in 1950 and 1951.

Birth weight lb:oz	*Birth order* 1	8
3:0–3:7	—	—
3:8–3:15	2	—
4:0–4:7	3	—
4:8–4:15	7	4
5:0–5:7	111	5
5:8–5:15	267	19
6:0–6:7	457	52
6:8–6:15	485	55
7:0–7:7	363	61
7:8–7:15	162	48

(*continued*)

Birth weight lb:oz	*Birth order* *1*	*8*
8:0–8:7	64	39
8:8–8:15	6	19
9:0–9:7	5	4
9:8–9:15	—	—
10:0–10:7	—	1
10:8–10:15	—	—
	1932	307

Which birth order appears to be accompanied by heavier infants? Is this difference significant? Can you conclude that birth order causes differences in birth weight? (Computational note: The variable should be coded as simply as possible.) Reanalyze, using the t-test, and verify that $t_s^2 = F_s$. ANS. $t_s = 11.016$.

8.4 The following cytochrome oxidase assessments of male *Periplaneta* roaches in cubic millimeters per ten minutes per milligram were taken from a larger study by Brown and Brown (1956):

	n	$\bar{Y}$	$s_{\bar{Y}}$
24 hours after methoxychlor injection	5	24.8	0.9
Control	3	19.7	1.4

Are the two means significantly different?

8.5 The following data are measurements of five random samples of domestic pigeons collected during the months of January, February, and March in Chicago in 1955. The variable is the length from the anterior end of the narial opening to the tip of the bony beak and is recorded in millimeters. Data from Olson and Miller (1958).

		Samples		
1	*2*	*3*	*4*	*5*
5.4	5.2	5.5	5.1	5.1
5.3	5.1	4.7	4.6	5.5
5.2	4.7	4.8	5.4	5.9
4.5	5.0	4.9	5.5	6.1
5.0	5.9	5.9	5.2	5.2
5.4	5.3	5.2	5.0	5.0
3.8	6.0	4.8	4.8	5.9
5.9	5.2	4.9	5.1	5.0
5.4	6.6	6.4	4.4	4.9
5.1	5.6	5.1	6.5	5.3
5.4	5.1	5.1	4.8	5.3
4.1	5.7	4.5	4.9	5.1
5.2	5.1	5.3	6.0	4.9
4.8	4.7	4.8	4.8	5.8
4.6	6.5	5.3	5.7	5.0
5.7	5.1	5.4	5.5	5.6
5.9	5.4	4.9	5.8	6.1
5.8	5.8	4.7	5.6	5.1
5.0	5.8	4.8	5.5	4.8
5.0	5.9	5.0	5.0	4.9

Are the five samples homogeneous?

8.6 P. E. Hunter (1959, detailed data unpublished) selected two strains of *Drosophila melanogaster*, one for short larval period (SL) and one for long larval periods (LL). A nonselected control strain (CS) was also maintained. At generation 42 the following data were obtained for the larval period (measured in hours). Analyze and interpret.

		Strain		
	SL	*CS*	LL	
n_i	80	69	33	
$\sum^{n_i} Y$	8070	7291	3640	$\sum^{3}\sum^{n_i} Y^2 = 1{,}994{,}650$

Note that part of the computation has already been performed for you. Perform a priori tests among the three means (short versus long larval periods and each against the control). Set 95% confidence limits to the observed differences of means for which these comparisons are made. ANS. $MS_{(\text{SL vs. LL})} =$ 2076.6697.

8.7 The following data were taken from a study of blood protein variations in deer (Cowan and Johnston, 1962). The variable is the mobility of serum protein fraction *II* expressed as 10^{-5} cm²/volt seconds.

	$\bar{Y}$	$s_{\bar{Y}}$
Sitka	2.8	0.07
California Blacktail	2.5	0.05
Vancouver Island Blacktail	2.9	0.05
Mule Deer	2.5	0.05
White Tail	2.8	0.07

$n = 12$ for each mean. Perform an analysis of variance and a multiple comparisons test, using the sums of squares *STP* procedure. ANS. $MS_{\text{within}} =$ 0.0416, maximal nonsignificant sets (at $P = 0.05$) are samples 1, 3, 5 and 2, 4 (numbered by order of magnitude of the means).

9

Two-Way Analysis of Variance

From the single classification anova of Chapter 8 we progress to the two-way anova of the present chapter by a single logical step. Individual items may be grouped into classes representing the different possible combinations of two treatments or factors. Thus the housefly wing lengths studied in earlier chapters, which yielded samples representing different medium formulations, might also be divided into males and females. We would like to know not only whether medium 1 induced a different wing length than medium 2 but also whether male houseflies differed in wing length from females. Obviously each combination of factors should be represented by a sample of flies. Thus for seven media and two sexes we need at least $7 \times 2 = 14$ samples. Similarly, the experiment testing five sugar treatments on pea sections (Box 9.4) might have been carried out at three different temperatures. This would have resulted in a *two-way analysis of variance* of the effects of sugars as well as of temperatures.

It is the assumption of this method of anova that a given temperature and a given sugar each contribute a certain amount to the growth of a pea section and that these two contributions add their effects without influencing each other. In Section 9.1 we shall see how departures from the assumption are measured; we shall also consider the expression for decomposing variates in a two-way anova.

The two factors in the present design may represent either Model I or Model II effects or one of each, in which case we talk of a *mixed model.*

The computation of a two-way anova for replicated subclasses (more than one variate per subclass or factor combination) is shown in Section 9.1, which also contains a discussion of the meaning of interaction as used in statistics. Significance testing in a two-way anova is the subject of Section 9.2. This is followed by a section (9.3) on two-way anova without replication, or with only a single variate per subclass. The well-known method of paired comparisons is a special case of a two-way anova without replication.

We will now proceed to illustrate the computation of a two-way anova. You will obtain closer insight into the structure of this design as we explain the computations.

9.1 Two-way anova with replication

We illustrate the computation of a two-way anova in a study of oxygen consumption by two species of limpets at three concentrations of seawater. Eight replicate readings were obtained for each combination of species and seawater concentration. We have continued to call the number of columns a, and are calling the number of rows b. The sample size for each cell (row and column combination) of the table is n. The cells are also called subgroups or subclasses.

The data are featured in Box 9.1. The computational steps labeled *Preliminary computations* provide an efficient procedure for the analysis of variance but we shall undertake several digressions to ensure that the concepts underlying this design are appreciated by the reader. We commence by considering the six subclasses as though they were six groups in a single classification anova. Each subgroup or sublcass represents eight oxygen consumption readings. Such an anova will test whether there is any variation among the six subgroups over and above the variance within the subgroups. For if there were no such added variation it would be unlikely that either species or salinity would significantly affect oxygen uptake. Steps **1** through **3** in Box 9.1 correspond to the identical steps in Box 8.1, although the symbolism has changed slightly since in place of a groups we now have ab subgroups. To complete the anova we need a correction term, which is labeled step **6** in Box 9.1. From these quantities we obtain SS_{total}, and SS_{within} in steps **7**, **8**, and **12**, corresponding to steps **5**, **6**, and **7** in the layout of Box 8.1. The results of this preliminary anova are featured in Table 9.1. They indicate clearly that there is considerable added variation among subgroups, making it likely that we shall find significant effects for at least one of the factors.

BOX 9.1

Two-way anova with replication.

Oxygen consumption rates of two species of limpets, *Acmaea scabra* and *A. digitalis,* at three concentrations of seawater. The variable measured is μl O_2/mg dry body weight/min at 22°C. There are eight replicates per combination of species and salinity ($n = 8$). This is a Model I anova.

Factor B: seawater concentrations ($b = 3$)	*Factor A: Species* ($a = 2$) *Acmaea scabra*		*Acmaea digitalis*		Σ
100%	7.16	8.26	6.14	6.14	
	6.78	14.00	3.86	10.00	
	13.60	16.10	10.40	11.60	
	8.93	9.66	5.49	5.80	
	$\Sigma = 84.49$		$\Sigma = 59.43$		143.92
75%	5.20	13.20	4.47	4.95	
	5.20	8.39	9.90	6.49	
	7.18	10.40	5.75	5.44	
	6.37	7.18	11.80	9.90	
	$\Sigma = 63.12$		$\Sigma = 58.70$		121.82
50%	11.11	10.50	9.63	14.50	
	9.74	14.60	6.38	10.20	
	18.80	11.10	13.40	17.70	
	9.74	11.80	14.50	12.30	
	$\Sigma = 97.39$		$\Sigma = 98.61$		196.00
Σ	245.00		216.74		461.74

Source: Unpublished study by F. J. Rohlf.

BOX 9.1 continued

Preliminary computations

1. Grand total $= \sum^{a}\sum^{b}\sum^{n} Y = 461.74$

2. Sum of the squared observations $= \sum^{a}\sum^{b}\sum^{n} Y^2 = (7.16)^2 + \cdots + (12.30)^2 = 5065.1530$

3. Sum of the squared subgroup (cell) totals, divided by the sample size of the subgroups

$$= \frac{\sum^{a}\sum^{b}\left(\sum^{n} Y\right)^2}{n} = \frac{[(84.49)^2 + \cdots + (98.61)^2]}{8} = 4663.6317$$

4. Sum of the squared column totals divided by the sample size of a column $= \dfrac{\sum^{a}\left(\sum^{b}\sum^{n} Y\right)^2}{bn}$

$$= \frac{[(245.00)^2 + (216.74)^2]}{(3 \times 8)} = 4458.3844$$

5. Sum of the squared row totals divided by the sample size of a row $= \dfrac{\sum^{b}\left(\sum^{a}\sum^{n} Y\right)^2}{an}$

$$= \frac{[(143.92)^2 + (121.82)^2 + (196.00)^2]}{(2 \times 8)} = 4623.0674$$

6. Grand total squared and divided by the total sample size = correction term CT

$$= \frac{\left(\sum^{a}\sum^{b}\sum^{n} Y\right)^2}{abn} = \frac{(\text{quantity } \mathbf{1})^2}{abn} = \frac{(461.74)^2}{(2 \times 3 \times 8)} = 4441.7464$$

7. $SS_{\text{total}} = \sum^{a}\sum^{b}\sum^{n} Y^2 - CT =$ quantity **2** − quantity **6** $= 5065.1530 - 4441.7464 = 623.4066$

8. $SS_{\text{subgr}} = \dfrac{\sum^{a}\sum^{b}\left(\sum^{n} Y\right)^2}{n} - CT =$ quantity **3** − quantity **6** $= 4663.6317 - 4441.7464 = 221.8853$

9. SS_A (SS of columns) $= \dfrac{\sum^{a}\left(\sum^{b}\sum^{n} Y\right)^2}{bn} - CT =$ quantity **4** − quantity **6** = 4458.3844 − 4441.7464 = 16.6380

10. SS_B (SS of rows) $= \dfrac{\sum^{b}\left(\sum^{a}\sum^{n} Y\right)^2}{an} - CT =$ quantity **5** − quantity **6** = 4623.0674 − 4441.7464 = 181.3210

11. $SS_{A \times B}$ (Interaction SS) $= SS_{\text{subgr}} - SS_A - SS_B =$ quantity **8** − quantity **9** − quantity **10**
= 221.8853 − 16.6380 − 181.3210 = 23.9263

12. SS_{within} (Within subgroups; error SS) $= SS_{\text{total}} - SS_{\text{subgr}} =$ quantity **7** − quantity **8** = 623.4066 − 221.8853 = 401.5213

As a check on your computations, ascertain that the following relations hold for some of the above quantities: $\mathbf{2} \geq \mathbf{3} \geq \mathbf{4} \geq \mathbf{6}$; $\mathbf{3} \geq \mathbf{5} \geq \mathbf{6}$.

Now fill in the anova table.

	Source of variation	*df*	*SS*	*MS*	*Expected MS (Model I)*
$\bar{Y} - \bar{\bar{Y}}$	Subgroups	$ab - 1$	**8**	$\dfrac{\mathbf{8}}{(ab-1)}$	
$\bar{Y}_A - \bar{\bar{Y}}$	A (columns)	$a - 1$	**9**	$\dfrac{\mathbf{9}}{(a-1)}$	$\sigma^2 + \dfrac{nb}{a-1}\sum^{a}\alpha^2$
$\bar{Y}_B - \bar{\bar{Y}}$	B (rows)	$b - 1$	**10**	$\dfrac{\mathbf{10}}{(b-1)}$	$\sigma^2 + \dfrac{na}{b-1}\sum^{b}\beta^2$
$\bar{Y} - \bar{Y}_A - \bar{Y}_B + \bar{\bar{Y}}$	$A \times B$ (interaction)	$(a-1)(b-1)$	**11**	$\dfrac{\mathbf{11}}{(a-1)(b-1)}$	$\sigma^2 + \dfrac{n}{(a-1)(b-1)}\sum^{ab}(\alpha\beta)^2$
$Y - \bar{Y}$	Within subgroups	$ab(n-1)$	**12**	$\dfrac{\mathbf{12}}{ab(n-1)}$	σ^2
$Y - \bar{\bar{Y}}$	Total	$abn - 1$	**7**		

Since the present example is a Model I anova for both factors the expected MS above are correct. Below are the corresponding expressions for other models.

BOX 9.1 continued

Source of variation	*Model II*	*Mixed Model (A fixed, B random)*
A	$\sigma^2 + n\sigma^2_{A\times B} + nb\sigma^2_A$	$\sigma^2 + n\sigma^2_{A\times B} + \frac{nb}{a-1}\sum^{a}\alpha^2$
B	$\sigma^2 + n\sigma^2_{A\times B} + na\sigma^2_B$	$\sigma^2 + na\sigma^2_B$
$A \times B$	$\sigma^2 + n\sigma^2_{A\times B}$	$\sigma^2 + n\sigma^2_{A\times B}$
Within subgroups	σ^2	σ^2

Anova table

Source of variation	*df*	*SS*	*MS*	F_s
Subgroups	5	221.8853	44.377	
A (columns; species)	1	16.6380	16.638	1.740 *ns*
B (rows; salinities)	2	181.3210	90.660	9.483**
$A \times B$ (interaction)	2	23.9263	11.963	1.251 *ns*
Within subgroups (error)	42	401.5213	9.560	
Total	47	623.4066		

$F_{.05[1,42]} = 4.07$ $F_{.05[2,42]} = 3.22$ $F_{.01[2,42]} = 5.15$

Since this is a Model I anova, all mean squares are tested over the error *MS*. For a discussion of significance tests see Section 9.2.

Conclusions.—Oxygen consumption does not differ significantly between the two species of limpets but differs with the salinity. At 50% seawater the O_2 consumption is increased. Salinity appears to affect the two species equally, for there is insufficient evidence of a species × salinity interaction.

TABLE 9.1
Preliminary anova of subgroups in two-way anova. Data from Box 9.1.

Source of variation		*df*		*SS*	*MS*
$\bar{Y} - \bar{\bar{Y}}$	Among subgroups	5	$ab - 1$	221.8853	44.377**
$Y - \bar{Y}$	Within subgroups	42	$ab(n - 1)$	401.5213	9.560
$Y - \bar{\bar{Y}}$	Total	47	$abn - 1$	623.4066	

The computation is continued by finding the sums of squares for rows and columns of the table. This is done by the general formula stated at the end of Section 8.1. Thus for columns, we square the column sums, sum these squares, and divide them by 24, the number of items per row. This is step **4** in Box 9.1. A similar quantity is computed for rows (step **5**). From these quotients we subtract the correction term, computed as quantity **6**. These subtractions are carried out as steps **9** and **10**, respectively. Since the rows and columns are based on equal sample sizes, we do not have to obtain separate quotients for the square of each row or column sum but carry out a single division after accumulating the squares of the sums.

Let us return for a moment to the preliminary analysis of variance in Table 9.1, which divided the total sum of squares into two parts, the sum of squares among the six subgroups and that within the subgroups, the error sum of squares. The new sums of squares pertaining to row and column effects clearly are not part of the error, but must contribute to the differences that comprise the sum of squares among the four subgroups. We therefore subtract row and column SS from the subgroup SS. The latter is 221.8853. The row SS is 181.3210, and the column SS is 16.6380. Together they add up to 197.9590, almost but not quite the value of the subgroup sum of squares. The difference represents a third sum of squares, called the *interaction sum of squares*, whose value in this case is 23.9263. We shall discuss the meaning of this new sum of squares presently. At the moment let us say only that it is almost always present (but not necessarily significant) and generally need not be independently computed but may be obtained as illustrated above—by the subtraction of the row SS and the column SS from the subgroup SS. This procedure is shown graphically in Figure 9.1, which illustrates the decomposition of the total sum of squares into the subgroup SS and error SS, the former being subdivided into the row SS, column SS, and interaction SS. The relative magnitudes of these sums of squares will differ from experiment to experiment. In Figure 9.1 they are not shown proportional to their actual values in the limpet experiment; otherwise the area representing the row SS would have to be about eleven times that allotted to the column SS. Before we can intelligently test for significance in this anova we must understand the meaning of interaction.

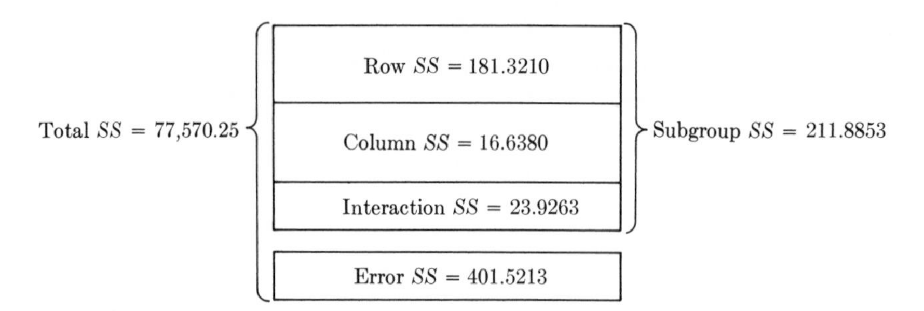

FIGURE 9.1

Diagrammatic representation of the partitioning of the total sums of squares in a two-way orthogonal anova. The areas of the subdivisions are not shown proportional to the magnitudes of the sums of squares.

We can best explain interaction in a two-way anova by means of an artificial illustration based on the limpet data we have just studied. If we interchange the readings for 75% and 50% for *A. digitalis* only, we obtain the data table shown in Table 9.2. Only the sums of the subgroups, rows, and columns are shown. We complete the analysis of variance in the manner presented above and note the results at the foot of Table 9.2. The total,

TABLE 9.2

An artificial example to illustrate the meaning of interaction. The readings for 75% and 50% seawater concentrations of *Acmaea digitalis* in Box 9.1 have been interchanged. Only subgroup and marginal totals are given below.

	Species		
Seawater concentration	*A. scabra*	*A. digitalis*	Σ
100%	84.49	59.43	143.92
75%	63.12	98.61	161.73
50%	97.39	58.70	156.09
Σ	245.00	216.74	461.74

Completed anova

Source of variation	*df*	*SS*	*MS*
Subgroups	5	221.8853	44.377**
Species	1	16.6380	16.638 *ns*
Salinities	2	10.3566	5.178 *ns*
Sp × Sal	2	194.8907	97.445**
Error	42	401.5213	9.560
Total	47	623.4066	

TABLE 9.3
Comparison of means of the data in Tables 9.1 and 9.2.

Seawater concentration	*Species*		
	A. scabra	*A. digitalis*	*Mean*
	Original data from Table 9.1		
100%	10.56	7.43	9.00
75%	7.89	7.34	7.61
50%	12.17	12.33	12.25
Mean	10.21	9.03	9.62
	Artificial data from Table 9.2		
100%	10.56	7.43	9.00
75%	7.89	12.33	10.11
50%	12.17	7.34	9.76
Mean	10.21	9.03	9.62

subgroup, and error *SS* are the same as before (Table 9.1). This should not be surprising, since we are using the same data. All that we have done is to interchange the contents of the lower two cells in the right-hand column of the table. When we partition the subgroup *SS* we do find some differences. We note that the *SS* between species (between columns) is unchanged. Since the change we made was within one column, the total for that column was not altered and consequently the column *SS* did not change. However, the sums of the second and third rows have been altered appreciably as a result of the interchange of the readings for 75% and 50% salinity in *A. digitalis*. The sum for 75% salinity is now very close to that for 50% salinity and the difference between the salinities, previously quite marked, is now no longer so. By contrast, the interaction *SS*, obtained by subtracting the sums of squares of rows and columns from the subgroup *SS*, is now a large quantity. Remember that the subgroup *SS* is the same in the two examples. In the first example we subtracted sums of squares due to the effects of both species and salinities, leaving only a tiny residual representing the interaction. In the second example these two *main effects* (species and salinities) account only for little of the subgroup sum of squares, leaving the interaction sum of squares as a substantial residual. What is the essential difference between these two examples?

In Table 9.3 we have shown the subgroup and marginal means for the original data from Table 9.1 and for the "doctored" data of Table 9.2. The original results are quite clear: at 75% salinity oxygen consumption is lower than at the other two salinities and this is true for both species. We note further that *A. scabra* consumes more oxygen than *A. digitalis* at two of the salinities. Thus our statements about differences due to species or to salinity can be made largely independent of each other. However, if we had to interpret the artificial data (lower half of Table 9.3), we would note that although *A. scabra* still consumes more oxygen than *A. digitalis* (since column sums have not changed), this difference depends greatly on the salinity. At 100% and 50% *A. scabra* consumes considerably more oxygen than *A. digitalis*, but at 75% this relationship is reversed. Thus we are no longer able to make an unequivocal statement about the amount of oxygen taken up by the two species. We have to qualify our statement by the seawater concentration at which they are kept. At 100% and 50% $\bar{Y}_{scabra} > \bar{Y}_{digitalis}$, but at 75% $\bar{Y}_{scabra} < \bar{Y}_{digitalis}$. If we examine the effects of salinity in the artificial example, we notice a mild increase in oxygen consumption 75%. However, again we have to qualify this statement by the species of the consuming limpet; *scabra* consumes least at 75%, while *digitalis* consumes most at this concentration.

This dependence of the effect of one factor on the level of another factor is called *interaction*. It is a common and fundamental scientific idea. It indicates that the effects of the two factors are not simply additive but that any given combination of levels of factors such as salinity combined with one species contributes a positive or negative increment to the level of expression of the variable. In common biological terminology a large positive increment of this sort is called *synergism*. When drugs act synergistically, the result of the interaction of the two drugs may be above and beyond the separate effects of each drug. When a combination of levels of two factors inhibit each other's effects, we call it *interference*. Synergism and interference will both tend to magnify the interaction *SS*.

Testing for interaction is an important procedure in analysis of variance. If the artificial data of Table 9.2 were real, it would be of little value to state that 75% salinity led to slightly greater consumption of oxygen. This statement would cover up the important differences in the data, which are that *scabra* consumes least at this concentration, while *digitalis* consumes most.

We are now able to write an expression symbolizing the decomposition of a single variate in a two-way analysis of variance in the manner of Expression (7.3) for single classification anova. The expression below assumes that both factors represent fixed treatment effects, Model I. This would seem reasonable, since species as well as salinity are fixed treatments. Variate Y_{ijk} is the kth item in the subgroup representing the ith group of treatment A and the jth group of treatment B. It is decomposed as follows:

$$Y_{ijk} = \mu + \alpha_i + \beta_j + (\alpha\beta)_{ij} + \epsilon_{ijk} \tag{9.1}$$

where μ equals the parametric mean of the population, α_i is the fixed treatment effect for the ith group of treatment A, β_j is the fixed treatment effect of the jth group of treatment B, $(\alpha\beta)_{ij}$ is the interaction effect in the subgroup representing the ith group of factor A and the jth group of factor B, and ϵ_{ijk} is the error term of the kth item in subgroup ij. We make the usual assumption that ϵ_{ijk} is normally distributed with a mean of 0 and a variance of σ^2. If one or both of the factors are Model II, we replace the α_i and β_j in the formula by A_i and/or B_j.

In previous chapters we have seen that each sum of squares represents a sum of squared deviations. What actual deviations does an interaction SS represent? We can see this easily by referring back to the anovas of Table 9.1. The variation among subgroups is represented by $\bar{Y} - \bar{\bar{Y}}$, where $\bar{Y}$ stands for the subgroup mean and $\bar{\bar{Y}}$ for the grand mean. When we subtract the deviations due to rows $\bar{R} - \bar{\bar{Y}}$ and columns $\bar{C} - \bar{\bar{Y}}$ from those of subgroups we obtain

$$\begin{aligned}(\bar{Y} - \bar{\bar{Y}}) - (\bar{R} - \bar{\bar{Y}}) - (\bar{C} - \bar{\bar{Y}}) &= \bar{Y} - \bar{\bar{Y}} - \bar{R} + \bar{\bar{Y}} - \bar{C} + \bar{\bar{Y}} \\ &= \bar{Y} - \bar{R} - \bar{C} + \bar{\bar{Y}}\end{aligned}$$

This somewhat involved expression is the deviation due to interaction. When we evaluate one such expression for each subgroup, square it, sum these squares, and multiply the sum by n, we obtain the interaction SS. This partition of the deviations also holds for their squares. This is so because the sums of the products of the separate terms cancel out.

A simple method for revealing the nature of the interaction present in the data is to inspect the means of the original data table. We can do this in Table 9.3. The original data, showing no interaction, would yield the following pattern of relative magnitudes:

	scabra	*digitalis*
100%		
	$\vee$	$\vee$
75%		
	$\wedge$	$\wedge$
50%		

The relative magnitudes of the means in the lower table yielding interaction can be summarized as follows:

	scabra	*digitalis*
100%		
	$\vee$	$\wedge$
75%		
	$\wedge$	$\vee$
50%		

When the pattern of signs expressing relative magnitudes is not uniform, as in the lower table, interaction is indicated. As long as the pattern of means is uniform, as in the upper table, interaction may not be present. However, interaction is often present without change in the *direction* of the differences; only the relative magnitudes may be affected. Thus visual inspection should not take the place of the statistical test.

In summary, when the effect of two treatments applied together cannot be predicted from the average responses of the separate factors, statisticians call this phenomenon interaction and test its significance by means of an interaction mean square. This is a very common scientific relationship. If we say that the effect of density on the fecundity or weight of a beetle depends on its genotype, we imply that a genotype × density interaction is present. If the geographic variation of a parasite depends on the nature of the host species it attacks, we speak of a host × locality interaction. If the effect of temperature on a metabolic process is independent of the effect of oxygen concentration, we say that temperature × oxygen interaction is absent.

Significance testing in a two-way anova will be deferred until the next section. However, we should point out that the computational steps **4** and **9** of Box 9.1 could have been shortened by employing the simplified formula for a sum of squares between *two* groups illustrated in Section 8.4. In an analysis with only two rows and two columns the interaction SS can be computed directly as

$$(\text{sum of one diagonal} - \text{sum of other diagonal})^2/abn.$$

9.2 Two-way anova: significance testing

Before we can test hypotheses about the sources of variation isolated in Box 9.1, we must become familiar with the expected mean squares for this design. In the anova table of Box 9.1 we first show the expected mean squares for Model I, both species differences and seawater concentrations being fixed treatment effects. Note that the within-subgroups or error MS again estimates the parametric variance of the items. The most important fact to remember about a Model I anova is that the mean square at each level of variation carries only the added effect due to that level of treatment; except for the parametric variance of the items, it does not contain any term from a lower line. Thus the expected MS of factor A contains only the parametric variance of the items plus the added term due to factor A, but does not also include interaction effects. In Model I the significance test is therefore simple and straightforward. Any source of variation is tested by the variance ratio of the appropriate mean square over the error MS. Thus for the appropriate tests we employ variance ratios $\boldsymbol{A}$/**Error**, $\boldsymbol{B}$/**Error** and $\boldsymbol{A} \times \boldsymbol{B})$/**Error**, where each boldface term signifies a mean square. Thus $\boldsymbol{A} = MS_A$, **Error** $= MS_{\text{within}}$.

When we do this in the example of Box 9.1 we find only factor B, salinity, significant. Neither factor A nor the interaction are significant. We conclude that the differences in oxygen consumption are induced by varying salinities (the two variables appear inversely related), and there does not appear to be

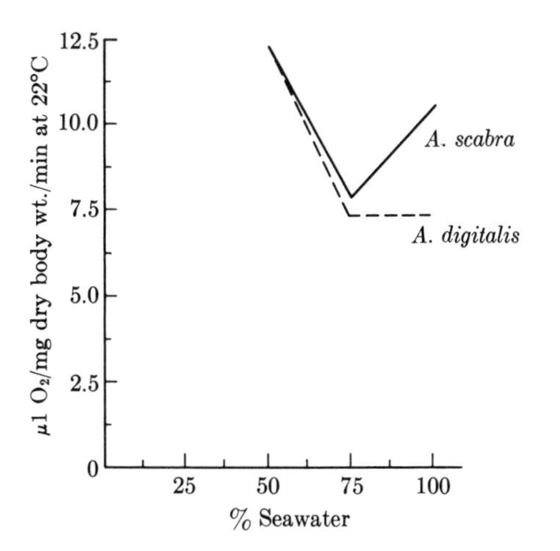

FIGURE 9.2

Oxygen consumption by two species of limpets at three salinities. Data from Box 9.1.

sufficient evidence for species differences in oxygen consumption. The tabulation of the relative magnitudes of the means in the previous section shows that the pattern of signs in the two lines is identical. However, this may be misleading, since the mean of *A. scabra* is far higher at 100% seawater than at 75%, but that of *A. digitalis* is only very slightly higher. Although the oxygen consumption curves of the two species when graphed appear far from parallel (see Figure 9.2), this suggestion of a species × salinity interaction cannot be shown to be significant when compared to the within-subgroups variance. Finding a significant difference among salinities does not conclude the analysis. The data suggest that at 75% salinity there is a real reduction in oxygen consumption. Whether this is really so could be tested by the methods of Section 8.6.

When we analyze the results of the artificial example in Table 9.2, we find only the interaction *MS* significant. Thus we would conclude that the response to salinity differs in the two species. This is brought out by inspection of the data, which show that at 75% salinity *A. scabra* consumes least, and *A. digitalis* consumes most oxygen.

In the last (artificial) example the mean squares of the two factors (main effects) are not significant in any case. However, many statisticians would not even test them once they found the interaction mean square to be significant, since in such a case an overall statement for each factor would have little meaning. A simple statement of response to salinity would be unclear. The presence of interaction makes us qualify our statements: the pattern of response to changes in salinity differed in the two species. We would consequently have to describe separate, nonparallel response curves for the two species. Occasionally it becomes important to test for overall significance in a Model I anova in spite of the presence of interaction. We may wish to demonstrate the significance of the effect of a drug, regardless of its significant interaction with age of the patient. To support this contention we might wish to test the mean square among drug concentrations (over the error MS), regardless of whether the interaction MS is significant.

Box 9.1 also lists expected mean squares for a Model II and a mixed model two-way anova. In the Model II note that the two main effects contain the variance component of the interaction as well as their own variance component. In a Model II we first test $(\boldsymbol{A} \times \boldsymbol{B})/\mathbf{Error}$. If the interaction is significant we continue testing $\boldsymbol{A}/(\boldsymbol{A} \times \boldsymbol{B})$ and $\boldsymbol{B}/(\boldsymbol{A} \times \boldsymbol{B})$, but when $\boldsymbol{A} \times \boldsymbol{B}$ is not significant some authors suggest computation of a pooled error $MS = (SS_{A\times B} + SS_{\text{within}})/(df_{A\times B} + df_{\text{within}})$ to test the significance of the main effects. The conservative position is to continue to test these over the interaction MS, and we shall follow this procedure in this book. Only one type of mixed model is shown in which factor A is assumed to be fixed and factor B to be random. If the situation is reversed the expected mean squares change accordingly. Here it is the mean square representing the fixed treatment that carries with it the variance component of the interaction, while the mean square representing the random factor contains only the error variance and its own variance component and does not include the interaction component. We therefore test the MS of the random main effect over the error but test the fixed treatment MS over the interaction.

9.3 Two-way anova without replication

In many experiments there will be no replication for each combination of factors represented by a cell in the data table. In such cases we cannot easily talk of "subgroups," since each cell contains a single reading only. Frequently it may be too difficult or too expensive to obtain more than one reading per cell, or the measurements may be known to be so repeatable that there is little point in estimating their error. As we shall see below, a two-way anova without replication can be properly applied only with certain assumptions. For some models and tests in anova we must assume that there is no interaction present.

Our illustration for this design is taken from limnology. In Box 9.2 we show temperatures of a lake taken at about the same time on four successive summer afternoons. These temperature measurements were made at ten different depths and were known to be very repeatable. Therefore only single readings were taken at each depth on any one day. What is the appropriate model for this anova? Clearly, the depths are Model I. The four days, however, are not likely to be of specific interest. It is improbable that an investigator would ask whether the water was colder on the 30th of July than on the 31st of July. A more meaningful way of looking at this problem would be to consider these four summer days as random samples that enable us to estimate the day-to-day variation of the temperature stratification in the lake during the stable midsummer period.

BOX 9.2

Two-way anova without replication.

Temperatures (°C) of Rot Lake on four early afternoons of the summer of 1952 at 10 depths. This is a mixed model anova.

Factor B: Depth in meters ($b = 10$)	Factor A: Days ($a = 4$) 29 July	30 July	31 July	1 August	Σ
0	23.8	24.0	24.6	24.8	97.2
1	22.6	22.4	22.9	23.2	91.1
2	22.2	22.1	22.1	22.2	88.6
3	21.2	21.8	21.0	21.2	85.2
4	18.4	19.3	19.0	18.8	75.5
5	13.5	14.4	14.2	13.8	55.9
6	9.8	9.9	10.4	9.6	39.7
9	6.0	6.0	6.3	6.3	24.6
12	5.8	5.9	6.0	5.8	23.5
15.5	5.6	5.6	5.5	5.6	22.3
Σ	148.9	151.4	152.0	151.3	603.6

Source: Data from Vollenweider and Frei (1953).

The four sets of readings are treated as replications (blocks) in this analysis. Depth is a fixed treatment effect, days are considered as random effects, hence this is a mixed model anova.

Preliminary computations

1. Grand total $= \sum^{a} \sum^{b} Y = 603.6$

2. Sum of the squared observations $= \sum^{a} \sum^{b} Y^2$

$$= (23.8)^2 + \cdots + (5.6)^2 = 11{,}230.78$$

3. Sum of squared column totals divided by sample size of a column

$$= \frac{\sum^{a} \left(\sum^{b} Y\right)^2}{b} = \frac{[(148.9)^2 + \cdots + (151.3)^2]}{10} = 9108.89$$

4. Sum of squared row totals divided by sample size of a row

$$= \frac{\sum^{b} \left(\sum^{a} Y\right)^2}{a} = \frac{[(97.2)^2 + \cdots + (22.3)^2]}{4} = 11{,}227.98$$

5. Grand total squared and divided by the total sample size = correction term

$$CT = \frac{\left(\sum^{a} \sum^{b} Y\right)^2}{ab} = \frac{(\text{quantity } \mathbf{1})}{ab} = \frac{(603.6)^2}{40} = 9108.32$$

BOX 9.2 continued

6. $SS_{\text{total}} = \sum^{a}\sum^{b} Y^2 - CT$
$= \text{quantity } \mathbf{2} - \text{quantity } \mathbf{5} = 11{,}230.78 - 9108.32 = 2122.46$

7. SS_A (*SS* of columns) $= \dfrac{\sum^{a}\left(\sum^{b} Y\right)^2}{b} - CT$
$= \text{quantity } \mathbf{3} - \text{quantity } \mathbf{5} = 9108.89 - 9108.32 = 0.57$

8. SS_B (*SS* of rows) $= \dfrac{\sum^{b}\left(\sum^{a} Y\right)^2}{a} - CT$
$= \text{quantity } \mathbf{4} - \text{quantity } \mathbf{5} = 11{,}227.98 - 9108.32 = 2119.66$

9. SS_{error} (remainder; discrepance) $= SS_{\text{total}} - SS_A - SS_B$
$= \text{quantity } \mathbf{6} - \text{quantity } \mathbf{7} - \text{quantity } \mathbf{8} = 2122.46 - 0.57 - 2119.66$
$= 2.23$

Anova table

Source of variation		*df*	*SS*	*MS*	F_s	*Expected MS*
$\bar{Y}_A - \bar{\bar{Y}}$	A (column; days)	3	0.57	0.190	2.30 *ns*	$\sigma^2 + b\sigma^2_A$
$\bar{Y}_B - \bar{\bar{Y}}$	B (rows; depths)	9	2119.66	235.5	2851.1**	$\sigma^2 + \sigma^2_{AB} + \frac{a}{b-1}\Sigma\beta^2$
$Y - \bar{Y}_A - \bar{Y}_B + \bar{\bar{Y}}$	Error (remainder; discrepance)	27	2.23	0.0826		$\sigma^2 + \sigma^2_{AB}$
$Y - \bar{\bar{Y}}$	Total	39	2122.46			

Conclusions.—Highly significant decreases in temperature occur as depth increases. For testing "days" we must assume interaction between days and depths to be zero. No added variance component among days can be demonstrated.

The computations are shown in Box 9.2. They are the same as those in Box 9.1 except that the expressions to be evaluated are considerably simpler. Since $n = 1$, much of the summation can be omitted. The subgroup sum of squares in this example is the same as the total sum of squares. If this is not immediately apparent, consult Figure 9.3, which, when compared with Figure 9.1, illustrates that the error sum of squares based on variation within subgroups is missing in this example. Thus, after we subtract the sum of squares for columns (factor A) and for rows (factor B) from the total *SS*, we are left with only a single sum of squares, which is the equivalent of the

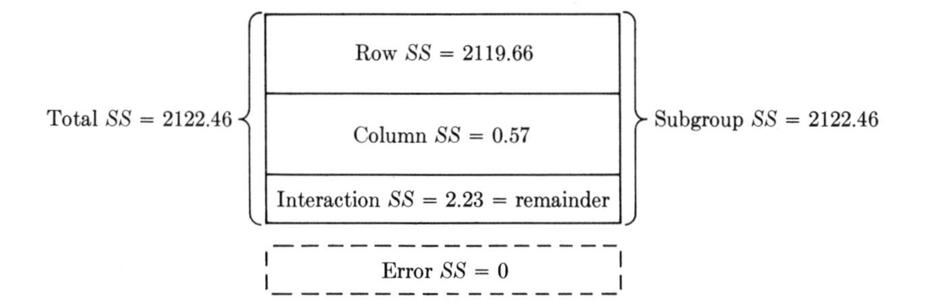

FIGURE 9.3

Diagrammatic representation of the partitioning of the total sums of squares in a two-way orthogonal anova without replication. The areas of the subdivisions are not shown proportional to the magnitudes of the sums of squares.

previous interaction SS but which is now the only source for an error term in the anova. This SS is known as the *remainder* SS or the *discrepance.*

If you will refer to the expected mean squares for the two-way anova in Box 9.1, you will discover why we made the statement earlier that for some models and tests in a two-way anova without replication we must assume that the interaction is not significant. If interaction is present, only a Model II anova can be entirely tested, while in a mixed model only the fixed level can be tested over the remainder mean square. But in a pure Model I, or for the random factor in a mixed model, it would be improper to test the main effects over the remainder unless we could reliably assume that no added effect due to interaction is present. General inspection of the data in Box 9.2 convinces us that the temperature trends present on any one day are faithfully reproduced on the other days. Thus, interaction is unlikely to be present. If, for example, a severe storm had churned up the lake on one day, changing the temperature relationships at various depths, interaction would have been apparent, and the test of the mean square among days carried out in Box 9.2 would not have been legitimate.

Since we assume no interaction, the row and column mean squares are tested over the error MS. The results are not surprising; casual inspection of the data would have predicted our findings. Added variance is not significant among days, but the differences in temperature due to depths are highly significant, yielding a value of $F_s = 2851.1$. It would be extremely unlikely that chance alone could produce such differences.

A common application of two-way anova without replication is the *repeated testing of the same individuals*. By this we mean that the same group of individuals is tested repeatedly over a period of time. The individuals are one factor (usually considered as random and serving as replication) and the time dimension is the second factor, a fixed treatment effect. For example, we might measure growth of a structure in ten individuals at regular intervals. When testing for the presence of an added variance component (due to the random factor) such a model again assumes that there is no interaction between time and the individuals; that is, the responses of the several individuals are parallel through time. Another use of this design is found in various physiological and psychological experiments in which we test the same group of individuals for the appearance of some response after treatment. Examples include increasing immunity after antigen inoculations, altered responses after conditioning, and measures of learning after a number of trials. Thus we may study the speed with which ten rats, repeatedly tested on the same maze, reach the end point. The fixed treatment effect would be the successive trials to which the rats have been subjected. The second factor, the ten rats, is random, presumably representing a random sample of rats from the laboratory population.

One special case, common enough to merit separate discussion, is randomized complete blocks in which there are only two treatments or groups ($a = 2$). This case is also known as *paired comparisons* because each observation for one treatment is paired with one for the other treatment. This pair is composed of the same individuals tested twice or of two individuals with common experiences so that we can legitimately arrange the data as a two-way anova.

Let us elaborate on this point. Suppose we test the muscle tone of a group of individuals, subject them to severe physical exercise, and measure their muscle tone once more. Since the same group of individuals will have been tested twice, we can arrange our muscle tone readings in pairs, each pair representing readings on one individual (before and after exercise). Such data are appropriately treated by a two-way anova without replication, which in this case would be a paired comparisons test because there are only two treatment classes. This "before and after treatment" comparison is a very frequent design leading to paired comparisons. Another design simply measures two stages in the development of a group of organisms, time being the treatment intervening between the two stages. The example in Box 9.3 is of this nature. It measures lower face width in a group of girls five years old and in the same group of girls when they are six years old. The paired comparison is for each individual girl, between her face width when she is five years old and her face width at six years.

Paired comparisons often result from dividing an organism or other individual unit so that half receives treatment 1 and the other half treatment 2, which may be the control. Thus if we wish to test the strength of two antigens or allergens we might inject one into each arm of a single individual and measure the diameter of the red area produced. It would not be wise from the point of view of experimental design to test antigen 1 on individual 1 and antigen 2 on individual 2. These individuals may be differentially susceptible to these antigens and we may learn little about the relative potency of the antigens, since this would be confounded by the differential responses of the subjects. A much better design would be to inject antigen 1 into the left arm and antigen 2 into the right arm of a group of n individuals and to analyze the data as a two-way anova without replication with n rows (individuals) and 2 columns (treatments). It is probably immaterial whether an antigen is injected into the right or left arm, but if we were designing such an experiment and knew little about the reaction of humans to antigens we might, as a precaution, randomly allocate antigen 1 to the left or right arm for different subjects, antigen 2 being injected into the opposite arm. A similar example is the testing of certain plant viruses by rubbing a certain concentration of the virus over the surface of a leaf and counting the resulting lesions. Since different leaves are differentially susceptible, a conventional way of measuring the strength of the virus is to wipe it over half of the leaf on one side of the midrib, rubbing the other half of the leaf with a control or standard solution.

Another design leading to paired comparisons is when the treatment is given to two individuals sharing a common experience, be this genetic or environmental. Thus a drug or a psychological test might be given to groups of twins or sibs, one of each pair receiving the treatment, the other one not.

Finally, the paired comparisons technique may be used when the two individuals to be compared share a single experimental unit and are thus subjected to common environmental experiences. If we have a set of rat cages, each of which holds two rats, and we are trying to compare the effect of a hormone injection with a control, we might inject one of each pair of rats with the hormone and use his cage mate as a control. This would yield a $2 \times n$ anova for n cages.

One reason for featuring the paired comparisons test separately is that it alone among the two-way anovas without replication has an equivalent, alternative method of analysis—the t-test for paired comparisons, which is the traditional method of solving it.

The paired comparisons case shown in Box 9.3 analyzes face widths of five- and six-year-old girls, as already mentioned. The question being asked is whether the faces of six-year-old girls are significantly wider than those of five-year-old girls. The data are shown in columns (1) and (2) of Box 9.3 for 15 individual girls. Column (3) features the row sums that are necessary for the analysis of variance. The computations for the two-way anova without replication in Box 9.3 are the same as those already shown for Box 9.2 and are not shown in detail. The anova table shows that there is a highly significant difference in face width between the two age groups. If interaction is assumed to be zero, there is a large added variance component among the individual girls, undoubtedly representing genetic as well as environmental differences.

BOX 9.3

Paired comparisons (randomized blocks with $a = 2$).

Lower face width (skeletal bigonial diameter in cm) for 15 North American white girls measured when 5 and again when 6 years old.

	(1)	*(2)*	*(3)*	*(4)*
Individuals	*5-year-olds*	*6-year-olds*	Σ	$D = Y_{i2} - Y_{i1}$ *(difference)*
1	7.33	7.53	14.86	0.20
2	7.49	7.70	15.19	.21
3	7.27	7.46	14.73	.19
4	7.93	8.21	16.14	.28
5	7.56	7.81	15.37	.25
6	7.81	8.01	15.82	.20
7	7.46	7.72	15.18	.26
8	6.94	7.13	14.07	.19
9	7.49	7.68	15.17	.19
10	7.44	7.66	15.10	.22
11	7.95	8.11	16.06	.16
12	7.47	7.66	15.13	.19
13	7.04	7.20	14.24	.16
14	7.10	7.25	14.35	.15
15	7.64	7.79	15.43	.15
ΣY	111.92	114.92	226.84	3.00
ΣY^2	836.3300	881.8304	3435.6992	0.6216

Source: From a larger study by Newman and Meredith (1956).

Two-way anova without replication

Anova table

Source of variation	*df*	*SS*	*MS*	F_s	*Expected MS*
Ages (columns factor A)	1	0.3000	0.3000	389.11**	$\sigma^2 + \sigma^2_{AB} + \frac{b}{a-1}\Sigma\alpha^2$
Individuals (rows factor B)	14	2.6367	0.1883	242.02**	$\sigma^2 + a\sigma^2_B$
Remainder	14	0.0108	0.000771		$\sigma^2 + \sigma^2_{AB}$
Total	29	2.9475			

$F_{.01[1,14]} = 8.86 \qquad F_{.01[12,14]} = 3.80$

Conclusions.—The variance ratio for ages is highly significant. We conclude that faces of 6-year-old girls are wider than those of 5-year-olds. If we are willing to assume that the interaction σ^2_{AB} is zero, we may test for an added variance component among individual girls and would find it significant.

BOX 9.3 continued

The t*-test for paired comparisons*

$$t_s = \frac{\bar{D} - (\mu_1 - \mu_2)}{s_{\bar{D}}}$$

where $\bar{D}$ is the mean difference between the paired observations

$$(\bar{D} = \sum D/b = 3.00/15 = 0.20)$$

and $s_{\bar{D}} = s_D/\sqrt{b}$ is the standard error of $\bar{D}$ calculated from the observed differences in column (4).

$$s_D = \sqrt{\frac{\sum D^2 - (\sum D)^2/b}{b - 1}} = \sqrt{\frac{0.6216 - (3.00^2/15)}{14}} = \sqrt{0.0216/14}$$

$$= \sqrt{0.001543} = 0.0392810$$

$$s_{\bar{D}} = \frac{s}{\sqrt{b}} = \frac{0.0392810}{\sqrt{15}} = 0.0101423$$

We assume that the true difference between the means of the two groups, $\mu_1 - \mu_2$, equals zero.

$$t_s = \frac{\bar{D} - 0}{s_{\bar{D}}} = \frac{0.20 - 0}{0.0101423} = 19.7194 \qquad \text{with } b - 1 = 14 \ df, \text{ and } P \ll 0.001.$$

$$t_s^2 = 388.85$$

which agrees with the previous F_s within an acceptable rounding error.

The other method of analyzing paired comparisons designs is the well-known t*-test for paired comparisons*. It is quite simple to apply and is illustrated in the second half of Box 9.3. It tests whether the mean of sample differences between pairs of readings in the two columns is significantly different from a hypothetical mean, which the null hypothesis puts at zero. The standard error over which this is tested is the standard error of the mean difference. The difference column has to be calculated and is shown in column (4) of the data table in Box 9.3. The computations are quite straightforward and the conclusions are the same as for the two-way anova. This is another instance in which we obtain the value of F_s within rounding error when we square the value of t_s.

Although the paired comparisons t-test is the traditional method of solving this type of problem, we prefer the two-way anova. Its computation is no more time-consuming, it avoids finding a square root, and it has the advantage of providing a measure of the variance component among the rows (blocks). This is useful knowledge because if there is no significant added variance component among blocks one might simplify the analysis and design of future, similar studies by employing a completely randomized anova.

Exercises 9

9.1 Swanson, Latshaw, and Tague (1921) determined soil *p*H electrometrically for various soil samples from Kansas. An extract of their data (acid soils) is shown below. Do subsoils differ in *p*H from surface soils (assume that there is no interaction between localities and depth for *p*H reading)? ANS. $F_s = 0.894$.

County	*Soil type*	*Surface pH*	*Subsoil pH*
Finney	Richfield silt loam	6.57	8.34
Montgomery	Summit silty clay loam	6.77	6.13
Doniphan	Brown silt loam	6.53	6.32
Jewell	Jewell silt loam	6.71	8.30
Jewell	Colby silt loam	6.72	8.44
Shawnee	Crawford silty clay loam	6.01	6.80
Cherokee	Oswego silty clay loam	4.99	4.42
Greenwood	Summit silty clay loam	5.49	7.90
Montgomery	Cherokee silt loam	5.56	5.20
Montgomery	Oswego silt loam	5.32	5.32
Cherokee	Bates silt loam	5.92	5.21
Cherokee	Cherokee silt loam	6.55	5.66
Cherokee	Neosho silt loam	6.53	5.66

9.2 The following data were extracted from a Canadian record book of purebred dairy cattle. Random samples of 10 mature (five-year-old and older) and 10 two-year-old cows were taken from each of five breeds (honor roll, 305-day class). The average butterfat percentages of these cows were recorded. This gave us a total of 100 butterfat percentages, broken down into five breeds and into two age classes. The 100 butterfat percentages are given below. Analyze and discuss your results. You will note that the tedious part of the calculation has been done for you.

	Breed									
	Ayshire		*Canadian*		*Guernsey*		*Holstein-Friesian*		*Jersey*	
	Mature	*2-yr*	*Mature*	*2-yr*	*Mature*	*2-yr*	*Mature*	*2-yr*	*Mature*	*2-yr*
	3.74	4.44	3.92	4.29	4.54	5.30	3.40	3.79	4.80	5.75
	4.01	4.37	4.95	5.24	5.18	4.50	3.55	3.66	6.45	5.14
	3.77	4.25	4.47	4.43	5.75	4.59	3.83	3.58	5.18	5.25
	3.78	3.71	4.28	4.00	5.04	5.04	3.95	3.38	4.49	4.76
	4.10	4.08	4.07	4.62	4.64	4.83	4.43	3.71	5.24	5.18
	4.06	3.90	4.10	4.29	4.79	4.55	3.70	3.94	5.70	4.22
	4.27	4.41	4.38	4.85	4.72	4.97	3.30	3.59	5.41	5.98
	3.94	4.11	3.98	4.66	3.88	5.38	3.93	3.55	4.77	4.85
	4.11	4.37	4.46	4.40	5.28	5.39	3.58	3.55	5.18	6.55
	4.25	3.53	5.05	4.33	4.66	5.97	3.54	3.43	5.23	5.72
$\sum Y$	40.03	41.17	43.66	45.11	48.48	50.52	37.21	36.18	52.45	53.40
$\bar{Y}$	4.003	4.117	4.366	4.511	4.848	5.052	3.721	3.618	5.245	5.340

$$\sum^{abn} Y^2 = 2059.6109$$

9.3 Blakeslee (1921) studied length/width ratios of second seedling leaves of two types of Jimson weed called globe (G) and nominal (N). Three seeds of each type were planted in 16 pots. Is there sufficient evidence to conclude that globe and nominal differ in length/width ratio?

Pot identification number	*Types*					
		G			*N*	
16533	1.67	1.53	1.61	2.18	2.23	2.32
16534	1.68	1.70	1.49	2.00	2.12	2.18
16550	1.38	1.76	1.52	2.41	2.11	2.60
16668	1.66	1.48	1.69	1.93	2.00	2.00
16767	1.38	1.61	1.64	2.32	2.23	1.90
16768	1.70	1.71	1.71	2.48	2.11	2.00
16770	1.58	1.59	1.38	2.00	2.18	2.16
16771	1.49	1.52	1.68	1.94	2.13	2.29
16773	1.48	1 44	1.58	1.93	1.95	2.10
16775	1.28	1.45	1.50	1.77	2.03	2.08
16776	1.55	1.45	1.44	2.06	1.85	1.92
16777	1.29	1.57	1.44	2.00	1.94	1.80
16780	1.36	1.22	1.41	1.87	1.87	2.26
16781	1.47	1.43	1.61	2.24	2.00	2.23
16787	1.52	1.56	1.56	1.79	2.08	1.89
16789	1.37	1.38	1.40	1.85	2.10	2.00

9.4 The following data were extracted from a more extensive study by Sokal and Karten (1964). The data represent mean dry weights (in mg) of three genotypes of beetles, *Tribolium castaneum*, reared at a density of 20 beetles per gram of flour. The four series of experiments represent replications.

	Genotypes		
Series	$++$	$+b$	bb
1	0.958	0.986	0.925
2	0.971	1.051	0.952
3	0.927	0.891	0.829
4	0.971	1.010	0.955

Test whether the genotypes differ in mean dry weight.

10

Assumptions of Analysis of Variance

We shall now examine the underlying assumptions of the analysis of variance, methods for testing whether these assumptions are valid, the consequences for an anova if the assumptions are violated, and steps to be taken if the assumptions cannot be met. We should stress that before carrying out any anova on an actual research problem you should reassure yourself that the assumptions listed in this chapter seem reasonable and, if they are not, you should carry out one of several possible alternative steps to remedy the situation.

The first section (10.1) lists briefly the various assumptions of analysis of variance, describes procedures for testing some of them, briefly states the consequences if the assumptions do not hold, and (if they do not) gives instructions on how to proceed. The assumptions include random sampling, independence, homogeneity of variances, normality, and additivity.

In many cases departure from the assumptions of analysis of variance can be rectified by transformation of the original data by use of a new scale. The rationale behind this and some of the common transformations is given in Section 10.2.

When transformations are unable to make the data conform to the assumptions of analysis of variance, other techniques of analysis, analogous to the intended anova, must be employed. These are the nonparametric techniques briefly mentioned in Section 2.6. As stated there, these techniques are sometimes used by preference even when the parametric method (anova in this case) can be legitimately employed. Rapidity of computation and a preference for the generally simple assumptions of the nonparametric analyses cause many research workers to turn to them. However, when the assumptions of anova are met, these methods are less efficient than anova. Section 10.3 examines three nonparametric methods in lieu of anova for two-sample cases only.

10.1 The assumptions of anova

Randomness. All anovas require that sampling of individuals be at random. Thus, in a study of the effects of three doses of a drug (plus a control) on five rats each, the five rats allocated to each treatment must be selected at random. If the five rats employed as controls are either the youngest or the smallest or the heaviest rats, while those allocated to some other treatment are selected in some other way, it is clear that the results are not apt to yield an unbiased estimate of the true treatment effects. Nonrandomness of sample selection may well be reflected in lack of independence of the items, in heterogeneity of variances, or in nonnormal distribution—all discussed in this section. Adequate safeguards to ensure random sampling during the design of an experiment, or when sampling from natural populations, are essential.

Independence. An assumption stated in each explicit expression for the expected value of a variate [for example, Expression (7.3) was $Y_{ij} = \mu + \alpha_i + \epsilon_{ij}$] is that the error term ϵ_{ij} is a random normal variable. In addition, for completeness we should also add the statement that it is assumed that the ϵ's are independently and identically (see below) distributed.

Thus, if you arrange the variates within any one group in some logical order independent of their magnitude (such as the order in which the measurements were obtained), we would expect the ϵ_{ij}'s to succeed each other in a random sequence. Consequently, we assume a long sequence of large positive values followed by an equally long sequence of negative values to be quite unlikely. We would also not expect positive and negative values to alternate with regularity.

How could departures from independence arise? An obvious example would be an experiment in which the experimental units were plots of ground laid out in a field. In such a case it is often found that adjacent plots of ground give rather similar yields. It would thus be important not to group all the plots containing the same treatment into an adjacent series of plots but rather to randomize the allocation of treatments among the experimental plots. The physical process of randomly allocating the treatments to the experimental plots ensures that the ϵ's will be independent.

Lack of independence of the ϵ's can result from correlation in time rather than space. In an experiment we might measure the effect of a treatment by recording weights of ten individuals. Our balance may suffer from a maladjustment that results in giving successive underestimates, compensated for by several overestimates. Conversely, compensation by the operator of the balance may result in regularly alternating over- and underestimates of the true weight. Here again randomization may overcome the problem of non-independence of errors. For example, we may determine the sequence in which individuals of the various groups are weighed according to some random procedure.

There is no simple adjustment or transformation to overcome the lack of independence of errors. The basic design of the experiment or the way in which it was performed must be changed. If the ϵ's are not independent, the validity of the usual F-test of significance can be seriously impaired.

Homogeneity of variances. In Section 8.4 and Box 8.2, in which we described the t-test for the difference between two means, you were told that the statistical test was valid only if we could assume that the variances of the two samples were equal. Although we have not stressed it so far, this assumption that the ϵ_{ij}'s have identical variances also underlies the equivalent anova test for two samples—and in fact any type of anova. *Equality of variances* in a group of samples is an important precondition for several statistical tests. Synonyms for this condition are *homogeneity of variances* or *homoscedasticity*. This term is coined from Greek roots meaning equal scatter; the converse condition (inequality of variances among samples) is called *heteroscedasticity*. Because we assume that each sample variance is an estimate of the same parametric error variance, the assumption of homogeneity of variances makes intuitive sense.

We have already seen how to test whether two samples are homoscedastic prior to a t-test of the differences between two means or a two-sample analysis of variance: we use an F-test for the hypotheses $H_0: \sigma_1^2 = \sigma_2^2$ and $H_1: \sigma_1^2 \neq \sigma_2^2$, as illustrated in Section 7.3 and Box 7.1. For more than two samples there is a "quick and dirty" method, preferred by many because of its simplicity. This is the $F_{\max}$-test. This test relies on the tabled cumulative probability distribution of a statistic that is the variance ratio of the largest to the smallest of several sample variances. This distribution is shown in Table **VI**. Let us assume that we have six anthropological samples of 10 bone lengths each, for which we wish to carry out an anova. The variances of the six samples range from 1.2 to 10.8. We compute the maximum variance ratio $s^2_{\max}/s^2_{\min} = 10.8/1.2 = 9.0$ and compare it with $F_{\max\,\alpha[a,\nu]}$, critical values of which are found in Table **VI**. For $a = 6$ and $\nu = n - 1 = 9$, $F_{\max}$ is 7.80 and 12.1 at the 5% and 1% levels, respectively. We conclude that the variances of the six samples are significantly heterogeneous.

What may cause such heterogeneity? In this case, we suspect that some of the populations are inherently more variable than others. Some races or species are relatively uniform for one character, while others are quite variable for the same character. In an anova representing the results of an experiment, it may well be that one sample was obtained under less standardized conditions than the others and hence has a greater variance. There are also many cases in which the heterogeneity of variances is a function of an improper choice of measurement scale. With some measurement scales, variances vary as functions of means. Thus differences among means bring about heterogeneous variances. For example, in variables following the Poisson distribution the variance is in fact equal to the mean, and populations with greater means will therefore have greater variances. Such departures from the assumption of homoscedasticity can often be easily corrected by a suitable transformation, as discussed later in this chapter.

A rapid first inspection for heteroscedasticity is to check for correlation between the means and variances or between the means and the ranges of the samples. If the variances increase with the means (as in a Poisson distribution), the ratios $s^2/\bar{Y}$ or $s/\bar{Y} = CV$ will be approximately constant for the samples. If means and variances are independent, these ratios will vary widely.

The consequences of moderate heterogeneity of variances are not too serious for the overall test of significance, but single degree of freedom comparisons may be far from accurate.

If transformation cannot cope with heteroscedasticity, nonparametric methods (Section 10.3) may have to be resorted to.

Normality. We have assumed that the error terms, ϵ_{ij}, of the variates in each sample will be independent, that the variances of the error terms of the several samples will be equal, and, finally, that the error terms are normally distributed. If there is serious question about the normality of the data, a graphic test as illustrated in Section 5.5 might be applied to each sample separately.

The consequences of nonnormality of error are not too serious. Only very skewed distribution would have a marked effect on the significance level of the F-test or on the efficiency of the design. The best way to correct for lack of normality is to carry out a transformation that will make the data normally distributed, as explained in the next section. If no simple transformation is satisfactory, a nonparametric test as carried out in Section 10.3 should be substituted for the analysis of variance.

Additivity. In two-way anova without replication it is necessary to assume that interaction is not present if one is to make tests of the main effects in a Model I anova. This assumption of no interaction in a two-way anova is sometimes also referred to as the assumption of additivity of the main

TABLE 10.1

Illustration of additive and multiplicative effects.

Factor B	Factor A			
	$\alpha_1 = 1$	$\alpha_2 = 2$	$\alpha_3 = 3$	
	2	3	4	Additive effects
$\beta_1 = 1$	1	2	3	Multiplicative effects
	0	0.30	0.48	Log of multiplicative effects
	6	7	8	Additive effects
$\beta_2 = 5$	5	10	15	Multiplicative effects
	0.70	1.00	1.18	Log of multiplicative effects

effects. By this we mean that any single observed variate can be decomposed into additive components representing the treatment effects of a particular row and column as well as a random term special to it. If interaction is actually present, then the F-test will be very inefficient and possibly misleading if the effect of the interaction is very large. A check of this assumption requires either more than a single observation per cell (so that an error mean square can be computed) or an independent estimate of the error mean square from previous *comparable* experiments.

Interactions can be due to a variety of causes. Most frequently it means that a given treatment combination, such as level 2 of factor A when combined with level 3 of factor B, makes a variable deviate from the expected value. Such a deviation is regarded as an inherent property of the natural system under study, as in examples of synergism or interference. Similar effects occur when a given replicate is quite aberrant, as may happen if an exceptional plot is included in an agricultural experiment, if a diseased individual is included in a physiological experiment, or if by mistake an individual from a different species is included in a biometric study. Finally, an interaction term will result if the effect of the two factors A and B on the response variable Y are multiplicative rather than additive. An example will make this clear.

In Table 10.1 we show the additive and multiplicative treatment effects in a hypothetical two-way anova. Let us assume that the expected population mean μ is zero. Then the mean of the sample subjected to treatment 1 of factor A and treatment 1 of factor B should be 2 by the conventional additive model. This is so because each factor at level 1 contributes unity to the mean. Similarly, the expected subgroup mean subjected to level 3 for factor A and level 2 for factor B is 8, the respective contributions to the mean being 3 and 5. However, if the process is multiplicative rather than additive, as occurs in a variety of physicochemical and biological phenomena, the expected values would be quite different. For treatment A_1B_1, the expected value equals 1, which is the product of 1 and 1. For treatment A_3B_2, the expected value is 15, the product of 3 and 5. If we were to analyze multiplicative data of this sort by a conventional anova, we would find that the interaction sum of squares would be greatly augmented because of the nonadditivity of the treatment effects. In this case, there is a simple remedy. By transforming the variable into logarithms (Table 10.1), we are able to restore the additivity of the data. The third item in each cell gives the logarithm of the expected value, assuming multiplicative relations. Notice that the increments are strictly additive again ($SS_{A \times B} = 0$). As a matter of fact, on a logarithmic scale we could simply write $\alpha_1 = 0$, $\alpha_2 = 0.30$, $\alpha_3 = 0.48$, $\beta_1 = 0$, $\beta_2 = 0.70$. Here is a good illustration of how transformation of scale, discussed in detail in Section 10.2, helps us meet the assumptions of analysis of variance.

10.2 Transformations

If the evidence indicates that the assumptions for an analysis of variance or a t-test cannot be maintained, two courses of action are open to us. We may carry out a different test not requiring the rejected assumptions, such as the distribution-free tests in lieu of anova, discussed in the next section. A second approach would be to transform the variable to be analyzed in such a manner that the resulting transformed variates meet the assumptions of the analysis. Let us look at a simple example of what transformation will do. A single variate of the simplest kind of anova (completely randomized, single classification, Model I) decomposes as follows: $Y_{ij} = \mu + \alpha_i + \epsilon_{ij}$. In this model the components are additive with the error term ϵ_{ij} normally distributed. However, we might encounter a situation in which the components were multiplicative in effect so that $Y_{ij} = \mu\alpha_i\epsilon_{ij}$, the product of these three terms. In such a case the assumptions of normality and of homoscedasticity would break down. The general parametric mean μ is constant in any one anova, but the treatment effect α_i differs from group to group. Clearly, the scatter among the variates Y_{ij} would double in a group in which α_i is twice as great as in another. Assume that $\mu = 1$, the smallest $\epsilon_{ij} = 1$, and the greatest 3; then if $\alpha_i = 1$, the range of the Y's will be $3 - 1 = 2$. However, when $\alpha_i = 4$, the corresponding range will be four times as wide from $4 \times 1 = 4$ to $4 \times 3 = 12$, a range of 8. Such data will be heteroscedastic. We can correct this situation simply by transforming our model into logarithms. We would therefore obtain $\log Y_{ij} = \log \mu + \log \alpha_i + \log \epsilon_{ij}$, which is additive and homoscedastic. The entire analysis of variance would then be carried out on the transformed variates.

At this point many of you will feel more or less uncomfortable about what we have done. Transformation seems too much like "data grinding." When you learn that often a statistical test may be made significant after transformation of a set of data, though it would not have been so without such a transformation, you may feel even more suspicious. What is the justification for transforming the data? It takes some getting used to the idea, but there is really no scientific necessity to employ the common linear or arithmetic scale to which we are accustomed. Teaching of the "new math" in elementary schools has done much to dispel the naive notion that the decimal system of numbers is the only "natural" one. It takes extensive experience in science and in the handling of statistical data to appreciate the fact that the linear scale, so familiar to all of us from our earliest experience, occupies a similar position with relation to other scales of measurement as does the decimal system of numbers with respect to the binary or octal numbering systems and others. If a system is multiplicative on a linear scale, it may be much more convenient to think of it as an additive system on a logarithmic scale. The square root of a variable is another frequent transformation. The square root of the surface area of an organism is often a more appropriate measure of the fundamental biological variable subjected to physiological and evolutionary forces than is the area. This is reflected in the normal distribution of the square root of the variable as compared to the skewed distribution of areas. In many cases experience has taught us to express experimental variables not in linear scale but as logarithms, square roots, reciprocals, or angles. Thus, *p*H values are logarithms and dilution series in microbiological titrations are expressed as reciprocals. As soon as you are ready to accept the idea that the scale of measurement is arbitrary, you simply have to look at the distributions of transformed variates to decide which transformation most closely satisfies the assumptions of the analysis of variance before carrying out an anova.

A fortunate fact about transformations is that very often several departures from the assumptions of anova are simultaneously cured by the same transformation to a new scale. Thus frequently, simply by making the data homoscedastic, we also make them approach normality and insure additivity of the treatment effects.

When a transformation is applied, tests of significance are performed on the transformed data, but estimates of means are usually given in the familiar untransformed scale. Since the transformations discussed in this chapter are nonlinear, confidence limits computed in the transformed scale and changed back to the original scale would be asymmetrical. Stating the standard error in the original scale would therefore be misleading. In reporting results of research with variables that require transformation, furnish means in the back-transformed scale followed by their (asymmetrical) confidence limits rather than by their standard errors.

An easy way to find out whether a given transformation will yield a distribution satisfying the assumptions of anova is to plot the cumulative distributions of the several samples on probability paper. By changing the scale of the second coordinate axis from linear to logarithmic, square root, or any other one, we can see whether a previously curved line, indicating skewness, straightens out to indicate normality (you may wish to refresh your memory on these graphic techniques studied in Section 5.5). We can look up upper class limits on transformed scales or employ a variety of available probability graph papers whose second axis is in logarithmic, angular, or other scale. Thus we not only test whether the data become more normal through transformation, but can also get an estimate of the standard deviation under transformation as measured by the slope of the fitted line. The assumption of homoscedasticity implies that the slopes for the several samples should be the same. If the slopes are very heterogeneous, homoscedasticity has not been achieved.

The logarithmic transformation. The most common transformation applied is conversion of all variates into logarithms, usually common logarithms. Whenever the mean is positively correlated with the variance (greater means are accompanied by greater variances), the logarithmic transformation is quite likely to remedy the situation and make the variance independent of the mean. Frequency distributions skewed to the right are often made more symmetrical by transformation to logarithmic scale. We saw in the previous section and in Table 10.1 that logarithmic transformation is also called for when effects are multiplicative.

The square root transformation. We shall use a square root transformation as a detailed illustration of transformation of scale. When the data are counts, as of insects on a leaf or blood cells in a hemacytometer, we frequently find the square root transformation of value. You will remember that such distributions are likely to be Poisson rather than normally distributed and that in a Poisson distribution the variance is the same as the mean. Therefore, the mean and variance cannot be independent but will vary identically. Transforming the variates to square roots will generally make the variances independent of the means. When the counts include zero values, it has been found desirable to code all variates by adding 0.5. The transformation then is $\sqrt{Y + \frac{1}{2}}$.

Table 10.2 shows an application of the square root transformation. The sample with the greater mean has a significantly greater variance prior to transformation. After transformation the variances are not significantly different. For reporting of means the transformed means are squared again and confidence limits are reported in lieu of standard errors.

The arcsine transformation. This transformation, also known as the *angular transformation,* is especially appropriate to percentages and proportions. You may remember from Section 4.2 that the standard deviation of a binomial distribution is $\sigma = \sqrt{pq/k}$. Since $\mu = p$, $q = 1 - p$, and k is constant for any one problem, it is clear that in a binomial distribution the variance would be a function of the mean. The arcsine transformation prevents this.

The transformation finds $\theta = \arcsin \sqrt{p}$, where p is a proportion. The term arcsin is synonymous with inverse sine or $\sin^{-1}$, which stands for the

TABLE 10.2

An application of the square root transformation. The data represent the number of adult *Drosophila* emerging from single pair cultures for two different medium formulations (medium A contained DDT).

(1) Number of flies emerging Y	(2) Square root of number of flies $\sqrt{Y}$	(3) Medium A f	(4) Medium B f
0	0.00	1	—
1	1.00	5	—
2	1.41	6	—
3	1.73	—	—
4	2.00	3	—
5	2.24	—	—
6	2.45	—	—
7	2.65	—	2
8	2.83	—	1
9	3.00	—	2
10	3.16	—	3
11	3.32	—	1
12	3.46	—	1
13	3.61	—	1
14	3.74	—	1
15	3.87	—	1
16	4.00	—	2
		15	15

	Medium A	Medium B
Untransformed variable		
$\bar{Y}$	1.933	11.133
s^2	1.495	9.410
Square root transformation		
$\overline{\sqrt{Y}}$	1.297	3.307
$s^2_{\sqrt{Y}}$	0.2630	0.2089

Tests of equality of variances

$$F_s = \frac{s^2_2}{s^2_1} = \frac{9.410}{1.495} = 6.294* \qquad F_{.025[14,14]} = 2.98 \qquad F_s = \frac{s^2_{\sqrt{Y_1}}}{s^2_{\sqrt{Y_2}}} = \frac{0.2630}{0.2089} = 1.259\ ns$$

	Medium A	Medium B
Back-transformed (squared) means		
$(\overline{\sqrt{Y}})^2$	1.682	10.936
95% confidence limits		
$L_1 = \overline{\sqrt{Y}} - t_{.05} s_{\overline{\sqrt{Y}}}$	$1.297 - 2.145\sqrt{\frac{0.2630}{15}}$ $= 1.013$	$3.307 - 2.145\sqrt{\frac{0.2089}{15}}$ $= 3.054$
$L_2 = \overline{\sqrt{Y}} + t_{.05} s_{\overline{\sqrt{Y}}}$	1.581	3.560
Back-transformed (squared) confidence limits		
L_1^2	1.026	9.327
L_2^2	2.500	12.674

Source: Data from unpublished study by R. R. Sokal.

angle whose sine is the given quantity. Thus, if we look up arcsin 0.431 we find 41.03°, the angle whose sine is 0.431. The arcsine transformation stretches out both tails of a distribution of percentages or proportions and compresses the middle. When the percentages in the original data fall between 30% and 70% it is generally not necessary to apply the arcsine transformation.

10.3 Nonparametric methods in lieu of anova

If none of the above transformations manage to make our data meet the assumptions of analysis of variance, we may resort to an analogous nonparametric method. These techniques are also called *distribution-free methods,* since they are not dependent on a given distribution (such as the normal in anova), but usually will work for a wide range of different distributions. They are called nonparametric methods because their null hypothesis is not concerned with specific parameters (such as the mean in analysis of variance) but only with the distribution of the variates. In recent years, nonparametric analysis of variance has become quite popular because it is simple to compute and permits freedom from worry about the assumptions of an anova. Yet we should point out that in cases where these assumptions hold entirely or even approximately, the analysis of variance is generally the more efficient statistical test for detecting departures from the null hypothesis.

We shall only discuss nonparametric tests for two samples in this section. For a design that would give rise to a t-test or anova with two classes, we employ the nonparametric *Mann-Whitney* U*-test* (Box 10.1). The null hypothesis is that the two samples come from populations having the same distribution. The data in Box 10.1 are morphological measurements on two samples of chigger nymphs. The Mann-Whitney U-test as illustrated in Box 10.1 is a semigraphical test and is quite simple to apply. It will be especially convenient when the data are already graphed and there are not too many items in each sample.

Note that this method does not really require that each individual observation represent a precise measurement. So long as you can order the observations you are able to perform these tests. Thus, for example, suppose you placed some meat out in the open and studied the arrival times of individuals of two species of blowflies. You could record exactly the time of arrival of each individual fly, starting from a point zero in time when the meat was set out. On the other hand, you might simply rank arrival times of the two species, noting that individual 1 of species B came first, 2 individuals from species A next, then 3 individuals of B, followed by the simultaneous arrival of one of each of the two species (a tie), and so forth. While such ranked or ordered data could not be analyzed by the parametric methods studied earlier, the techniques of Box 10.1 are entirely applicable.

The method of calculating the sample statistic U_s for the Mann-Whitney and the Wilcoxon tests is straightforward, as shown in Box 10.1. The critical

BOX 10.1

Mann-Whitney *U*-test for two samples, ranked observations, not paired.

Two samples of nymphs of the chigger *Trombicula lipovskyi* (unpublished data by D. A. Crossley). Variate measured is length of cheliceral base stated as micrometer units. Actual data are not shown. Only a graph of the two samples is needed for the test as shown in step **1**. Designate sample size of larger sample as n_1 and that of the smaller sample as n_2. In this case, $n_1 = 16$, $n_2 = 10$. If the two samples are of equal size, it does not matter which is designated as 1.

1. Graph the two samples as shown below. Indicate the ties by placing asterisks or crosses one above the other.

Sample B

Sample A

100 110 120 130

Micrometer units

2. For each observation in one sample (it is convenient to use the smaller sample), count the number of observations in the other sample which are lower in value (to the left). Count $\frac{1}{2}$ for each tied observation. For example, there are zero observations in Sample A less than the first observation in Sample B; 1 observation less than the second, third, fourth, and fifth observations in Sample B; 2 observations in A less than the sixth in B; 4 observations in A less than the seventh in B, but one is equal (tied) with it, so we count $4\frac{1}{2}$. Continuing in a similar manner, we obtain counts of 8, $8\frac{1}{2}$, and $9\frac{1}{2}$. The sum of these counts $C = 36\frac{1}{2}$. The Mann-Whitney statistic U_s is the greater of the two quantities C and $(n_1 n_2 - C)$, in this case $36\frac{1}{2}$ and $[(16 \times 10) - 36\frac{1}{2}] = 123\frac{1}{2}$.

3. We compare the value of $U_s = 123\frac{1}{2}$ with the critical values for $U_{\alpha[n_1,n_2]}$ in Table **XIII.** The null hypothesis is rejected if the observed value is too large. Since $U_{.025[16,10]} = 118$ and $U_{.01[16,10]} = 124$, the two samples are significantly different at $0.05 > P > 0.02$ (we double the probabilities because this is a two-tailed test).

In cases where $n_1 > 20$, calculate the following quantity

$$t_s = \frac{\left(U_s - \dfrac{n_1 n_2}{2}\right)}{\sqrt{\dfrac{n_1 n_2 (n_1 + n_2 + 1)}{12}}}$$

which is approximately normally distributed. The denominator 12 is a constant. Look up the significance of t_s in Table **III** against critical values of $t_{\alpha[\infty]}$ for a

BOX 10.1 continued

one-tailed test or two-tailed test as required by the hypothesis. When tied values occur, the above formula is modified as follows:

$$t_s = \frac{\left(U_s - \frac{n_1 n_2}{2}\right)}{\sqrt{\left(\frac{n_1 n_2}{(n_1 + n_2)(n_1 + n_2 - 1)}\right)\left(\frac{(n_1 + n_2)^3 - (n_1 + n_2) - \sum^m T_j}{12}\right)}}$$

In this expression T_j is a function of the t_j, the number of variates tied in the jth group of ties. (This t has no relation to Student's t.) The function is $T_j = t_j^3 - t_j$, computed easiest as $(t_j - 1)t_j(t_j + 1)$. Since in most cases the tied group will range from $t = 2$ to $t = 10$ ties, we give a small table of T over this range; the summation of T_j is over the m different ties.

t_j	2	3	4	5	6	7	8	9	10
T_j	6	24	60	120	210	336	504	720	990

For example, if we had to calculate $\sum^m T_j$ for the above problem (not necessary since $n_1 < 20$), we would compute $t_1 = 2$ for the first two tied variates (107 micrometer units). Similarly, we could construct a table for all values of t_j and T_j in this problem.

t_j	2	2	2	2	2	2	3
T_j	6	6	6	6	6	6	24

$$\sum^m T_j = 6 + 6 + \ldots + 24 = 60$$

values for $U_{\alpha[n_1,n_2]}$ are shown in Table **XIII**, which is adequate for cases in which the larger sample size $n_1 \leq 20$. The probabilities in Table **XIII** assume a one-tailed test. For a two-tailed test you should double the value of the probability shown in that table. When $n_1 > 20$, compute the expression shown near the bottom of Box 10.1. Since this expression is distributed as a normal deviate, consult the table of t (Table **III**), for $t_{\alpha[\infty]}$ using one- or two-tailed probabilities, depending on your hypothesis. A further complication arises from tied observations, which require the more elaborate formula shown at the bottom of Box 10.1. However, it takes a substantial number of ties to affect the outcome of the test appreciably. Corrections for ties increase the t_s-value slightly; hence the uncorrected formula is more conservative.

It is desirable to obtain an intuitive understanding of the rationale behind this test. In the Mann-Whitney test we can conceive of two extreme situations: in one case the two samples overlap and coincide entirely; in the other they are quite separate. In the latter case, if we take the sample with the lower-valued variates there will be no points of the contrasting sample to the left of it; that is, we can go through every observation in the lower-valued sample without having any items of the higher-valued one to the left of it. Conversely, all the points of the lower-valued sample would be to the left of every point of the higher-valued one had we started out with the latter. Our total count would therefore be the total count of one sample multiplied by every observation in the second sample, which yields n_1n_2. Thus, since we are told to take the greater of the two values, the sum of the counts C or $n_1n_2 - C$, our result in this case would be n_1n_2. On the other hand, if the two samples coincide completely, then for each point in one sample we would have those points below it plus a half point for the tied value representing that observation in the second sample which is at exactly the same level as the observation under consideration. A little experimentation will show this value to be $[n(n - 1)/2] + (n/2) = n^2/2$. Clearly the range of possible U-values must be between this and n_1n_2 and the critical value must be somewhere within this range.

Our conclusion as a result of the tests in Box 10.1 is that the two samples differ significantly in the distribution of cheliceral base length. It is obvious that the chiggers in sample A have longer cheliceral bases than those of sample B.

Finally we shall present a nonparametric method for the paired comparisons design, discussed in Section 9.3 and illustrated in Box 9.3. The most widely used method is that of *Wilcoxon's signed-ranks test,* illustrated in Box 10.2. The example to which it is applied has not yet been encountered. It records mean litter size in two strains of guinea pigs kept in large colonies during the years 1916 through 1924. Each of these values is the average of a large number of litters. Note the parallelism in the changes in the variable in the two strains. During 1917 and 1918 (war years for the U.S.), a shortage of caretakers and of food resulted in a decrease in the number of offspring per litter. As soon as better conditions returned, the mean litter size increased again. Notice that a subsequent drop in 1922 is again mirrored in both lines, suggesting that these fluctuations are environmentally caused. It is therefore quite appropriate that the data be treated as paired comparisons, with years as replications and the strain differences as the fixed treatments to be tested. Column (3) in Box 10.2 lists the differences on which a conventional paired comparisons t-test could be performed. For Wilcoxon's test these differences are ranked *without regard to sign* so that the smallest absolute difference is ranked 1, and the largest absolute difference (of the nine differences) is ranked 9. Tied ranks are computed as averages of the ranks; thus if the fourth and fifth difference have the same absolute magnitude they would both be assigned rank 4.5. After the ranks have been computed the original sign of each difference is assigned to the corresponding rank. The sum of the positive or of the negative ranks, whichever one is smaller in absolute value, is then computed (it is labeled T_s) and is compared with the critical value T in Table **XIV** for the corresponding sample size. In view of the significance of the rank sum, it is clear that strain B has a litter size different from that of strain 13. This is a

BOX 10.2

Wilcoxon's signed-ranks test for two groups, arranged as paired observations.

Mean litter size of two strains of guinea pigs, compared over $n = 9$ years.

Year	(1) *Strain B*	(2) *Strain 13*	(3) D	(4) *Rank* (R)
1916	2.68	2.36	+0.32	+9
1917	2.60	2.41	+0.19	+8
1918	2.43	2.39	+0.04	+2
1919	2.90	2.85	+0.05	+3
1920	2.94	2.82	+0.12	+7
1921	2.70	2.73	−0.03	−1
1922	2.68	2.58	+0.10	+6
1923	2.98	2.89	+0.09	+5
1924	2.85	2.78	+0.07	+4
	Absolute sum of negative ranks			1
	Sum of positive ranks			44

Source: Data by S. Wright.

Procedure

1. Compute the differences between the n pairs of observations. These are entered in column (3), labeled D.

2. Rank these differences from the smallest to the largest *without regard to sign.*

3. Assign to the ranks the original signs of the differences.

4. Sum the positive and negative ranks separately. The sum that is smaller in absolute value, T_s, is compared with the values in Table **XIV** for $n = 9$.

Since $T_s = 1$, which is equal to or less than the entry for one-tailed $\alpha = 0.005$ in the table, our observed difference is significant at the 1% level. Litter size in strain B is significantly different from that of strain 13.

For large samples ($n > 50$) compute

$$t_s = \frac{T_s - \dfrac{n(n+1)}{4}}{\sqrt{\dfrac{n(n+\frac{1}{2})(n+1)}{12}}}$$

where T_s is as defined in step 4 above. Compare the computed value with $t_{\alpha[\infty]}$ in Table **III**.

very simple test to carry out, but it is, of course, not as efficient as the corresponding parametric t-test, which should be preferred if the necessary assumptions hold. Note that one needs minimally six differences in order to carry out Wilcoxon's signed-ranks test. In six paired comparisons, all differences must be of like sign for the test to be significant at the 5% level.

For a large sample an approximation to the normal curve is available, which is given in Box 10.2. Note that the absolute magnitudes of these differences play a role only so far as they affect the ranks of the differences.

A still simpler test is the *sign test*, which counts the number of positive and negative signs among the differences (omitting all differences of zero). We then test the hypothesis that the n plus and minus signs are sampled from a population in which the two kinds of signs are present in equal proportions, as might be expected if there were no true difference between the two paired samples. Such sampling should follow the binomial distribution, and the test of the hypothesis that the parametric frequency of the plus signs is $\hat{p} = 0.5$ can be made in a number of ways. Let us learn these by applying the sign test to the guinea pig data of Box 10.2. There are nine differences, of which eight are positive and one is negative. We could follow the methods of Section 4.2 (illustrated in Table 4.3) in which we calculate the expected probability of sampling one minus sign in a sample of nine on the assumption of $\hat{p} = \hat{q} = 0.5$. The probability of such an occurrence and all "worse" outcomes equals 0.0195. Since we have no a priori notions that one strain should have a greater litter size than the other, this is a two-tailed test and we double the probability to 0.0390. Clearly, this is an improbable outcome and we reject the null hypothesis that $\hat{p} = \hat{q} = 0.5$.

Since the computation of the exact probabilities may be quite tedious if no table of cumulative binomial probabilities is at hand, a second approach is to make use of Table **IX**, which furnishes confidence limits for $\hat{p}$ for various sample sizes and sampling outcomes. Looking up sample size 9 and $Y = 1$ (number showing the property), we find the 95% confidence limits to be 0.0028 and 0.4751 by interpolation, thus excluding the value $\hat{p} = \hat{q} = 0.5$ postulated by the null hypothesis. At least at the 5% significance level we can conclude that it is unlikely that the number of plus and minus signs is equal. The confidence limits imply a two-tailed distribution; if we intend a one-tailed test, we can infer a 0.025 significance level from the 95% confidence limits and a 0.005 level from the 99% limits. Obviously, such a one-tailed test would be carried out only if the results were in the direction of the alternative hypothesis. Thus, if the alternative hypothesis were that strain 13 in Box 10.2 had greater litter size than strain B, we would not have bothered testing this example at all, since the observed proportion of years showing this relation was less than half. For larger samples, we can use the normal approximation to the binomial distribution as follows: $t_s = (Y - \mu)/\sigma_Y = (Y - kp)/\sqrt{kpq}$, substituting the mean and standard deviation of the binomial distribution learned in Section 4.2. In our case, we let n stand for k and assume that $p = q = 0.5$. Therefore, $t_s = (Y - \frac{1}{2}n)/\sqrt{\frac{1}{4}n} = (Y - \frac{1}{2}n)/\frac{1}{2}\sqrt{n}$. The value of t_s is then compared with $t_{\alpha[\infty]}$ in Table **III**, using one tail or two tails of the distribution as warranted. When the sample size $n \geq 12$, this is a satisfactory approximation.

A third approach would be to test the departure from expectation $\hat{p} = \hat{q} = 0.5$ by one of the methods of Chapter 13.

Exercises 10

10.1 In a study of flower color in Butterflyweed (*Asclepias tuberosa*), Woodson (1964) obtained the following results:

Geographic region	$\bar{Y}$	*n*	*s*
C1	29.3	226	4.59
SW2	15.8	94	10.15
SW3	6.3	23	1.22

The variable recorded was a color score (ranging from 1 for pure yellow to 40 for deep orange red) obtained by matching flower petals to sample colors in Maerz and Paul's *Dictionary of Color*. Test whether the samples are homoscedastic.

10.2 Allee and Bowen (1932) studied survival time of goldfish (in minutes) when placed in colloidal silver suspensions. Experiment No. 9 involved 5 replications, and experiment No. 10 involved 10 replicates. Do the results of the two experiments differ? Addition of urea, NaCl, and Na_2S to a third series of suspensions apparently prolonged the life of the fish.

Colloidal silver		
Experiment No. 9	*Experiment No. 10*	*Urea and salts added*
210	150	330
180	180	300
240	210	300
210	240	420
210	240	360
	120	270
	180	360
	240	360
	120	300
	150	120

Analyze and interpret. Test equality of variances. Compare anova results with those obtained using the Mann-Whitney U-test for the two comparisons under study. To test the effect of urea it might be best to pool Experiments 9 and 10, if they prove not to differ significantly. ANS. $U_s = 33$ between Experiments 9 and 10.

10.3 Number of bacteria in 1 cc of milk from three cows counted at three periods (data from Park, Williams, and Krumwiede, 1924).

	At time of milking	*After 24 hours*	*After 48 hours*
Cow No. 1	12,000	14,000	57,000
2	13,000	20,000	65,000
3	21,500	31,000	106,000

(a) Calculate means and variances for the three periods and examine the relation between these two statistics. Transform the variates to logarithms and compare means and variances based on the transformed data. Discuss.
(b) Carry out an anova on transformed and untransformed data. Discuss your results.

10.4 Test for a difference in surface and subsoil *p*H in the data of Exercise 9.1, using Wilcoxon's signed-ranks test. ANS. $T_s = 38$; $P > 0.10$.

11

Regression

Our studies so far have dealt with only one variable at a time. However, we frequently measure two or more variables on each individual and we consequently would like to be able to express more precisely the nature of the relationships between these variables. This brings us to the subjects of *regression* and *correlation*. In regression we estimate the relationship of one variable with another by expressing the one in terms of a linear (or a more complex) function of the other. In correlation analysis, which is sometimes confused with regression, we estimate the degree to which two variables vary together. Chapter 12 deals with correlation, and we shall postpone our effort to clarify the relation and distinction between regression and correlation until then. The variables involved in regression and correlation are continuous or, if meristic, they are treated as though they were continuous. If the variables are qualitative (that is, attributes), then the methods of regression and correlation cannot be used.

In Section 11.1, we review the notion of mathematical functions and introduce the new terminology required for regression analysis. This is followed in Section 11.2 by a discussion of the appropriate statistical models for regression analysis. The basic computations in simple linear regression are shown in Section 11.3 for the case of one dependent variate for each independent variate. The case with several dependent variates for each independent variate is treated in Section 11.4. Tests of significance and computation of confidence intervals for regression problems are discussed in Section 11.5.

Section 11.6 serves as a summary of regression and discusses the various uses of regression analysis in biology. Finally, Section 11.7 shows how transformation of scale can straighten out curvilinear relationships for ease of analysis.

11.1 Introduction to regression

Much scientific thought concerns the relations between pairs of variables hypothesized to be in a cause-and-effect relationship. We shall be content with establishing the form and significance of *functional relationships* between two variables, leaving the demonstration of cause-and-effect relationships to the established procedures of the scientific method. A *function* is a mathematical relationship enabling us to predict what values of a variable Y correspond to given values of a variable X. Such a relationship, generally written as $Y = f(X)$, is familiar to all of us.

A typical linear regression is of the form shown in Figure 11.1, which illustrates the effect of two drugs on the blood pressures of two species of animals. The relationships depicted in this graph can be expressed by the formula $Y = a + bX$. Clearly, Y is a function of X. We call the variable Y the *dependent variable*, while X is called the *independent variable*. The magnitude of blood pressure Y depends on the amount of the drug X and can therefore be predicted from the independent variable, which presumably is free to vary. Although a cause would always be considered an independent variable and an effect a dependent variable, a functional relationship observed in nature may not actually be a cause-and-effect relationship. The highest line is of the relationship $Y = 20 + 15X$, which represents the effect of drug A on animal P. The quantity of drug is measured in micrograms, the blood pressure in millimeters mercury. Thus, after 4 μg of the drug have been given, the blood pressure would be $Y = 20 + (15)(4) = 80$ mm Hg. The independent variable X is multiplied by a coefficient b, the slope factor. In the example chosen, $b = 15$; that is, for an increase in one microgram of the drug the blood pressure is raised by 15 mm.

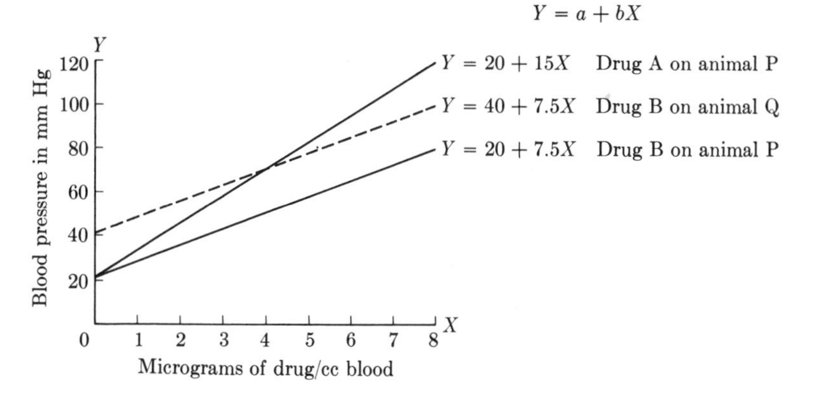

FIGURE 11.1

Blood pressure of an animal in mm Hg as a function of drug concentration in μg per cc of blood.

In biology, such a relationship can clearly be appropriate over only a limited range of values of X. Negative values of X are meaningless in this case and it is unlikely that the blood pressure will continue to increase at a uniform rate. Quite probably the slope of the functional relationship will flatten out as the drug level rises. But, for a limited portion of the range of variable X (micrograms of the drug), the linear relationship $Y = a + bX$ may be an adequate description of the functional dependence of Y on X.

By this formula, when the independent variable equals zero, the dependent variable equals a. This point is the intersection of the function line with the Y-axis. It is called the *Y-intercept.* In Figure 11.1, when $X = 0$, the function just studied will yield a blood pressure of 20 mm, which is the normal blood pressure of animal P in the absence of the drug.

The two other functions in Figure 11.1 show the effects of varying both a, the Y-intercept, and b, the slope. In the lowest line, $Y = 20 + 7.5X$, the Y-intercept remains the same, but the slope has been halved. We visualize this as the effect of a different drug B on the same organism P. Obviously, when no drug is administered, the blood pressure should be at the same Y-intercept, since the identical organism is being studied. However, a different drug is likely to exert a different hypertensive effect, as reflected by the different slope. The third relationship also describes the effect of drug B, which is assumed to remain the same, but the experiment is carried out on a different species Q, whose normal blood pressure is assumed to be 40 mm Hg. Thus, the equation for the effect of drug B on species Q is written as $Y = 40 + 7.5X$. This line is parallel to the one studied previously.

From your knowledge of analytical geometry you will have recognized the slope factor b as the *slope* of the function $Y = a + bX$, generally symbolized by m. In calculus, b is the *derivative* of that same function ($dY/dX = b$). In biostatistics, b is called the *regression coefficient* and the function is called a *regression equation*. When we wish to stress that the regression coefficient is of variable Y on variable X, we write $b_{Y \cdot X}$.

11.2 Models in regression

In any real example observations would not lie perfectly along a regression line because of random error in measuring and because of the effects of unpredictable environmental factors. Thus in regression a functional relationship does not mean that given an X the value of Y must be $a + bX$, but rather that the mean (or expected value) of Y is $a + bX$.

Significance tests in regression are based on the following two models. The more common of these, *Model I regression*, is especially suitable in experimental situations. It is based on four assumptions.

1. The independent variable X is measured without error. We therefore say that the X's are "fixed." We mean by this that only Y, the dependent variable, is a random variable; X does not vary at random, but is under the control of the investigator. Thus in the example of Figure 11.1 we varied dose of drug at will and studied the response of the random variable blood pressure. We can manipulate X in the same way that we were able to manipulate the treatment effect in a Model I anova. As a matter of fact, as you shall see later, there is a very intimate relationship between Model I anova and Model I regression.

2. The expected value for the variable Y for any given value X is described by the linear function $\mu_Y = \alpha + \beta X$. This is the same relation we have encountered before, but we use Greek letters for a and b, since we are describing a parametric relationship. Another way of stating this assumption is that the parametric means μ_Y of the values of Y are a function of X and lie on a straight line described by this equation.

3. For any given value X_i, the Y_i's are independently and normally distributed. This can be represented by the equation $Y_{ij} = \alpha + \beta X_i + \epsilon_{ij}$, where ϵ_{ij} is assumed to be a normally distributed error term with a mean of zero. Figure 11.2 illustrates this concept with a regression line similar to the ones in Figure 11.1. A given experiment can be repeated several times. Thus, for instance, we could administer 2, 4, 6, 8, and 10 μg of the drug to each of 20 individuals of an animal species and obtain a frequency distribution of blood pressure responses Y to the independent variates $X = 2, 4, 6, 8$, and 10 μg. In view of the inherent variability of biological material, it is obvious that the responses to each dosage would not be the same in every individual; you would obtain a frequency distribution of values of Y (blood pressure) around the expected value. Assumption 3 states that these sample values

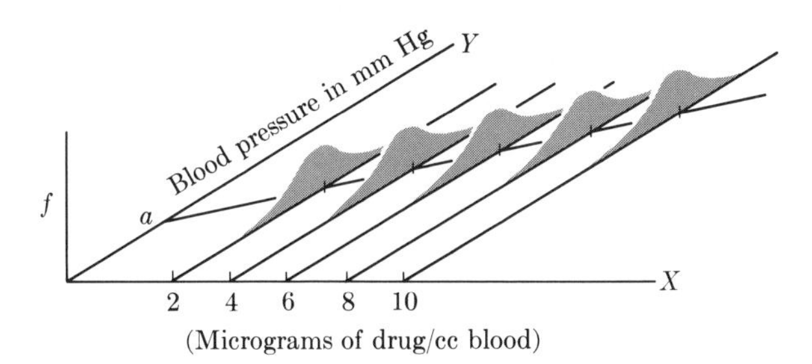

FIGURE 11.2

Blood pressure of an animal in mm Hg as a function of drug concentration in μg per cc of blood. Repeated sampling for a given drug concentration.

would be independently and normally distributed. This is indicated by the normal curves, which are superimposed about several points in the regression line in Figure 11.2. A few are shown to give you an idea of the scatter about the regression line. In actuality there is, of course, a continuous scatter, as though these separate normal distributions were stacked right next to each other, there being, after all, an infinity of intermediate values of X between any two dosages. In those rare cases in which the independent variable is discontinuous, the distributions of Y would be physically separate from each other and would occur only along those points of the abscissa corresponding to independent variates. An example of such a case would be weight of offspring (Y) as a function of number of offspring (X) in litters of mice. There may be three or four offspring per litter, but there would be no intermediate value of X representing 3.25 mice per litter.

Not every experiment will have more than one replicate of Y for each value of X. In fact, the basic computations we shall learn in the next section are for only one value of Y per X, this being the more common case. However, you should realize that even in such instances the basic assumption of Model I regression is that the single variate of Y corresponding to the given value of X is a sample from a population of independently and normally distributed variates.

4. The final assumption is a familiar one. We assume that these samples along the regression line are homoscedastic; that is, they have a common variance, σ^2, which is the variance of the ϵ's in the above expression. Thus we assume that the variance around the regression line is constant and independent of the magnitude of X or Y.

In *Model II regression* the independent variable is also measured with error. We do not consider X to be fixed and at the control of the investigator. Suppose you sample a population of female flies and measure a number of ovarioles and total weight of each individual. The distributions of these variables will probably differ. You might be interested in studying a number of ovarioles as a function of weight. In this case the weight, which you assume to be the independent variable, is not fixed and certainly not the "cause" of ovarian development. Although, as will be discussed in the next chapter, cases of this sort are much better analyzed by the methods of correlation analysis, we sometimes wish to describe the functional relationship between such variables. To do so, we need to resort to the special techniques of Model II regression. In this book we shall limit ourselves to a treatment of Model I regression.

11.3 The basic computations (single Y for each value of X)

To learn the basic computations necessary to carry out a Model I linear regression, we shall choose an example with only one Y-value per independent variate X, as this is computationally simpler. The extension to replicated values of Y per single value of X is shown in the next section. The computations illustrated in Table 11.1 are shown for pedagogical reasons to facilitate an understanding of the meaning of regression. Simple computational formulas are presented at the end of this section.

The data on which we shall learn regression come from a study of water loss in *Tribolium confusum*, the confused flour beetle. Nine batches of 25 beetles were weighed (individual beetles could not be weighed with available equipment), kept at different relative humidities, and weighed again after six days of starvation. Weight loss in milligrams was computed for each batch. This is clearly a Model I regression in which the weight loss is the dependent variable Y and the relative humidity is the independent variable X, a fixed treatment effect under the control of the experimenter. The purpose of the analysis is to establish whether the relationship between relative humidity and weight loss can be adequately described by a linear regression of the general form $Y = a + bX$. The original data are shown in columns (1) and (2) of Table 11.1. They are plotted in Figure 11.3, from which it appears that a negative relationship exists between weight loss and humidity; as the humidity increases, the weight loss decreases. The means of weight loss and relative humidity, $\bar{Y}$ and $\bar{X}$, respectively, are marked along the coordinate axes. The average humidity is 50.39%, and the average weight loss is 6.022 mg. How can we fit a regression line to these data, permitting us to estimate a value of Y for a given value of X? Unless the actual observations lie exactly on a straight line, we will need a criterion for determining the best possible placing of the regression line. Statisticians have generally followed the principle of least squares, which we first encountered in Chapter 3 when learning about the arithmetic mean and the variance. If we were to draw a horizontal line through $\bar{X}$, $\bar{Y}$ (that is, a line parallel to the X-axis at the level of $\bar{Y}$), then deviations to that line drawn parallel to the Y-axis would represent the deviations from the mean for these observations with respect to variable Y (see Figure 11.4). We learned in Chapter 3 that the sum of these observations, $\sum(Y - \bar{Y}) = \sum y = 0$. The sum of squares of these deviations, $\sum(Y - \bar{Y})^2 = \sum y^2$, is less than that from any other horizontal line. Another way of saying this is that the arithmetic mean of Y represents the least squares horizontal line. Any horizontal line drawn through the data

TABLE 11.1

Basic computations in regression. Weight loss (in mg) of nine batches of 25 *Tribolium* beetles after six days of starvation at nine different humidities.

	(1)	(2)	(3)	(4)	(5)	(6)	(7)	(8)	(9)	(10)	(11)	(12)
	Percent relative humidity	*Weight loss in mg*	$x =$	$y =$					$d_{Y \cdot X} =$		$\hat{y} =$	
	X	Y	$(X - \bar{X})$	$(Y - \bar{Y})$	x^2	xy	y^2	$\hat{Y}$	$Y - \hat{Y}$	$d^2_{Y \cdot X}$	$\hat{Y} - \bar{Y}$	$\hat{y}^2$
	0	8.98	−50.39	2.958	2539.1521	−149.0536	8.7498	8.7038	0.2762	0.0763	2.6818	7.1921
	12	8.14	−38.39	2.118	1473.7921	− 81.3100	4.4859	8.0652	0.0748	0.0056	2.0432	4.1747
	29.5	6.67	−20.89	0.648	436.3921	− 13.5367	0.4199	7.1338	−0.4638	0.2151	1.1118	1.2361
	43	6.08	− 7.39	0.058	54.6121	− 0.4286	0.0034	6.4153	−0.3353	0.1124	0.3933	0.1547
	53	5.90	2.61	−0.122	6.8121	− 0.3184	0.0149	5.8831	0.0169	0.0003	−0.1389	0.0193
	62.5	5.83	12.11	−0.192	146.6521	− 2.3251	0.0369	5.3776	0.4524	0.2047	−0.6444	0.4153
	75.5	4.68	25.11	−1.342	630.5121	− 33.6976	1.8010	4.6857	−0.0057	0.0000	−1.3363	1.7857
	85	4.20	34.61	−1.822	1197.8521	− 63.0594	3.3197	4.1801	0.0199	0.0004	−1.8419	3.3926
	93	3.72	42.61	−2.302	1815.6121	− 98.0882	5.2992	3.7543	−0.0343	0.0012	−2.2677	5.1425
Sum	453.5	54.20	− 0.01	0.002	8301.3889	−441.8176	24.1307	54.1989	0.0011	0.6160	0.0009	23.5130
Mean	50.39	6.022						6.022				
Sum/$(n - 1)$					1037.6736	− 55.2272	3.0163			0.0880[a]		

Source: Nelson (1964).
[a] Sum divided by $n - 2$.

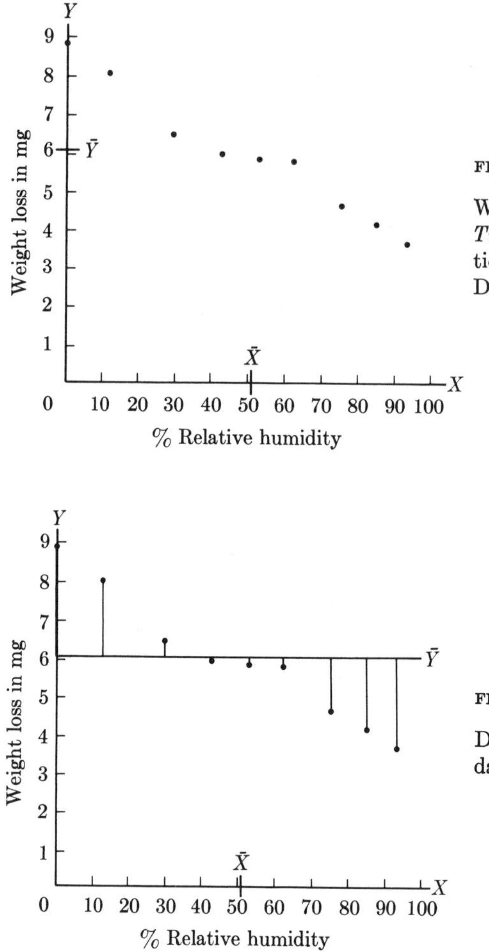

FIGURE 11.3

Weight loss (in mg) of nine batches of 25 *Tribolium* beetles after six days of starvation at nine different relative humidities. Data from Table 11.1 after Nelson (1964).

FIGURE 11.4

Deviations from the mean (of Y) for the data of Figure 11.3.

at a point other than $\bar{Y}$ would yield a sum of deviations other than zero and a sum of deviations squared greater than $\sum y^2$. Therefore a mathematically correct but impractical method for finding the mean of Y would be to draw a series of horizontal lines across a graph, calculate the sum of squares of deviations from it, and choose that line yielding the smallest sum of squares.

In linear regression, we still draw a straight line through our observations, but it is no longer necessarily horizontal. A sloped regression line will indicate for each value of the independent variable X_i an estimated value of the dependent variable. We should distinguish the estimated value of Y_i, which we shall hereafter designate as $\hat{Y}_i$ (read: Y-hat or Y-caret), and the observed values, conventionally designated as Y_i. The regression equation therefore should read

$$\hat{Y} = a + bX \tag{11.1}$$

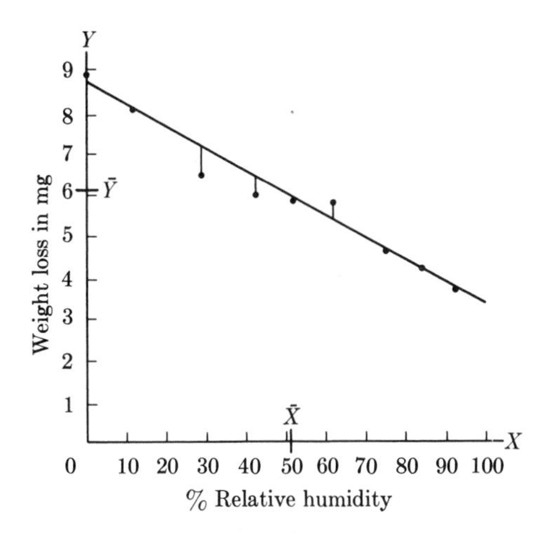

FIGURE 11.5

Deviations from the regression line for the data of Figure 11.3.

which indicates that for given values of X, this equation calculates estimated values $\hat{Y}$ (as distinct from the observed values Y in any actual case). The deviation of an observation Y_i from the regression line is $(Y_i - \hat{Y}_i)$ and is generally symbolized as $d_{Y \cdot X}$. These deviations can still be drawn parallel to the Y-axis, but since the regression line is sloped, they meet it at an angle (see Figure 11.5). The sum of these deviations is again zero ($\sum d_{Y \cdot X} = 0$), and the sum of their squares yields a quantity $\sum(Y - \hat{Y})^2 = \sum d^2_{Y \cdot X}$ analogous to the sum of squares $\sum y^2$. For reasons that will become clear later, $\sum d^2_{Y \cdot X}$ is called the unexplained sum of squares. The least squares *linear regression line* through a set of points is defined as that straight line which results in $\sum d^2_{Y \cdot X}$ being at a minimum. Geometrically, the basic idea is that one would prefer using a line that is in some sense close to as many points as possible. For purposes of regression analysis, it is most useful to define closeness in terms of the vertical distances from the points to a line and to use the line making the sum of these squared deviations as small as possible. A convenient consequence of this criterion is that the line must pass through the point $\bar{X}$, $\bar{Y}$. Again, it would be feasible but impractical to calculate the correct regression slope by pivoting a ruler around the point $\bar{X}$, $\bar{Y}$, and calculating the unexplained sum of squares, $\sum d^2_{Y \cdot X}$, for each of the innumerable possible positions. Whichever position gave the minimal value of $\sum d^2_{Y \cdot X}$ would be the least squares regression line.

The formula for the slope of a line based on a minimal value of $\sum d^2_{Y \cdot X}$ is obtained by means of the calculus. It is

$$b_{Y \cdot X} = \frac{\sum xy}{\sum x^2} \tag{11.2}$$

Let us calculate $b = \sum xy / \sum x^2$ for our weight loss data.

We first compute the deviations from the respective means of X and Y as shown in columns (3) and (4) of Table 11.1. The sums of these deviations, $\sum x$ and $\sum y$, are slightly different from their expected value of zero because of rounding errors. The squares of these deviations yield sums of squares and variances in columns (5) and (7). In column (6) we have computed the products xy, which in this example are all negative because the deviations are of unlike sign. An increase in humidity results in a decrease in weight loss. The sum of these products $\sum^{n} xy$ is a new quantity, called the *sum of products*. This is a poor, but well-established term, referring to $\sum xy$, the sum of the products of the deviations rather than $\sum XY$, the sum of the products of the variates. You will recall that $\sum y^2$ is called the sum of squares, while $\sum Y^2$ is the sum of the squared variates. The sum of products is analogous to the sum of squares. When divided by the degrees of freedom, it yields the *covariance* by analogy with the variance resulting from a similar division of the sum of squares. You may recall first having encountered covariances in Section 7.4. Note that the sum of products can be negative as well as positive. If it is negative, this indicates a negative slope of the regression line: as X increases, Y decreases. In this respect it differs from a sum of squares, which can only be positive. From Table 11.1 we find that $\sum xy = -441.8176$, $\sum x^2 = 8301.3889$, and $b = \sum xy / \sum x^2 = -0.05322$. Thus, for a one-unit increase in X, there is a decrease of 0.05322 units of Y. Relating it to our actual example, we can say that for a 1% increase in relative humidity, there is a reduction of 0.05322 mg in weight loss.

How can we complete the equation $Y = a + bX$? We have stated that the regression line will go through the point $\bar{X}, \bar{Y}$. Therefore, when X is at its mean, we estimate that Y should also be at its mean. Whether this is actually so in our data we cannot tell, since we do not have an X variate exactly at the mean. Even if we did, it would not be very likely that the observed value of Y would be exactly at the mean. Remember, after all, that the Y-values of our observations are only a sample from a population centering around $\mu_{\hat{Y}}$. At $\bar{X} = 50.39$, $\hat{Y} = 6.022$; that is, we use $\bar{Y}$, the observed mean of Y, as an estimate $\hat{Y}$ of the mean. We can substitute these means into Expression (11.1):

$$\hat{Y} = a + bX$$

$$\bar{Y} = a + b\bar{X}$$

$$a = \bar{Y} - b\bar{X}$$

$$a = 6.022 - (-0.05322)(50.39)$$

$$= 8.7038$$

Therefore

$$\hat{Y} = 8.7038 - 0.05322X.$$

This is the equation that relates weight loss to relative humidity. Note that when X is zero (humidity zero), the estimated weight loss is greatest. It is then equal to $a = 8.7038$ mg. But as X increases to a maximum of 100, the weight loss would decrease to 3.3818 mg.

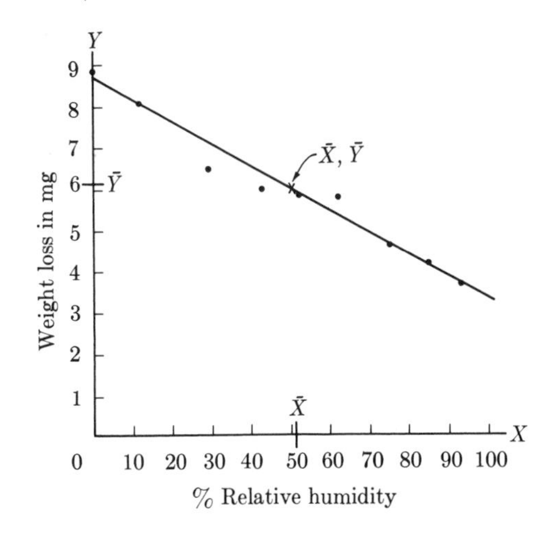

FIGURE 11.6

Linear regression fitted to data of Figure 11.3.

We can use the regression formula to draw the regression line: simply estimate $\hat{Y}$ at two convenient points of X such as $X = 0$ and $X = 100$ and draw a straight line between them. This line has been added to the observed data and is shown in Figure 11.6. Note that it goes through the point $\bar{X}$, $\bar{Y}$. In fact, for drawing the regression line, we frequently use the intersection of the two means and one other point.

Since

$$a = \bar{Y} - b\bar{X}$$

we can write Expression (11.1), $\hat{Y} = a + bX$, as

$$\hat{Y} = (\bar{Y} - b\bar{X}) + bX$$
$$= \bar{Y} + b(X - \bar{X})$$

Therefore

$$\hat{Y} = \bar{Y} + bx$$

Also,

$$\hat{Y} - \bar{Y} = bx$$

$$\hat{y} = bx \tag{11.3}$$

where $\hat{y}$ is defined as the deviation $\hat{Y} - \bar{Y}$. Next, using Expression (11.1), we estimate $\hat{Y}$ for every one of our given values of X. The estimated values $\hat{Y}$ are shown in column (8) of Table 11.1. Compare them with the observed values of Y in column (2). Overall agreement between the two columns of values is good. Note that except for rounding errors, $\sum\hat{Y} = \sum Y$ and hence $\bar{\hat{Y}} = \bar{Y}$. However, our actual Y-values usually are different from the estimated values $\hat{Y}$. This is due to individual variation around the regression line. Yet, the regression line is a better base from which to compute deviations than the arithmetic average $\bar{Y}$, since the value of X has been taken into account in constructing it.

When we compute deviations of each observed Y-value from its estimated value $(Y - \hat{Y}) = d_{Y \cdot X}$ and list these in column (9), we notice that these deviations exhibit one of the properties of deviations from a mean: they sum to zero except for rounding errors. Thus $\sum d_{Y \cdot X} = 0$, just as $\sum y = 0$. Next, we compute in column (10) the squares of these deviations of observed values of Y from values estimated by regression and sum them to give a new sum of squares, $\sum d_{Y \cdot X}^2 = 0.6160$. When we compare $\sum(Y - \bar{Y})^2 = \sum y^2 = 24.1307$ with $\sum(Y - \hat{Y})^2 = \sum d_{Y \cdot X}^2 = 0.6160$, we note that the new sum of squares is much less than the previous old one. What has caused this reduction? The regression line is a set of averages (one for each value of X) as compared with the single arithmetic mean of Y. Allowing for different magnitudes of X has eliminated most of the variance of Y from the sample. Remaining is the *unexplained sum of squares* $\sum d_{Y \cdot X}^2$, which expresses that portion of the total SS of Y that is not accounted for by differences in X. It is unexplained with respect to X. The difference between the total SS, $\sum y^2$, and the unexplained SS, $\sum d_{Y \cdot X}^2$, is not surprisingly called the *explained sum of squares*, $\sum \hat{y}^2$, and is based on the deviations $\hat{y} = \hat{Y} - \bar{Y}$. The computation of this deviation and its square is shown in columns (11) and (12). Note that $\sum \hat{y}$ approximates zero and that $\sum \hat{y}^2 = 23.5130$. Add the unexplained $SS = 0.6160$ to this and you obtain $\sum y^2 = \sum \hat{y}^2 + \sum d_{Y \cdot X}^2 = 24.1290$, which is equal (except for rounding errors) to the independently calculated value of 24.1307 in column (7). We shall return to the meaning of the unexplained and explained sums of squares in later sections.

We now proceed to demonstrate an efficient computational method for a regression equation in data with single values of Y for each value of X.

The regression coefficient $\sum xy / \sum x^2$ can be rewritten as

$$b_{Y \cdot X} = \frac{\sum^{n} (X - \bar{X})(Y - \bar{Y})}{\sum^{n} (X - \bar{X})^2} \tag{11.4}$$

The denominator of this expression is the sum of squares of X. Its computational formula as first encountered in Section 3.9 is $\sum x^2 = \sum X^2 - (\sum X)^2/n$. We shall now learn an analogous formula for the numerator of Expression (11.4), the sum of products. The customary formula is

$$\sum^{n} xy = \sum^{n} XY - \frac{\left(\sum^{n} X\right)\left(\sum^{n} Y\right)}{n} \tag{11.5}$$

The quantity $\sum XY$ is simply the accumulated product of the two variables. Expression (11.5) is derived in Appendix A1.6. The actual computations for a regression equation (single value of Y per value of X) are illustrated in Box 11.1, employing the weight loss data of Table 11.1.

Box 11.1 also shows how to compute the explained sum of squares $\sum \hat{y}^2 = \sum(\hat{Y} - \bar{Y})^2$ and the unexplained sum of squares $\sum d_{Y \cdot X}^2 = \sum(Y - \hat{Y})^2$.

BOX 11.1

Computation of regression statistics. Single value of Y for each value of X.

Data from Table 11.1.

Weight loss in mg (Y)	8.98	8.14	6.67	6.08	5.90	5.83	4.68	4.20	3.72
Percent relative humidity (X)	0	12.0	29.5	43.0	53.0	62.5	75.5	85.0	93.0

Basic computations

1. Compute sample size, sums, sums of the squared observations, and the sum of the XY's.

$$n = 9 \qquad \sum X = 453.5 \qquad \sum Y = 54.20$$

$$\sum X^2 = 31{,}152.75 \qquad \sum Y^2 = 350.5350 \qquad \sum XY = 2289.260$$

2. The means, sums of squares, and sum of products are

$$\bar{X} = 50.389 \qquad \bar{Y} = 6.022$$

$$\sum x^2 = 8301.3889 \qquad \sum y^2 = 24.1306$$

$$\sum xy = \sum XY - \frac{(\sum X)(\sum Y)}{n}$$

$$= 2289.260 - \frac{(453.5)(54.20)}{9} = -441.8178$$

3. The regression coefficient is

$$b_{Y \cdot X} = \frac{\sum xy}{\sum x^2} = \frac{-441.8178}{8301.3889} = -0.05322$$

4. The Y-intercept is

$$a = \bar{Y} - b_{Y \cdot X}\bar{X} = 6.022 - (-0.05322)(50.389) = 8.7037$$

5. The explained sum of squares is

$$\sum \hat{y}^2 = \frac{(\sum xy)^2}{\sum x^2} = \frac{(-441.8178)^2}{8301.3889} = 23.5145$$

6. The unexplained sum of squares is

$$\sum d^2_{Y \cdot X} = \sum y^2 - \sum \hat{y}^2 = 24.1306 - 23.5145 = 0.6161$$

That

$$\sum d_{Y \cdot X}^2 = \sum y^2 - \frac{(\sum xy)^2}{\sum x^2} \tag{11.6}$$

is demonstrated in Appendix A1.7. From this demonstration it also becomes obvious that the explained sum of squares is

$$\sum \hat{y}^2 = \sum b^2 x^2 = b^2 \sum x^2 = \frac{(\sum xy)^2}{(\sum x^2)^2} \sum x^2 \tag{11.7}$$

$$\sum \hat{y}^2 = \frac{(\sum xy)^2}{\sum x^2}$$

11.4 More than one value of Y for each value of X

We now take up Model I regression as originally defined in Section 11.2 and illustrated by Figure 11.2. For each value of the treatment X we sample Y repeatedly, obtaining a sample distribution of Y-values at each of the chosen points of X. We have selected an experiment from the laboratory of one of us (Sokal) in which *Tribolium* beetles were reared from eggs to adulthood at four different densities. The percentage survival to adulthood was calculated for varying numbers of replicates at these densities. Following Section 10.2, these percentages were given arcsine transformations, which are listed in Box 11.2. These transformed values are more likely to be normal and homoscedastic than percentages. The arrangement of these data is very much like that of a single classification Model I anova. There are four different densities and several replicated survival values at each density. We now would like to determine whether there are differences in survival among the four groups, and also whether we can establish a regression of survival on density.

A first approach, therefore, is to carry out an analysis of variance, using the methods of Section 8.3 and Table 8.1. Our aim in doing this is illustrated diagrammatically in Figure 11.7. If the analysis of variance were not significant, this would indicate that the means are not significantly different from each other as shown in Figure 11.7A, and it would be unlikely that a regression line fitted to these data would have a slope significantly different from zero. Occasionally, when the means increase or decrease slightly as X increases, they may not be different enough for the mean square among groups to be significant by anova, yet a significant regression can be found. These are usually cases on the borderline of statistical significance. When we find a marked regression of the means on X, as shown in Figure 11.7B, we usually will find a significant difference among the means by an anova. However, we cannot turn this argument around and say that a significant difference among means as shown by an anova necessarily indicates that a significant linear regression can be fitted to these data. In Figure 11.7C, the means follow a **U**-shaped function (a parabola). Though the means would

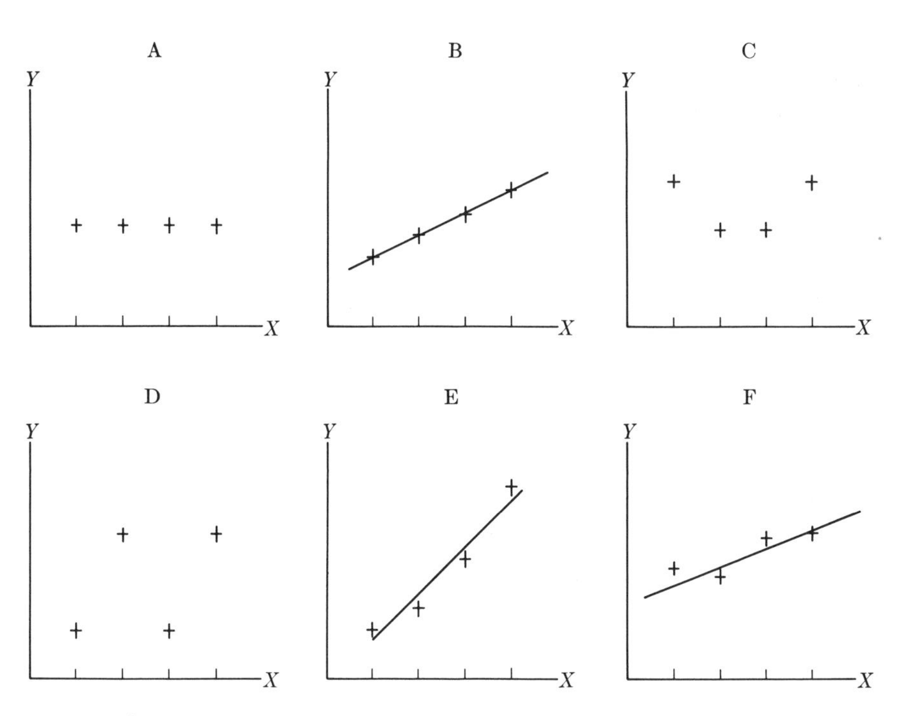

FIGURE 11.7

Differences among means and linear regression. General trends only are indicated by these figures. Significance of any of these would depend on the outcomes of appropriate tests. (For further explanation see text.)

likely be significantly different from each other, clearly a straight line fitted to these data would be a horizontal line halfway between the upper and the lower points. In such a set of data, *linear* regression can explain only little of the variation of the dependent variable. However, a curvilinear parabolic regression would fit these data and remove most of the variance of Y. A similar case is shown in Figure 11.7D, in which the means describe a periodically changing phenomenon, rising and falling alternatingly. Again the regression line for these data has slope zero. A curvilinear (cyclical) regression could also be fitted to such data, but our main purpose in showing this example is to indicate that there could be heterogeneity among the means of Y apparently unrelated to the magnitude of X. Remember that in real examples you will rarely ever get a regression as clear-cut as the linear case in 11.7B, or the curvilinear one in 11.7C, nor will you necessarily get heterogeneity of the type shown in 11.7D, in which any straight line fitted to the data would be horizontal and halfway between the upper and lower points. You are more likely to get data in which linear regression can be demonstrated, but which will not fit a straight line well. The residual deviations of the means around linear regression might be removed by changing from linear to curvilinear regression (as is suggested by the pattern of points in Figure 11.7E) or they remain as inexplicable residual heterogeneity around the regression line, as indicated in Figure 11.7F.

We carry out the computations following the by now familiar outline for analysis of variance and obtain the anova table shown in Box 11.2. The three degrees of freedom among the four groups yield a mean square that is highly significant when tested over the within-groups mean square. It therefore is clearly worthwhile to continue the analysis to see whether a significant regression of survival on density exists. The additional steps for the regression analysis follow in Box 11.2. We compute the sum of squares of X, the sum of products of X and Y, the explained sum of squares of Y, and the unexplained sum of squares of Y. The formulas will look unfamiliar because of the complication of the several Y's per value of X. The computations for the sum of squares of X involve the multiplication of X by the number of items in the study. Thus, though there may appear to be only four densities, there are, in fact, as many densities (although of only four magnitudes) as there are values of Y in the study. Having completed the computations, we

BOX 11.2

Computation of regression with more than one value of Y per value of X.

The variates Y are arcsine transformations of the percentage survival of the beetle *Tribolium castaneum* at 4 densities (X = number of eggs per gram of flour medium).

	Density = X ($a = 4$)			
	5/g	*20*/g	*50*/g	*100*/g
Survival; in degrees	61.68	68.21	58.69	53.13
	58.37	66.72	58.37	49.89
	69.30	63.44	58.37	49.82
	61.68	60.84		
	69.30			
$\sum^{n_i} Y$	320.33	259.21	175.43	152.84
n_i	5	4	3	3
$\bar{Y}_i$	64.07	64.80	58.48	50.95

$\sum^{a} n_i = 15$

$\sum^{a}\sum^{n_i} Y = 907.81$

Source: Data by Sokal (1967).

The anova computations are carried out as in Table 8.1.

BOX 11.2 continued

Anova table

Source of variation	*df*	*SS*	*MS*	F_s
$\bar{Y} - \bar{\bar{Y}}$ Among groups	3	423.7016	141.2339	11.20**
$Y - \bar{Y}$ Within groups	11	138.6867	12.6079	
$Y - \bar{\bar{Y}}$ Total	14	562.3883		

The groups differ significantly.

We proceed to test whether the differences among the survival values can be accounted for by linear regression on density. If no significant differences among groups had been found, it would be possible, although unlikely, for linear regression to be significant. If $SS_{\text{groups}} < (MS_{\text{within}} \times F_{[1, \sum^a n_i - a]})$, it is impossible for regression to be significant.

Computations for regression analysis

1. Sum of X weighted by sample size $= \sum^a n_i X$

$$= 5(5) + 4(20) + 3(50) + 3(100) = 555$$

2. Sum of X^2 weighted by sample size $= \sum^a n_i X^2$

$$= 5(5)^2 + 4(20)^2 + 3(50)^2 + 3(100)^2 = 39{,}225$$

3. Sum of products of X and $\bar{Y}$ weighted by sample size $= \sum^a n_i X \bar{Y}$

$$= \sum^a X\left(\sum^{n_i} Y\right) = 5(320.33) + \cdots + 100(152.84) = 30{,}841.35$$

4. Correction term for $X = CT_X = \dfrac{\left(\sum^a n_i X\right)^2}{\sum^a n_i}$

$$= \frac{(\text{quantity } \mathbf{1})^2}{\sum^a n_i} = \frac{(555)^2}{15} = 20{,}535.00$$

5. Sum of squares of $X = \sum x^2 = \sum^a n_i X^2 - CT_X$

$$= \text{quantity } \mathbf{2} - \text{quantity } \mathbf{4} = 39{,}225 - 20{,}535 = 18{,}690$$

6. Sum of products $= \sum xy$

$$= \sum^n X\left(\sum^{n_i} Y\right) - \frac{\left(\sum^a n_i X\right)\left(\sum^a \sum^{n_i} Y\right)}{\sum^a n_i}$$

$$= \text{quantity } \mathbf{3} - \frac{\text{quantity } \mathbf{1} \times \sum^a \sum^{n_i} Y}{\sum^a n_i}$$

$$= 30{,}841.35 - \frac{(555)(907.81)}{15} = -2747.62$$

BOX 11.2 continued

7. Explained sum of squares $= \sum \hat{y}^2 = \dfrac{(\sum xy)^2}{\sum x^2}$

$$= \frac{(\text{quantity } \mathbf{6})^2}{\text{quantity } \mathbf{5}} = \frac{(-2747.62)^2}{18{,}690} = 403.9281$$

8. Unexplained sum of squares $= \sum d^2_{Y \cdot X} = SS_{\text{groups}} - \sum \hat{y}^2$

$$= SS_{\text{groups}} - \text{quantity } \mathbf{7} = 423.7016 - 403.9281 = 19.7735$$

Completed anova table with regression

Source of variation	*df*	*SS*	*MS*	F_s
$\bar{Y} - \bar{\bar{Y}}$ Among densities (groups)	3	423.7016	141.2339	11.20**
$\hat{Y} - \bar{\bar{Y}}$ Linear regression	1	403.9281	403.9281	40.86*
$\bar{Y} - \hat{Y}$ Deviations from regression	2	19.7735	9.8868	<1 *ns*
$Y - \bar{Y}$ Within groups	11	138.6867	12.6079	
$Y - \bar{\bar{Y}}$ Total	14	562.3883		

In addition to the familiar mean squares, MS_{groups} and MS_{within}, we now have the mean square due to linear regression, $MS_{\hat{Y}}$, and the mean square for deviations from regression, $MS_{Y \cdot X}$ $(= s^2_{Y \cdot X})$. To test if the deviations from linear regression are significant, compare the ratio $F_s = MS_{Y \cdot X}/MS_{\text{within}}$ with $F_{\alpha[a-2, \sum^a n_i - a]}$. Since we find $F_s < 1$, we accept the null hypothesis that the deviations from linear regression are zero.

To test for the presence of linear regression, we therefore tested $MS_{\hat{Y}}$ over the mean square of deviations from regression $s^2_{Y \cdot X}$ and, since $F_s = 403.9281/9.8868 = 40.86$ is greater than $F_{.05[1,2]} = 18.5$, we clearly reject the null hypothesis that there is no regression, or that $\beta = 0$.

9. Regression coefficient (slope of regression line) $= b_{Y \cdot X} = \dfrac{\sum xy}{\sum x^2}$

$$= \frac{\text{quantity } \mathbf{6}}{\text{quantity } \mathbf{5}} = \frac{-2747.62}{18{,}690} = -0.14701$$

10. Y-intercept $= a = \bar{\bar{Y}} - b_{Y \cdot X}\bar{X}$

$$= \frac{\sum^a \sum^{n_i} Y}{\sum^a n_i} - \frac{\text{quantity } \mathbf{9} \times \text{quantity } \mathbf{1}}{\sum^a n_i}$$

$$= \frac{907.81}{15} - \frac{(-0.14701)555}{15} = 60.5207 + 5.4394 = 65.9601$$

Hence, the regression equation is $\hat{Y} = 65.9601 - 0.14701X$

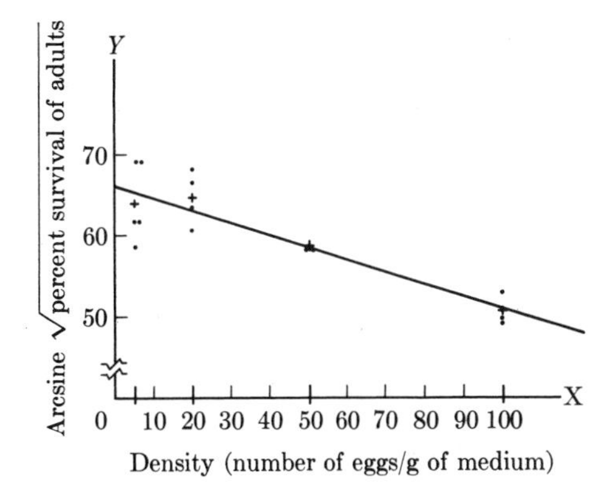

FIGURE 11.8

Linear regression fitted to data of Box 11.2. Sample means are identified by + signs.

again present the results in the form of an anova table, as shown in Box 11.2. Note that the major quantities in this table are the same as in a single classification anova, but in addition we now have a sum of squares representing linear regression, which is always based on one degree of freedom. This sum of squares is subtracted from the SS among groups, leaving a residual sum of squares (of two degrees of freedom in this case) representing the deviations from linear regression.

We should become quite certain what these sources of variation represent. The SS due to linear regression represents that portion of the SS among groups that is explained by linear regression on X. The SS due to deviations from regression represents the residual variation or scatter around the regression line as illustrated by the various examples in Figure 11.7. The SS within groups is a measure of the variation of the items around each group mean.

We first test whether the mean square for deviations from regression ($MS_{Y \cdot X} = s^2_{Y \cdot X}$) is significant by computing the variance ratio of $MS_{Y \cdot X}$ over the within-groups MS. In our case, the deviations from regression are clearly not significant, since the mean square for deviations is less than that within groups. We now test the mean square for regression, $MS_{\hat{Y}}$, over the mean square for deviations from regression and find it to be significant. Thus linear regression on density has clearly removed a significant portion of the variation of survival values. Significance of the mean square for deviations from regression could mean either that Y is a curvilinear function of X or that there is a large amount of random heterogeneity around the regression line (as already discussed in connection with Figure 11.7; actually a mixture of both conditions may prevail).

We complete the computation of the regression coefficient and regression equation as shown at the end of Box 11.2. Our conclusions are that as density increases, survival decreases and that this relationship can be expressed by a significant linear regression of the form $\hat{Y} = 65.9601 - 0.14701X$, where X is density per gram and $\hat{Y}$ is the arcsine transformation of percentage survival. This relation is graphed in Figure 11.8. The sums of products and regression slopes of both examples discussed so far have been negative and you may begin to believe that this is always so. However, it is only an accident of choice of these two examples. In the homework exercises a positive regression coefficient will be encountered.

When we have equal sample sizes of Y-values for each value of X, the computations become simpler. First we carry out the anova in the manner of Box 8.1. Steps **1** through **8** in Box 11.2 become simplified because the unequal sample sizes n_i are replaced by a constant sample size n, which can generally be factored out of the various expressions. Also, $\sum^{a} n_i = an$. Significance tests applied to such cases are also simplified.

11.5 Tests of significance in regression

We have so far interpreted regression as a method for providing an estimate, $\hat{Y}_1$, given a value of X_1. Another interpretation is as a method for explaining some of the variation of the dependent variable Y in terms of the variation of the independent variable X. The SS of a sample of Y-values, $\sum y^2$, is computed by summing and squaring deviations $y = Y - \bar{Y}$. In Figure 11.9 we can see that the deviation y can be decomposed into two parts, $\hat{y}$ and $d_{Y \cdot X}$. It is also clear from Figure 11.9 that the deviation $\hat{y} = \hat{Y} - \bar{Y}$ represents the deviation of the estimated value $\hat{Y}$ from the mean of Y. The height of $\hat{y}$ is clearly a function of x. We have already seen that $\hat{y} = bx$ [Expression (11.3)]. In analytical geometry this is called the point-slope form of the equation. If b, the slope of the regression line, were steeper, $\hat{y}$ would be relatively longer for a given value of x. The remaining portion of the deviation y is $d_{Y \cdot X}$. It represents the residual variation of the variable Y after the explained variation has been subtracted. We can see that $y = \hat{y} + d_{Y \cdot X}$ by writing out these deviations explicitly as $Y - \bar{Y} = (\hat{Y} - \bar{Y}) + (Y - \hat{Y})$.

For each of these deviations we can compute a corresponding sum of squares. Appendix A1.7 gives the computational formula for the unexplained sum of squares,

$$\sum d^2_{Y \cdot X} = \sum y^2 - \frac{(\sum xy)^2}{\sum x^2}$$

Transposed, this yields

$$\sum y^2 = \frac{(\sum xy)^2}{\sum x^2} + \sum d^2_{Y \cdot X}$$

Of course, $\sum y^2$ corresponds to y, $\sum d^2_{Y \cdot X}$ to $d_{Y \cdot X}$, and

$$\sum \hat{y}^2 = \frac{(\sum xy)^2}{\sum x^2}$$

to $\hat{y}$ (as shown in the previous section). Thus we are able to partition the

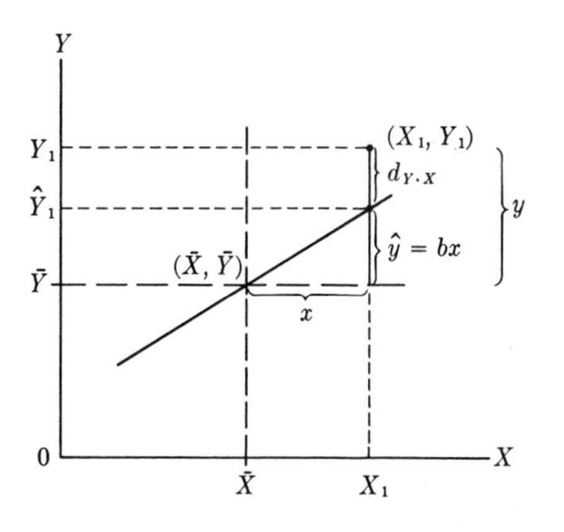

FIGURE 11.9

Schematic diagram to show relations involved in partitioning the sum of squares of the dependent variable.

sum of squares of the dependent variable in regression in an analogous way to the partition of the total SS in analysis of variance. You may wonder how the additive relation of the deviations can be matched by an additive relation of their squares without the presence of any cross products. Some simple algebra in Appendix A1.8 will show that the cross products cancel out. The magnitude of the unexplained deviation $d_{Y \cdot X}$ is independent of the magnitude of the explained deviation $\hat{y}$ just as in anova the magnitude of the deviation of an item from the sample mean is independent of the magnitude of the deviation of the sample mean from the grand mean. This relationship between regression and analysis of variance can be carried further. We can undertake an analysis of variance of the partitioned sums of squares as follows:

Source of variation	*df*	*SS*	*MS*
$\hat{Y} - \bar{Y}$ Explained (estimated Y from mean of Y)	1	$\sum \hat{y}^2 = \dfrac{(\sum xy)^2}{\sum x^2}$	$s^2_{\hat{Y}}$
$Y - \hat{Y}$ Unexplained, error (observed Y from estimated Y)	$n - 2$	$\sum d^2_{Y \cdot X} = \sum y^2 - \sum \hat{y}^2$	$s^2_{Y \cdot X}$
$Y - \bar{Y}$ Total (observed Y from mean of Y)	$n - 1$	$\sum y^2 = \sum Y^2 - \dfrac{(\sum Y)^2}{n}$	s^2_Y

The *explained mean square*, or *mean square due to linear regression*, measures the amount of variation in Y accounted for by variation of X. It is tested over the *unexplained mean square*, which measures the residual variation and is used as an error MS. The mean square due to linear regression, $s^2_{\hat{Y}}$. is based on one degree of freedom, and consequently $n - 2$ *df* remain for the error MS since the total sum of squares possesses $n - 1$ degrees of freedom. The test is of the null hypothesis $H_0: \beta = 0$. When we carry out such an anova on the weight loss data of Box 11.1, we obtain the following results:

Source of variation	df	SS	MS	F_s
Explained—due to linear regression	1	23.5145	23.5145	267.18**
Unexplained—error around regression line	7	0.6161	0.08801	
Total	8	24.1306		

The significance test is $F_s = s_{\hat{Y}}^2/s_{Y \cdot X}^2$. It is clear from the observed value of F_s that a large and significant portion of the variance of Y has been explained by regression on X.

We now proceed to the standard errors for various regression statistics, their employment in tests of hypotheses, and the computation of confidence limits. Box 11.3 lists these standard errors in two columns. The right-hand column is for the case with a single Y-value for each value of X. The first row of the table considers the *standard error of the regression coefficient,* which is simply the square root of the quotient of the unexplained variance divided by the sum of squares of X. Note that the unexplained variance $s_{Y \cdot X}^2$ is a fundamental quantity that is a part of all standard errors in regression. The standard error of the regression coefficient permits us to test various hypotheses and to set confidence limits to our sample estimate of b. The computation of s_b is illustrated in step **1** of Box 11.4, using the weight loss example of Box 11.1.

The significance test illustrated in step **2** tests the *"significance" of the regression coefficient;* that is, it tests the null hypothesis that the sample value of b comes from a population with a parametric value $\beta = 0$ for the regression coefficient. This is a t-test, the appropriate degrees of freedom being $n - 2 = 7$. If we cannot reject the null hypothesis, there is no evidence that the regression is significantly deviant from zero in either the positive or negative direction. Our conclusions for the weight loss data are that a highly significant negative regression is present. We saw earlier (Section 8.4) that $t^2 = F$. When we square $t_s = -16.345$ from Box 11.4, we obtain 267.16, which (within rounding error) equals the value of F_s found in the anova earlier in this section. The significance test in step **2** of Box 11.4 could, of course, also have been used to test whether b is significantly different from a parametric value β other than zero.

Setting confidence limits to the regression coefficient presents no new features. The computation is shown in step **3** of Box 11.4. In view of the small magnitude of s_b, the confidence interval is quite narrow. The confidence limits are shown in Figure 11.10 as dotted lines representing the 95% bounds of the slope. Note that the regression line as well as its confidence limits pass through the means for X and Y. Variation in b therefore rotates the regression line about the point $\bar{X}$, $\bar{Y}$.

Next, we calculate a *standard error to the observed sample mean* $\bar{Y}$. You will recall from Section 6.1 that $s_{\bar{Y}}^2 = s_Y^2/n$. However, now that we have

BOX 11.3

Standard errors of regression statistics and their degrees of freedom.

For explanation of this box, see Section 11.5: ν identifies degrees of freedom; a = number of values of X when there are n_i Y-values for each X; n = sample size when there is a single Y-value for each value of X.

Statistic	s	*More than one Y-value for each value of X*		*Single Y-value for each value of X*	
$b_{Y \cdot X}$ (regression coefficient)	s_b	$\sqrt{\frac{s^2_{Y \cdot X}}{\Sigma x^2}}$	$\nu = a - 2$	$\sqrt{\frac{s^2_{Y \cdot X}}{\Sigma x^2}}$	$\nu = n - 2$
		At any value X_i		At $\bar{X}$	
$\bar{Y}_i$ (sample mean)	$s_{\bar{Y}}$	$\sqrt{\frac{MS_{\text{within}}}{n_i}}$	$\nu = \Sigma n_i - a$	$\sqrt{\frac{s^2_{Y \cdot X}}{n}}$	$\nu = n - 2$
$\hat{Y}_i$ (estimated Y for a given value X_i)	$s_{\hat{Y}}$	$\sqrt{s^2_{Y \cdot X}\left[\frac{1}{\Sigma n_i} + \frac{(X_i - \bar{X})^2}{\Sigma x^2}\right]}$	$\nu = a - 2$ ($\nu = \Sigma n_i - 2$ if pooled $s^2_{Y \cdot X}$ is employed)	$\sqrt{s^2_{Y \cdot X}\left[\frac{1}{n} + \frac{(X_i - \bar{X})^2}{\Sigma x^2}\right]}$	$\nu = n - 2$

BOX 11.4

Significance tests and computation of confidence limits of regression statistics. Single value of Y for each value of X.

Based on standard errors and degrees of freedom of Box 11.3; using example of Box 11.1.

$$n = 9 \qquad \bar{X} = 50.389 \qquad \bar{Y} = 6.022$$

$$b_{Y \cdot X} = -0.05322 \qquad \sum x^2 = 8301.3889$$

$$s^2_{Y \cdot X} = \frac{\sum d^2_{Y \cdot X}}{(n-2)} = \frac{0.6161}{7} = 0.08801$$

1. Standard error of the regression coefficient

$$s_b = \sqrt{\frac{s^2_{Y \cdot X}}{\sum x^2}} = \sqrt{\frac{0.08801}{8301.3889}} = \sqrt{0.000{,}010{,}602} = 0.003{,}2561$$

2. Testing significance of the regression coefficient

$$t_s = \frac{(b - 0)}{s_b} = \frac{-0.05322}{0.003{,}2561} = -16.345$$

$$t_{.001[7]} = 5.408 \qquad P < 0.001$$

3. 95% confidence limits for regression coefficient

$$t_{.05[7]}s_b = 2.365(0.003{,}2561) = 0.00770$$

$$L_1 = b - t_{.05[7]}s_b = -0.05322 - 0.00770 = -0.06092$$

$$L_2 = b + t_{.05[7]}s_b = -0.05322 + 0.00770 = -0.04552$$

4. Standard error of the sampled mean $\bar{Y}$ (at $\bar{X}$)

$$s_{\bar{Y}} = \sqrt{\frac{s^2_{Y \cdot X}}{n}} = \sqrt{\frac{0.08801}{9}} = 0.098{,}8883$$

5. 95% confidence limits for the mean μ_Y corresponding to $\bar{X}(\bar{Y} = 6.022)$

$$t_{.05[7]}s_{\bar{Y}} = 2.365(0.098{,}8883) = 0.233871$$

$$L_1 = \bar{Y} - t_{.05[7]}s_{\bar{Y}} = 6.022 - 0.2339 = 5.7881$$

$$L_2 = \bar{Y} + t_{.05[7]}s_{\bar{Y}} = 6.022 + 0.2339 = 6.2559$$

6. Standard error of $\hat{Y}$, an estimated Y for a given value of X_i

$$s_{\hat{Y}} = \sqrt{s^2_{Y \cdot X}\left[\frac{1}{n} + \frac{(X_i - \bar{X})^2}{\sum x^2}\right]}$$

for example, for $X_i = 100\%$ relative humidity,

$$s_{\hat{Y}} = \sqrt{0.08801\left[\frac{1}{9} + \frac{(100 - 50.389)^2}{8301.3889}\right]}$$

$$= \sqrt{0.08801[0.40760]} = \sqrt{0.035{,}873} = 0.18940$$

7. 95% confidence limits for μ_{Y_i} corresponding to the estimate $\hat{Y}_i = 3.3817$ at $X_i = 100\%$ relative humidity

BOX 11.4 continued

$t_{.05[7]}s_{\hat{Y}} = 2.365(0.18940) = 0.44793$

$L_1 = \hat{Y}_i - t_{.05[7]}s_{\hat{Y}} = 3.3817 - 0.4479 = 2.9338$

$L_2 = \hat{Y}_i + t_{.05[7]}s_{\hat{Y}} = 3.3817 + 0.4479 = 3.8296$

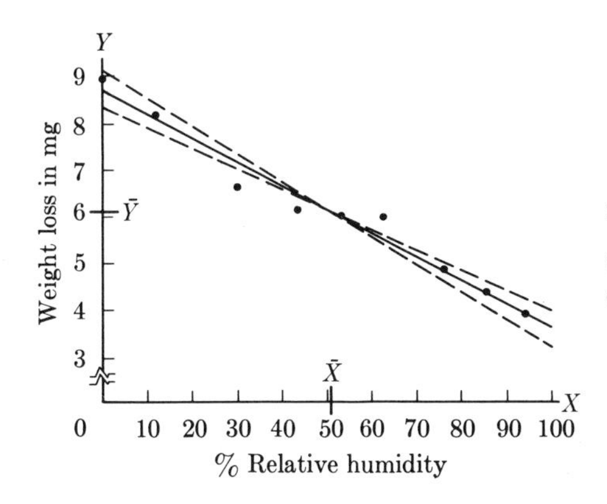

FIGURE 11.10

95% confidence limits to regression line of Figure 11.6.

regressed Y on X, we are able to account for some of the variation of Y in terms of the variation of X. The variance of Y around the point $\bar{X}$, $\bar{Y}$ on the regression line is less than s_Y^2; it is $s_{Y \cdot X}^2$. At $\bar{X}$ we may therefore compute confidence limits of $\bar{Y}$, using as a standard error of the mean $s_{\bar{Y}} = \sqrt{s_{Y \cdot X}^2/n}$ with $n - 2$ degrees of freedom. This standard error is computed in step **4** of Box 11.4, and 95% confidence limits for the sampled mean $\bar{Y}$ at $\bar{X}$ are calculated in step **5**. These limits (5.7881 – 6.2559) are considerably narrower than the confidence limits for the mean based on the conventional standard error $s_{\bar{Y}}$, which would be from 4.687 to 7.357. Clearly, differences in relative humidity explain much of the variation in weight loss.

The standard error for $\bar{Y}$ is only a special case of the *standard error for any estimated value $\hat{Y}$ along the regression line.* A new factor now enters the error variance, whose magnitude is in part a function of the distance of a given value X_i from its mean $\bar{X}$. Thus, the farther away X_i is from its mean, the greater will be the error of estimate. This factor is seen in the third row of Box 11.3 as the deviation $X_i - \bar{X}$, squared and divided by the sum of squares of X. The standard error for an estimate $\hat{Y}_i$ for a relative humidity $X_i = 100\%$ is given in step **6** of Box 11.4. The 95% confidence limits for $\mu_{\hat{Y}_i}$, the parametric value corresponding to the estimate $\hat{Y}_i$, are shown in step **7** of that box. Note that the width of the confidence interval is $3.8296 - 2.9338 = 0.8958$, considerably wider than the confidence interval at $\bar{X}$ calculated in step **5**, which was $6.2559 - 5.7881 = 0.4678$. This illustrates the point that confidence limits are wider away from the mean than at the mean. If we calculate a series of confidence limits for different values of X_i, we obtain a

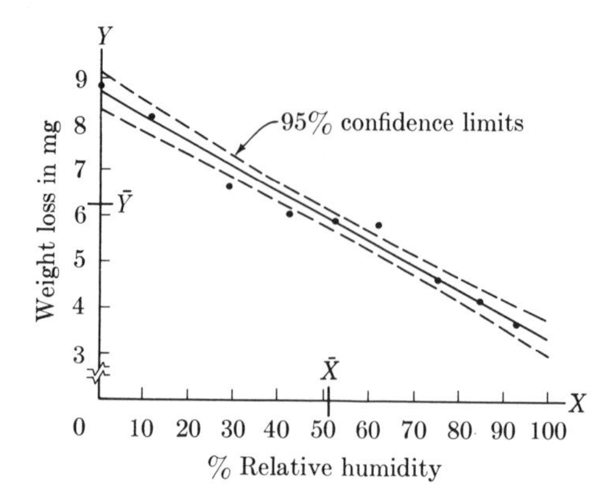

FIGURE 11.11

95% confidence limits to regression estimates for data of Figure 11.6.

biconcave confidence belt as shown in Figure 11.11. The farther we get away from the mean, the less reliable are our estimates of Y because of the uncertainty about the true slope, β, of the regression line.

Furthermore, the linear regressions that we fit are often only rough approximations to the more complicated functional relationships between biological variables. Very often there is an approximately linear relation along a certain range of the independent variable, beyond which range the slope changes rapidly. For example, heartbeat of a poikilothermic animal will be directly proportional to temperature over a range of tolerable temperatures, but beneath and above this range the heartbeat will eventually decrease as the animal freezes or suffers heat prostration. Hence common sense indicates that one should be very cautious about extrapolating from a regression equation if one has any doubts about the linearity of the relationship.

The confidence limits for α, the parametric value of a, are a special case of that for $\mu_{\hat{Y}_i}$ at $X_i = 0$.

Tests of significance in regression analyses where there is more than one variate Y per value of X are carried out in a manner similar to that of Box 11.4, except that the standard errors in the left-hand column of Box 11.3 are employed.

Another significance test in regression is a test of the differences between two regression lines. Why would we be interested in testing differences between regression slopes? We might find that different toxicants yield different dosage-mortality curves or that different drugs yield different relationships between dosage and response (see, for example, Figure 11.1). Or, genetically differing cultures might yield different responses to increasing density, an important fact in understanding the effect of natural selection in these cultures. The regression slope of one variable on another is as fundamental a statistic of a sample as is the mean or the standard deviation, and in comparing samples it may be as important to compare regression coefficients as it is to compare these other statistics.

The test for the difference between two regression coefficients can be carried out as an F-test. We compute

$$F_s = \frac{(b_1 - b_2)^2}{\dfrac{\sum x_1^2 + \sum x_2^2}{(\sum x_1^2)(\sum x_2^2)} \bar{s}_{Y \cdot X}^2}$$

where $\bar{s}_{Y \cdot X}^2$ is the weighted average $s_{Y \cdot X}^2$ of the two groups. Its formula is

$$\bar{s}_{Y \cdot X}^2 = \frac{(\sum d_{Y \cdot X}^2)_1 + (\sum d_{Y \cdot X}^2)_2}{\nu_2}$$

For one Y per value of X, $\nu_2 = n_1 + n_2 - 4$, but when there is more than one variate Y per value of X, $\nu_2 = a_1 + a_2 - 4$. Compare F_s with $F_{\alpha[1,\nu_2]}$. Since there is a single degree of freedom in the numerator, $t_s = \sqrt{F_s}$.

11.6 The uses of regression

We have been so busy learning the mechanics of regression analysis that we have not had time to give much thought to the various applications of regression. We shall take up four more or less distinct applications in this section. All are discussed in terms of Model I regression.

First, we might mention the *study of causation*. If we wish to know whether variation in a variable Y is caused by changes in another variable X, we manipulate X in an experiment and see whether we can obtain a significant regression of Y on X. The idea of causation is a complex, philosophical one that we shall not go into here. You have undoubtedly been cautioned from your earliest scientific experience not to confuse concomitant variation with causation. Variables may vary together, yet this covariation may be accidental or both may be functions of a common cause affecting them. The latter cases are usually Model II regression with both variables varying freely. When we manipulate one variable and find that such manipulations affect a second variable, we generally are satisfied that the variation of the independent variable X is the cause of the variation of the dependent variable Y (not the cause of the variable!). However, even here it is best to be cautious. When we find that heartbeat rate in a cold-blooded animal is a function of ambient temperature, we may conclude that temperature is one of the causes of differences in heartbeat rate. There may well be other factors affecting rate of heartbeat. A possible mistake is to invert the cause-and-effect relationship. It is unlikely that anyone would suppose that heartbeat rate affects the temperature of the general environment, but we might be mistaken about the cause-and-effect relationships between two chemical substances in the blood, for instance. Despite these cautions, regression analysis is a commonly used device for screening out causal relationships. While a significant regression of Y on X does not prove that changes in X are the cause of variations of Y, the converse statement is true. When we find no significant regression of Y on X, we can in all but the most complex cases infer quite safely (allowing for the possibility of type II error) that deviations of X do not affect Y.

The *description of scientific laws* and *prediction* are a second general area of application of regression analysis. Science aims at mathematical description of relations between variables in nature, and Model I regression analysis permits us to estimate functional relationships between variables, one of which is subject to error. These functional relationships do not always have clearly interpretable biological meaning. Thus, in many cases it may be difficult to assign a biological interpretation to the statistics a and b, or their corresponding parameters α and β. When we can do so, we speak of a *structural mathematical model*, one whose component parts have clear scientific meaning. However, mathematical curves that are not structural models are also of value in science. Most regression lines are *empirically fitted curves*, in which the functions simply represent the best mathematical fit (by a criterion such as least squares) to an observed set of data.

Comparison of dependent variates is another application of regression. As soon as it is established that a given variable is a function of another one, as in Box 11.2 where we found survival of beetles to be a function of density, one is bound to ask to what degree any observed difference in survival between two samples of beetles is a function of the density at which they have been raised. It would be unfair to compare beetles raised at very high density (and expected to have low survival) with those raised under optimal conditions of low density. This is the same point of view that makes us disinclined to compare the mathematical knowledge of a fifth-grader with that of a college student. Since we could undoubtedly obtain a regression of mathematical knowledge on years of schooling in mathematics, we should be comparing how far a given individual deviates from his expected value based on such a regression. Thus, relative to his classmates and age group, the fifth-grader may be far better than is the college student relative to his peer group. This suggests that we calculate *adjusted Y-values* that allow for the magnitude of the independent variable X. A conventional way of calculating such adjusted Y-values is as the overall mean of the population $\bar{Y}$ plus the deviation $d_{Y \cdot X} = (Y_i - \hat{Y}_i)$ of the given Y-value Y_i from the regression line. As you know, such deviations may be either positive or negative; therefore adjusted Y-values can be above or below the mean. We shall formally define an adjusted Y-value as

$$Y_{\text{adj}} = \bar{Y} + d_{Y \cdot X} = Y - bx \tag{11.8}$$

Statistical control is an application of regression that is not widely known among biologists and represents a scientific philosophy that is not well established in biology outside agricultural circles. Biologists frequently categorize work as either descriptive or experimental, with the implication that only the latter can be analytical. However, statistical approaches applied to descriptive work can, in a number of instances, take the place of experimental techniques quite adequately—occasionally they are even to be preferred. These approaches are attempts to substitute statistical manipulation of a concomitant variable for control of the variable by experimental means. An example will clarify this technique.

Let us assume that we are studying the effects of various diets on blood pressure in rats. We find that the variability of blood pressure in our rat population is considerable, even before we introduce differences in diet. Further study reveals that the variability is largely due to differences in age among the rats of the experimental population. This can be demonstrated by a significant linear regression of blood pressure on age. To reduce the variability of blood pressure in the population, we should keep the age of the rats constant. The reaction of most biologists at this point will be to repeat the experiment using rats of only one age group; this is a valid, commonsense approach, which is part of the experimental method. An alternative approach is superior in some cases, when it is not practical or too costly to hold the variable constant. We might continue to use rats of variable ages and simply record the age of each rat as well as its blood pressure. Then we regress blood pressure on age and use an adjusted mean as the basic blood pressure reading for each individual. We can now evaluate the effect of differences in diet on these adjusted means. Or we can analyze the effects of diet on unexplained deviations, $d_{Y \cdot X}$, after the experimental blood pressures have been regressed on age (which amounts to the same thing).

What are the advantages of such an approach? Often it will be impossible to secure adequate numbers of individuals all of the same age. By using regression we are able to utilize all the individuals in the population. The use of statistical control assumes that it is relatively easy to record the independent variable X and, of course, that this variable can be measured without error, which would be generally true of such a variable as age of a laboratory animal. Statistical control may also be preferable because we obtain information over a wider range of both Y and X and also because we obtain added knowledge about the relations between these two variables, which would not be so if we restricted ourselves to a single age group.

11.7 Transformations in regression

In transforming either or both variables in regression, we aim at simplifying a curvilinear relationship to a linear one. As a by-product of such a procedure the proportion of the variance of the dependent variable explained by the independent variable is generally increased and the distribution of the deviations of points around the regression line tends to become normal and homoscedastic. Rather than fit a complicated curvilinear regression to points plotted on an arithmetic scale, it is far more expedient to compute a simple linear regression for variates plotted on a transformed scale. A general test of whether transformation will improve linear regression is to graph the points to be fitted on ordinary graph paper as well as on other graph paper in a scale suspected to improve the relationship. If the function straightens out and the systematic deviation of points around a visually fitted line is reduced, the transformation is worthwhile.

We shall briefly discuss a few of the transformations commonly applied in regression analysis. Square root and arcsine transformations (Section 10.2) are not mentioned below, but they are also effective in regression cases involving data suited to such transformations.

The *logarithmic transformation* is the most frequently used. Anyone doing statistical work is therefore well advised to keep a supply of semilog paper handy. Most frequently we transform the dependent variable Y. This transformation is indicated when percentage changes in the dependent variable vary directly with changes in the independent variable. Such a relationship is indicated by the equation $\hat{Y} = ae^{bX}$, where a and b are constants and e is the base of the natural logarithm. After the transformation, we obtain $\log \hat{Y} = \log a + b(\log e)X$. In this expression $\log e$ is a constant which when multiplied times b yields a new constant factor b', which is equivalent to a regression coefficient. Similarly, $\log a$ is a new Y-intercept, a'. We can then simply regress $\log Y$ on X to obtain the function $\log \hat{Y} = a' + b'X$ and obtain all our prediction equations and confidence intervals in this form. Figure 11.12 shows an example of transforming the dependent variate to logarithmic form, which results in considerable straightening of the response curve.

A logarithmic transformation of the independent variable in regression is effective when proportional changes in the independent variable produce linear responses in the dependent variable. An example might be the decline in weight of an organism as density increases, where the successive increases in density need to be in a constant ratio in order to effect equal decreases in weight. This belongs to a well-known class of biological phenomena, another example of which is the Weber-Fechner law in physiology and psychology, which states that a stimulus has to be increased by a constant proportion in order to produce a constant increment in response. Figure 11.13 illustrates how logarithmic transformation of the independent variable results in the straightening of the regression line. For computations one would transform X into logarithms.

Logarithmic transformation for both variables is applicable in situations in which the true relationship can be described by the formula $\hat{Y} = aX^b$. The regression equation is rewritten as $\log \hat{Y} = \log a + b \log X$ and the computation is done in the conventional manner. Examples are the greatly disproportionate growth of various organs in some organisms, such as the sizes of antlers of deer or horns of stag beetles, with respect to their general body sizes. A double logarithmic transformation is indicated when plotting on log-log graph paper results in a straight line graph.

Reciprocal transformation. Many rate phenomena (a given performance per unit of time or per unit of population) such as wing beats per second or

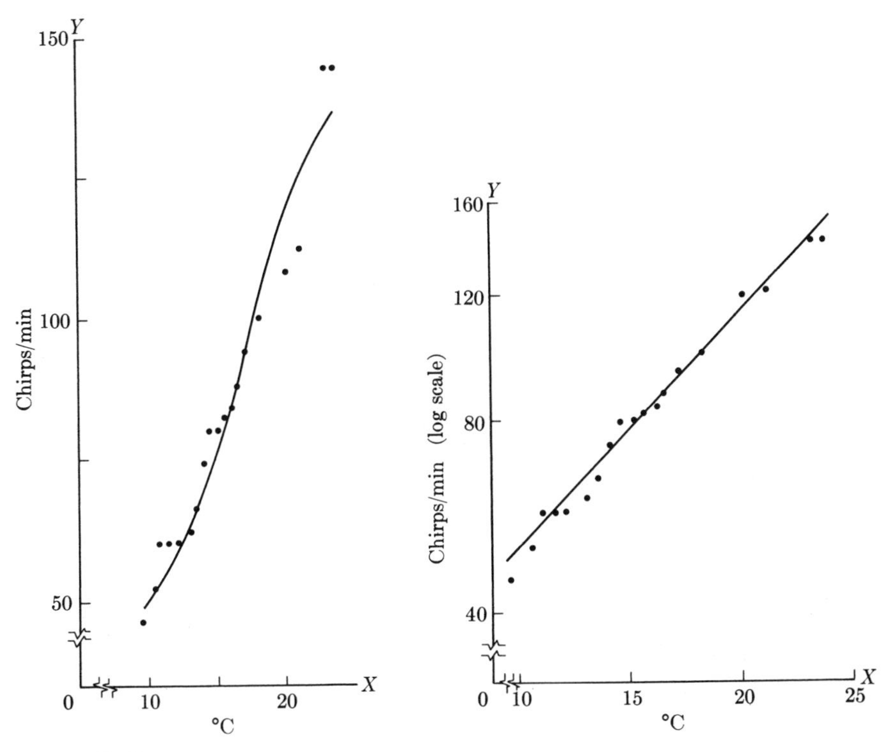

FIGURE 11.12

Logarithmic transformation of a dependent variable in regression. Chirp-rate as a function of temperature in males of the tree cricket *Oecanthus fultoni*. Each point represents the mean chirp-rate/min for all group observations at a given temperature in °C. Original data in left panel, Y plotted on logarithmic scale in right panel. (Data from Block, 1966.)

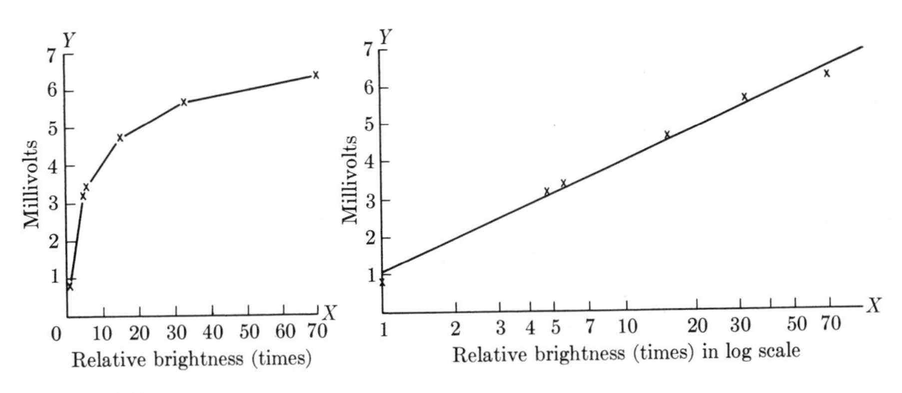

FIGURE 11.13

Logarithmic transformation of the independent variable in regression. This illustrates size of electrical response to illumination in cephalopod eye. Ordinate millivolts; abscissa relative brightness of illumination. A proportional increase in X (relative brightness) produces a linear electrical response Y. (Data in Fröhlich, 1921.)

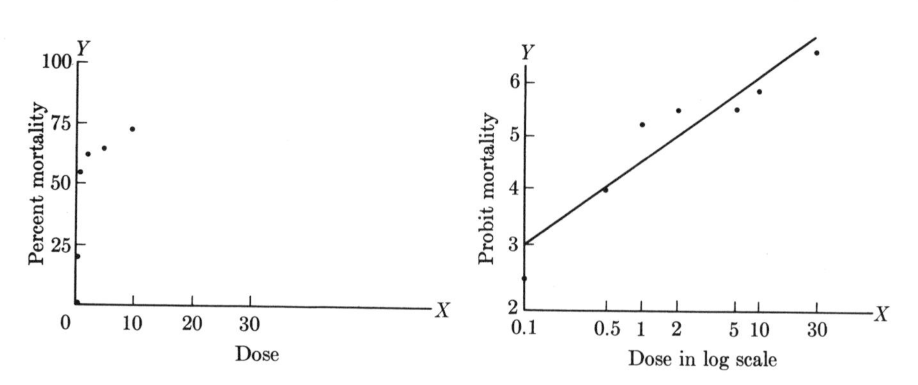

FIGURE 11.14

Dosage mortality data illustrating an application of the probit transformation. Data are mean mortalities for two replicates. Twenty *Drosophila melanogaster* per replicate were subjected to seven doses of an "unknown" insecticide in a class experiment. The point at dose 0.1 which yielded 0% mortality has been assigned a probit value of 2.5 in lieu of $-\infty$, which cannot be plotted.

number of eggs laid per female, will yield hyperbolic curves when plotted in original measurement scale. Thus, they form curves described by the general mathematical equations $bXY = 1$ or $(a + bX)Y = 1$. From these we can derive $1/Y = bX$ or $1/Y = a + bX$. By transforming the dependent variable into its reciprocal, we can frequently obtain straight-line regressions.

Finally, some cumulative curves can be straightened by the *probit transformation.* Refresh your memory on the cumulative normal curve shown in Figure 5.5. Remember that by changing the ordinate of the cumulative normal into probability scale we were able to straighten out this curve. We do the same thing here except that we graduate the probability scale in standard deviation units. Thus, the 50% point becomes 0 standard deviations, the 84.13% point becomes +1 standard deviation, and the 2.27% point becomes −2 standard deviations. Such standard deviations, corresponding to a cumulative percentage are called *normal equivalent deviates* (*N.E.D.*). If we use ordinary graph paper and mark the ordinate in *N.E.D.* units, we would obtain a straight line when plotting the cumulative normal curve against it. *Probits* are simply normal equivalent deviates coded by the addition of 5.0, which will avoid negative values for most deviates. Thus, the probit value 5.0 corresponds to a cumulative frequency of 50%, the probit value 6.0 corresponds to a cumulative frequency of 84.13%, and the probit value 3.0 corresponds to a cumulative frequency of 2.27%.

Figure 11.14 shows an example of mortality percentages for increasing doses of an insecticide. These represent differing points of a cumulative frequency distribution. With increasing dosages an ever greater proportion of the sample dies until at a high enough dose the entire sample is killed. It is often found that if the doses of toxicants are transformed into logarithms, the tolerances of many organisms to these poisons are approximately normally distributed. These transformed doses are often called *dosages.* Increasing dosages lead to a cumulative normal distribution of mortalities, often called *dosage-mortality curves.* These curves are the subject matter of an entire field of biometric analysis, *bioassay,* to which we can only refer in passing here. The most common technique in this field is *probit analysis.* The results of such analysis are plotted on *probit paper,* which is probability graph paper in which the abscissa has been transformed into logarithmic scale. A regression line is fitted to dosage-mortality data graphed on probit paper (see Figure 11.14). From the regression line the 50% lethal dose is estimated by a process of inverse prediction, that is, we estimate the value of X (dosage) corresponding to a kill of probit 5.0, which is equivalent to 50%.

Exercises 11

11.1 The following temperatures (Y) were recorded in a rabbit at various times (X) after being inoculated with rinderpest virus (data from Carter and Mitchell, 1958).

Time after injection (hrs)	*Temperature (°F)*
24	102.8
32	104.5
48	106.5
56	107.0
72	103.9
80	103.2
96	103.1

Graph the data. Clearly the last three data points represent a different phenomenon from the first four pairs. *For the first four points:* (a) Calculate b. (b) Calculate the regression equation and draw in the regression line. (c) Test the hypothesis that $\beta = 0$ and set 95% confidence limits. (d) Set 95% confidence limits to your estimate of the rabbit's temperature 50 hours after the injection. ANS. $b = 0.1300$, $\hat{Y}_{50} = 106.5$.

11.2 The following table is extracted from data by Sokoloff (1955). Adult weights of female *Drosophila persimilis* reared at 24°C are affected by their density as larvae. Carry out an anova among densities. Then calculate the regression of weight on density and partition the sums of squares among groups into that explained and unexplained by linear regression. Graph the data with the regression line fitted to the means. Interpret your results.

Larval density	*Mean weight of adults (in mg)*	*s of weights (not $s_{\bar{Y}}$)*	n
1	1.356	0.180	9
3	1.356	0.133	34
5	1.284	0.130	50
6	1.252	0.105	63
10	0.989	0.130	83
20	0.664	0.141	144
40	0.475	0.083	24

11.3 Using the complete data given in Exercise 11.1, calculate the regression equation and compare it with the one obtained earlier. Discuss the effect of the inclusion of the last three points in the analysis.

11.4 Davis (1955) reported the following results in a study of the amount of energy metabolized by the English sparrow, *Passer domesticus,* under various constant temperature conditions and a ten-hour photoperiod.

Temperature (°*C*)	*Calories* $\bar{Y}$	n	s
0	24.9	6	1.77
4	23.4	4	1.99
10	24.2	4	2.07
18	18.7	5	1.43
26	15.2	7	1.52
34	13.7	7	2.70

Analyze and interpret.

11.5 The following results were obtained in a study of oxygen consumption (microliters/mg dry weight/hr) in *Heliothis zea* by Phillips and Newsom (1966) under controlled temperatures and photoperiods.

Temperature (°*C*)	*Photoperiod* (*hrs*)	
	10	14
18	0.51	1.61
21	0.53	1.64
24	0.89	1.73

Compute regression for each photoperiod separately and test for homogeneity of slopes. ANS. $s^2_{Y \cdot X} = 0.019267$ and 0.00060 for 10- and 14-hour photoperiods.

11.6 Length of developmental period (in days) of the potato leafhopper, *Empoasca fabae,* from egg to adult at various constant temperatures (Kouskolekas and Decker, 1966). The original data were weighted means, but for purposes of this analysis we shall consider them as though they were single observed values.

Temperature °*F*	*Mean length of developmental period in days* $\bar{Y}$
59.8	58.1
67.6	27.3
70.0	26.8
70.4	26.3
74.0	19.1
75.3	19.0
78.0	16.5
80.4	15.9
81.4	14.8
83.2	14.2
88.4	14.4
91.4	14.6
92.5	15.3

Analyze and interpret. Compute deviations from the regression line ($\bar{Y}_i - \hat{Y}_i$) and plot against temperature.

11.7 The experiment cited in Exercise 11.4 was repeated using a 15-hour photoperiod, and the following results were obtained.

Temperature (°C)	*Calories* $\bar{Y}$	n	s
0	24.3	6	1.93
10	25.1	7	1.98
18	22.2	8	3.67
26	13.8	10	4.01
34	16.4	6	2.92

Test for the equality of slopes of the regression lines for the 10-hour and 15-hour photoperiod. ANS. $F_s = 0.003$.

12

Correlation

In this chapter we continue our discussion of bivariate statistics. The chapter on regression dealt with the functional relation of one variable upon the other; the present chapter treats the measurement of the amount of association between two variables. This general topic is called correlation analysis.

It is not always obvious which type of analysis—regression or correlation—one should employ in a given problem. There has been considerable confusion in the minds of investigators and also in the literature on this topic. We shall try to make the distinction between these two approaches clear at the outset in Section 12.1. In the following section (12.2) you will be introduced to the product-moment correlation coefficient, the common correlation coefficient of the literature. We shall derive a formula for this coefficient and give you something of its theoretical background. The close mathematical relationship between regression and correlation analysis will be examined in this section. We shall also compute a product-moment correlation coefficient in this section. Various tests of significance involving correlation coefficients are treated in the next section (12.3). Having learned something about correlation coefficients, it is time to discuss their applications in Section 12.4.

Section 12.5 contains a nonparametric method that tests for association. It is to be used in those cases in which the necessary assumptions for tests involving correlation coefficients do not hold, or where quick but less than fully efficient tests are preferred for reasons of speed in computation or for convenience.

12.1 Correlation and regression

There has been much confusion on the subject matter of correlation and regression. Quite frequently correlation problems are treated as regression in the scientific literature, and the converse is equally true. There are several reasons for this confusion. First of all, the mathematical relations between the two methods of analysis are quite close and mathematically one can easily move from one to the other. Hence, the temptation to do so is great. Second, earlier texts had not made the distinction between the two approaches sufficiently clear, and this liability has still not been entirely overcome. At least one current text synonymizes the two, a step that we feel can only compound the confusion. Finally, while the approach chosen by an investigator may be correct in terms of his intentions, the data available for analysis may be such as to make one or the other of the techniques inappropriate.

Let us examine these points at some length. The many and close mathematical relations between regression and correlation will be detailed in the following section. It suffices for now to state that for any given problem the majority of the computational steps are the same whether one carries out a regression or a correlation analysis. You will recall that the fundamental quantity required for regression analysis is the sum of products. This is the very same quantity that serves as the base for the computation of the correlation coefficient. There are some simple mathematical relations between regression coefficients and their corresponding correlation coefficients. Thus the temptation exists to compute a correlation coefficient corresponding to a given regression coefficient. Yet, as we shall see below, this would be wrong unless our intention at the outset had been to study association and the data were appropriate for such a computation.

Thus a correlation coefficient computed from data that were properly analyzed by Model I regression, is meaningless as an estimate of any population correlation coefficient. Conversely, we can evaluate a regression coefficient of one variable on another in data that have been properly computed as correlations. Not only would construction of such a functional dependence for these variables not meet our intentions, but we should point out that a conventional regression coefficient computed from data in which both variables are measured with error furnishes biased estimates of the functional relation.

Let us then look at the intentions or purposes behind the two types of analyses. In regression we intend to describe the dependence of a variable Y on an independent variable X. As we have seen, we employ regression equations to lend support to hypotheses regarding the possible causation of changes in Y by changes in X; for purposes of prediction, of Y in terms of X; and for purposes of explaining some of the variation of Y by X, by using the latter variable as a statistical control. Studies of the effects of temperature on heartbeat rate, nitrogen content of soil on growth rate in a plant, age of an animal on blood pressure, or dose of an insecticide on mortality of the insect population are all typical examples of regression for the purposes named above.

In correlation, by contrast, we are concerned largely whether two variables are interdependent or *covary*—that is, vary together. We do not express one as a function of the other. There is no distinction between independent and dependent variables. It may well be that of the pair of variables whose correlation is studied, one is the cause of the other, but we neither know nor assume this. A more typical (but not essential) assumption is that the two variables are both effects of a common cause. What we wish to estimate is the degree to which these variables vary together. Thus we might be interested in the correlation between arm length and leg length in a population of mammals or between body weight and egg production in female blowflies or between days to maturity and number of seeds in a weed. Reasons why we would wish to demonstrate and measure association between pairs of variables need not concern us yet. We shall take this up in Section 12.5. It suffices for now to state that when we wish to establish the degree of association between pairs of variables in a population sample, correlation analysis is the proper approach.

Even if we attempt the correct method in line with our purposes we may run afoul of the nature of the data. Thus we may wish to establish cholesterol content of blood as a function of weight, and to do so we may take a random sample of men of the same age group, obtain each individual's cholesterol content and weight, and regress the former on the latter. However, both these variables will have been measured with error. Individual variates of the supposedly independent variable X were not deliberately chosen or controlled by the experimenter. The underlying assumptions of Model I regression do not hold, and fitting a Model I regression to the data is not legitimate, although you will have no difficulty finding instances of such improper practices in the published research literature. If it is really an equation describing the dependence of Y on X that we are after, we should carry out a Model II regression. However, if it is the degree of association between the variables (interdependence) that is of interest, then we should carry out a correlation analysis, for which these data are suitable. The converse difficulty is trying to obtain a correlation coefficient from data that are properly computed as a regression—that is, when X is fixed. An example would be heartbeats of a

TABLE 12.1

The relations between correlation and regression. This table indicates the correct computation for any combination of purposes and variables, as shown.

Purpose of investigator	Nature of the two variables	
	Y random, X fixed	*Y_1, Y_2 both random*
Establish and estimate dependence of one variable upon another	Model I regression.	**Model II regression. (Not treated in this book.)**
Establish and estimate association (interdependence) between two variables	Meaningless for this case. If desired, an estimate of the proportion of the variation of Y explained by X can be obtained as the square of the correlation coefficient between X and Y.	Correlation coefficient. (Significance tests entirely appropriate only if Y_1, Y_2 are distributed as bivariate normal variables.)

poikilotherm as a function of temperature, where several temperatures have been applied in an experiment. Such a correlation coefficient is easily obtained mathematically but would simply be a numerical value, not an estimate of a parametric measure of correlation. There is an interpretation that can be given to the square of the correlation coefficient that has some relevance to a regression problem. However, it is not in any way an estimate of a parametric correlation.

This discussion is summarized in Table 12.1, which shows the relations between correlation and regression. The two columns of the table indicate the two conditions of the pair of variables: in one case one random and measured with error, the other variable fixed; in the other case, both variables random. We have departed from the usual convention of labeling the pair of variables Y and X or X_1, X_2 for both correlation and regression analyses. In regression we continue the use of Y for the dependent variable and X for the independent variable, but in correlation both of the variables are in fact random variables, which we have throughout the text designated as Y. We therefore refer to the two variables as Y_1 and Y_2. The rows of the table indicate the intention of the investigator in carrying out his analysis and the four quadrants of the table indicate the appropriate procedures for a given combination of intention of investigator and nature of the pair of variables.

12.2 The product-moment correlation coefficient

There are numerous correlation coefficients in statistics. The most common of these is called the *product-moment correlation coefficient,* which in its current formulation is due to Karl Pearson. We shall derive its formula through an intuitive approach.

You have seen that the sum of products is a measure of covariation and it is, therefore, likely that this will be the basic quantity from which to obtain a formula for the correlation coefficient. We shall label the variables whose correlation is to be estimated as Y_1 and Y_2. Their sum of products will, therefore, be $\sum y_1y_2$ and their covariance $[1/(n - 1)]\sum y_1y_2 = s_{12}$. The latter quantity is analogous to a variance, that is, a sum of squares divided by its degrees of freedom.

A standard deviation is expressed in original measurement units such as inches, grams, or cubic centimeters. Similarly, a regression coefficient is expressed as so many units of Y per unit of X, such as 5.2 grams/day. However, a measure of association should be independent of the original scale of measurement so that we can compare the degree of association in one pair of variables with that in another. One way to accomplish this is to divide the covariance by the standard deviations of variables Y_1 and Y_2. This results in dividing each deviation y_1 and y_2 by its proper standard deviation and making it into a standardized deviate. The expression now becomes the sum of the products of standardized deviates divided by $n - 1$

$$r_{Y_1Y_2} = \frac{\sum y_1y_2}{(n-1)s_{Y_1}s_{Y_2}} \tag{12.1}$$

This is the formula for the product-moment correlation coefficient $r_{Y_1Y_2}$ between variables Y_1 and Y_2. We shall simplify the symbolism to

$$r_{12} = \frac{\sum y_1y_2}{(n-1)s_1s_2} = \frac{s_{12}}{s_1s_2} \tag{12.2}$$

Expression (12.2) can be rewritten in another common form. Since

$$s\sqrt{n-1} = \sqrt{s^2(n-1)} = \sqrt{\frac{\sum y^2}{n-1}(n-1)} = \sqrt{\sum y^2}$$

Expression (15.2) can be rewritten as

$$r_{12} = \frac{\sum y_1y_2}{\sqrt{\sum y_1^2 \sum y_2^2}} \tag{12.3}$$

which is often preferable for computation. To state Expression (12.2), more generally for variables Y_j and Y_k, we can write it as

$$r_{jk} = \frac{\sum y_jy_k}{(n-1)s_js_k} \tag{12.4}$$

The correlation coefficient r_{jk} can range from $+1$ for perfect association to -1 for perfect negative association. This is intuitively obvious when we consider the correlation of a variable Y_j with itself. Expression (12.4) would then yield $r_{jj} = \sum y_jy_j/\sqrt{\sum y_j^2 \sum y_j^2} = \sum y_j^2/\sum y_j^2 = 1$, which yields a perfect correlation of $+1$. If deviations in one variable were paired with opposite but

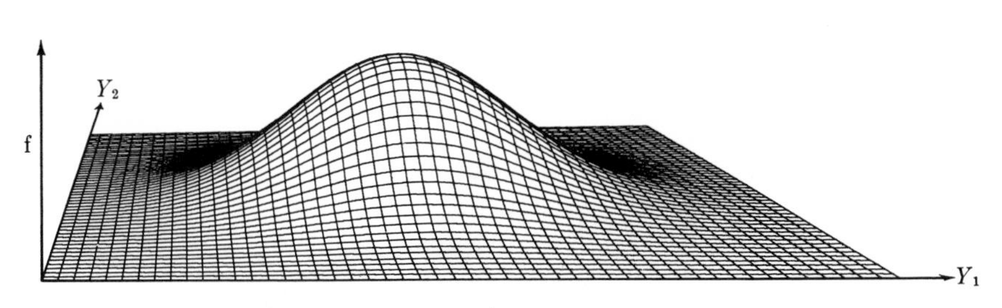

FIGURE 12.1

Bivariate normal frequency distribution. The parametric correlation ρ between variables Y_1 and Y_2 equals zero. The frequency distribution may be visualized as a bell-shaped mound.

equal deviations representing another variable, this would yield a correlation of -1 because the sum of products in the numerator would be negative. Proof that the correlation coefficient is bounded by $+1$ and -1 will be given shortly.

If the variates follow a specified distribution, the *bivariate normal distribution,* the correlation coefficient r_{jk} will estimate a parameter of that distribution symbolized by ρ_{jk}.

Let us approach the distribution empirically. Suppose you have sampled a hundred items and measured two variables on each item, obtaining two samples of 100 variates in this manner. If you plot these 100 items on a graph in which the variables Y_1 and Y_2 are the coordinates, you will obtain a scattergram of points as in Figure 12.3A. Let us assume that both variables, Y_1 and Y_2, are normally distributed and that they are quite independent of each other, so that the fact that one individual happens to be greater than the mean in character Y_1 has no effect whatsoever on its value for variable Y_2. Thus this same individual may be greater or less than the mean for variable Y_2. If there is absolutely no relation between Y_1 and Y_2 and if the two variables are standardized to make their scales comparable, you would find that the outline of the scattergram is roughly circular. Of course, for a sample of 100 items, the circle would be only imperfectly outlined; but the larger the sample, the more clearly could one discern a circle with the central area around the intersection $\bar{Y}_1$, $\bar{Y}_2$ heavily darkened because of the aggregation there of many points. If you keep sampling, you will have to super-impose new points upon previous points, and if you visualize these points in a physical sense, such as grains of sand, a mound peaked in a bell-shaped fashion would gradually accumulate. This is a three-dimensional realization of a normal distribution shown in perspective in Figure 12.1. Regarded from either coordinate axis, the mound would present a two-dimensional appearance, and its outline would be that of a normal distribution curve, the two perspectives giving the distributions of Y_1 and Y_2, respectively.

If we assume that the two variables Y_1 and Y_2 are not independent but are positively correlated to some degree, then if a given individual has a large value of Y_1, it is more likely than not to have a large value of Y_2 as well.

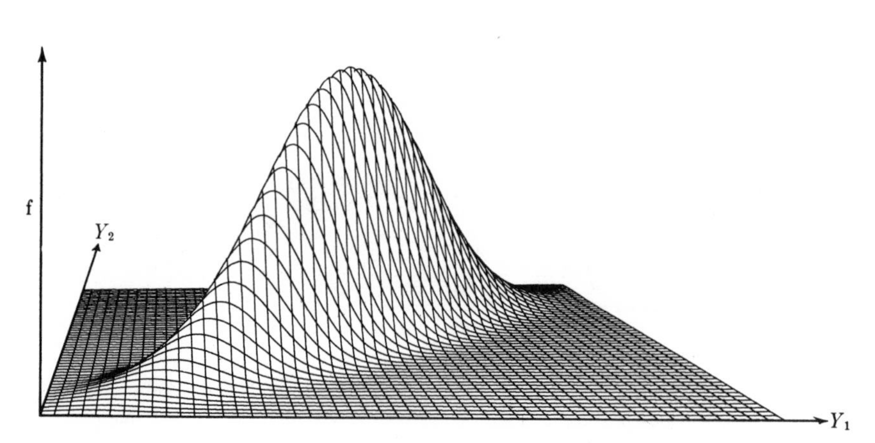

FIGURE 12.2

Bivariate normal frequency distribution. The parametric correlation ρ between variables Y_1 and Y_2 equals 0.9. The bell-shaped mound of Figure 12.1 has become elongated.

Similarly, a small value of Y_1 will likely be associated with a small value of Y_2. Were you to sample items from such a population, the resulting scattergram (shown in Figure 12.3D) would become elongated in the form of an ellipse. This is so because those parts of the circle that formerly included individuals high for one variable and low for the other (and vice versa), are now scarcely represented. Continued sampling (with the sand grain model) yields a three-dimensional elliptic mound shown in Figure 12.2. If correlation is perfect, all the data would fall along a single regression line (the identical line would describe the regression of Y_1 and Y_2 and of Y_2 on Y_1), and if we let them pile up in a physical model, they would result in a flat, essentially two-dimensional normal curve lying on this regression line.

The circular or elliptical shape of the outline of the scattergram and of the resulting mound is clearly a function of the degree of correlation between the two variables, and this is the parameter ρ_{jk} of the bivariate normal distribution. By analogy with Expression (12.2), the parameter ρ_{jk} can be defined as

$$\rho_{jk} = \sigma_{jk}/\sigma_j\sigma_k \tag{12.5}$$

where σ_{jk} is the parametric covariance of variables Y_j and Y_k, and σ_j and σ_k are the parametric standard deviations of variables Y_j and Y_k, as before. When two variables are distributed according to the bivariate normal, a sample correlation coefficient r_{jk} estimates the parametric correlation coefficient ρ_{jk}. We can make some statements about the sampling distribution of ρ_{jk} and set confidence limits to it.

Regrettably, the elliptical shape of scattergrams of correlated variables is not usually very clear unless either very large samples have been taken or the parametric correlation ρ_{jk} is very high. To illustrate this point, we show in Figure 12.3 several graphs illustrating scattergrams resulting from samples

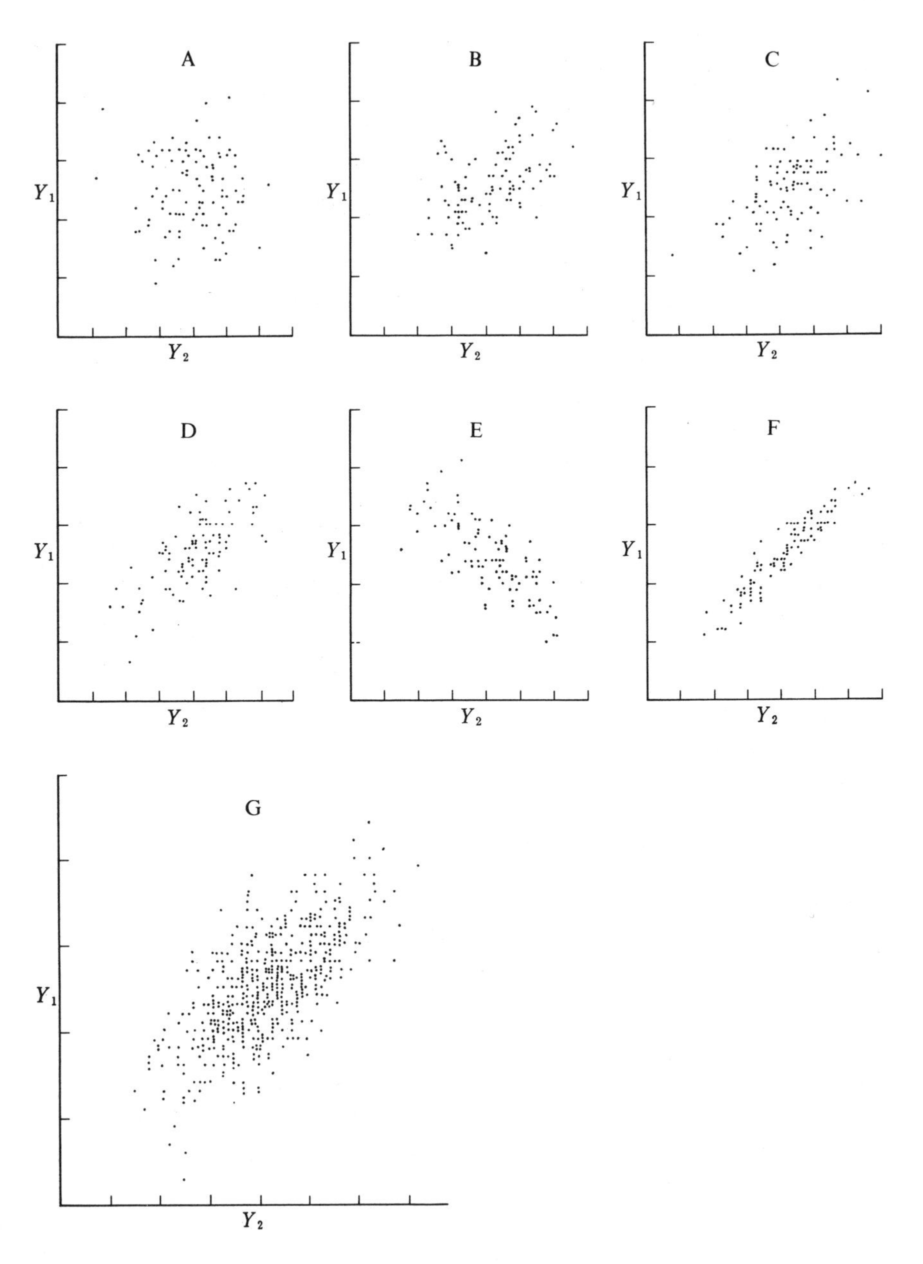

FIGURE 12.3

Random samples from bivariate normal distributions with varying values of the parametric correlation coefficient ρ. Sample sizes $n = 100$ in all graphs except G, which has $n = 500$. A. $\rho = 0.4$. B. $\rho = 0.3$. C. $\rho = 0.5$. D. $\rho = 0.7$. E. $\rho = -0.7$. F. $\rho = 0.9$. G. $\rho = 0.5$.

of 100 items from bivariate normal populations with differing values of ρ_{jk}. Note that in the first graph (Figure 12.3A), with $\rho_{jk} = 0$, the circular distribution is only very vaguely outlined. A far greater sample is required to demonstrate the circular shape of the distribution more clearly. No substantial difference is noted in Figure 12.3B, based on $\rho_{jk} = 0.3$. Knowing that this depicts a positive correlation, one can visualize a positive slope in the scattergram; but without prior knowledge this would be difficult to detect visually. The next graph (Figure 12.3C, based on $\rho_{jk} = 0.5$) is somewhat clearer, but still does not exhibit an unequivocal trend. In general, correlation cannot be inferred from inspection of scattergrams based on samples from populations with ρ_{jk} between -0.5 and $+0.5$ unless the sample is very numerous. This point is illustrated in the last graph (Figure 12.3G), also sampled from a population with $\rho_{jk} = 0.5$ but based on a sample of 500. Here, the positive slope and elliptical outline of the scattergram are quite evident. Figure 12.3D, based on $\rho_{jk} = 0.7$ and $n = 100$, shows the trend more clearly. Note that the next graph (Figure 12.3E), based on the same magnitude of ρ_{jk} but representing negative correlation, also shows the trend but is more strung out than Figure 12.3D. The difference in shape of the ellipse has no relation to the negative nature of the correlation; it is simply a function of sampling error, and the comparison of these two figures should give you some idea of the variability to be expected on random sampling from a bivariate normal distribution. Finally, Figure 12.3F, representing a correlation of $\rho_{jk} = 0.9$, shows a tight association between the variables and a reasonable approximation to an ellipse of points.

Now let us return to the expression for the sample correlation coefficient shown in Expression (12.3). Squaring this expression results in

$$r_{12}^2 = \frac{(\sum y_1 y_2)^2}{\sum y_1^2 \sum y_2^2}$$

$$= \frac{(\sum y_1 y_2)^2}{\sum y_1^2} \cdot \frac{1}{\sum y_2^2}$$

Look at the left term of the last expression. It is the square of the sum of products of variables Y_1 and Y_2, divided by the sum of squares of Y_1. If this were a regression problem, this would be the formula for the explained sum of squares of variable Y_2 on variable Y_1, $\sum \hat{y}_2^2$. In the symbolism of the regression chapter, it would be $\sum \hat{y}^2 = (\sum xy)^2/\sum x^2$. Thus, we can write

$$r_{12}^2 = \frac{\sum \hat{y}_2^2}{\sum y_2^2} \tag{12.6}$$

The square of the correlation coefficient, therefore, is the ratio of the explained sum of squares of variable Y_2 divided by the total sum of squares of variable Y_2 or, equivalently,

$$r_{12}^2 = \sum \hat{y}_1^2 / \sum y_1^2 \tag{12.6a}$$

which could be derived as easily. Remember that since we are not really regressing one variable on the other, it is just as legitimate to have Y_1 explained by Y_2 as the other way around. This ratio is a proportion between zero and one. This becomes obvious after a little contemplation of the meaning of this formula. The explained sum of squares of any variable must be smaller than its total sum of squares or, maximally, if all the variation of a variable has been explained, it can be as great as the total sum of squares, but certainly no greater. Minimally, it will be zero if none of the variable can be explained by the other variable with which the covariance has been computed. Thus, we obtain an important measure of the proportion of the variation of one variable determined by the variation of the other. This quantity, the square of the correlation coefficient, r_{12}^2, is called the *coefficient of determination*. It ranges from zero to 1 and must be positive regardless of whether the correlation coefficient is negative or positive. Incidentally, here is proof that the correlation coefficient cannot vary beyond -1 and $+1$. Since its square is the coefficient of determination, and we have just shown that the bounds of the latter are zero to 1, it is obvious that the bounds of its square root will be ± 1.

The coefficient of determination is useful also when one is considering the relative importance of correlations of different magnitudes. As can be seen by a re-examination of Figure 12.3, the rate at which the scatter diagrams go from a distribution with a circular outline to that of an ellipse seems to be more directly proportional to r^2 than to r itself. Thus, in Figure 12.3B, with $\rho^2 = 0.09$, it is difficult to detect the correlation visually. However, by the time we reach Figure 12.3D, with $\rho^2 = 0.49$, the presence of correlation is very apparent.

The coefficient of determination is a quantity that may be useful in regression analysis also. You will recall that in a regression we used anova to partition the total sum of squares into explained and unexplained sums of squares. Once such an analysis of variance has been carried out, one can obtain the ratio of the explained sums of squares over the total SS as a measure of the proportion of the total variation that has been explained by the regression. However, as already discussed in Section 12.1, it would not be meaningful to take the square root of such a coefficient of determination and consider it as an estimate of the parametric correlation of these variables.

We shall now take up a mathematical relation between the coefficients of correlation and regression. At the risk of being repetitious, we should stress again that though we can easily convert one coefficient into the other, this does not mean that the two types of coefficients can be used interchangeably on the same sort of data. One important relationship between the correlation coefficient and the regression coefficient can be derived as follows from Expression (12.3):

$$r_{12} = \frac{\sum y_1 y_2}{\sqrt{\sum y_1^2 \sum y_2^2}} = \frac{\sum y_1 y_2}{\sqrt{\sum y_1^2}} \cdot \frac{1}{\sqrt{\sum y_2^2}}$$

Multiplying numerator and denominator of this expression by $\sqrt{\sum y_1^2}$, we obtain

$$r_{12} = \frac{\sum y_1 y_2}{\sqrt{\sum y_1^2}\sqrt{\sum y_1^2}} \cdot \frac{\sqrt{\sum y_1^2}}{\sqrt{\sum y_2^2}} = \frac{\sum y_1 y_2}{\sum y_1^2} \cdot \frac{\sqrt{\sum y_1^2}}{\sqrt{\sum y_2^2}}$$

Dividing numerator and denominator of the right term of this expression by $\sqrt{n-1}$, we obtain

$$r_{12} = \frac{\sum y_1 y_2}{\sum y_1^2} \cdot \frac{\sqrt{\dfrac{\sum y_1^2}{n-1}}}{\sqrt{\dfrac{\sum y_2^2}{n-1}}} = b_{2\cdot 1}\frac{s_1}{s_2} \tag{12.7}$$

Similarly, we could demonstrate that

$$r_{12} = b_{1\cdot 2}\frac{s_2}{s_1} \tag{12.7a}$$

and hence

$$b_{2\cdot 1} = r_{12}\frac{s_2}{s_1} \qquad b_{1\cdot 2} = r_{12}\frac{s_1}{s_2} \tag{12.7b}$$

In these expressions $b_{2\cdot 1}$ is the regression coefficient for variable Y_2 on Y_1. We see, therefore, that the correlation coefficient is the regression slope multiplied by the ratio of the standard deviations of the variables. The correlation coefficient may thus be regarded as a standardized regression coefficient. If the two standard deviations are identical, both regression coefficients and the correlation coefficient will be identical in value.

Now that we know about the coefficient of correlation, some of the earlier work on paired comparisons (see Section 9.3) can be put into proper perspective. In Appendix A1.9 we show for the corresponding parametric expressions that the variance of a sum of two variables is

$$s^2_{(Y_1+Y_2)} = s_1^2 + s_2^2 + 2r_{12}s_1s_2 \tag{12.8}$$

where s_1 and s_2 are standard deviations of Y_1 and Y_2, respectively, and r_{12} is the correlation coefficient between these variables. Similarly, for a difference between two variables, we obtain

$$s^2_{(Y_1-Y_2)} = s_1^2 + s_2^2 - 2r_{12}s_1s_2 \tag{12.9}$$

What Expression (12.8) indicates is that if we make a new composite variable that is the sum of two other variables, the variance of this new variable will be the sum of the variances of the variables of which it is composed plus an added term, which is a function of the standard deviations of these two variables and of the correlation between them. It is shown in Appendix A1.9 that this added term is twice the covariance of Y_1 and Y_2. When the two variables being summed are uncorrelated, this added covariance term will also be zero, and the variance of the sum will simply be the sum of the variances of the two variables. This is the reason why, in an anova or in a t-test of the difference between the two means, we had to assume the independence of the two variables to permit us to add their variances. Otherwise we would have had to allow for a covariance term. By contrast, in the paired comparisons technique we expect correlation between the variables, since each pair shares a common experience. The paired comparisons test automatically subtracts a covariance term, resulting in a smaller standard error and consequently in a larger value of t_s, since the numerator of the ratio remains the same. Thus, whenever correlation between two variables is positive, the variance of their differences will be considerably smaller than the sum of their variances; this is the reason why the paired comparisons test has to be used in place of the t-test for difference of means. These considerations are equally true for the corresponding analyses of variance, single classification, and two-way anova.

The computation of a product-moment correlation coefficient is quite simple. The basic quantities needed are the same six required for computation of the regression coefficient (Section 11.3). Box 12.1 illustrates how the coefficient should be computed. The example is based on a sample of twelve crabs in which gill weight Y_1 and body weight Y_2 have been recorded. We wish to

BOX 12.1

Computation of the product-moment correlation coefficient.

Relationships between gill weight and body weight in the crab *Pachygrapsus crassipes*. $n = 12$.

(1) Y_1 *Gill weight in milligrams*	*(2)* Y_2 *Body weight in grams*
159	14.40
179	15.20
100	11.30
45	2.50
384	22.70
230	14.90
100	1.41
320	15.81
80	4.19
220	15.39
320	17.25
210	9.52

Source: Unpublished data by L. Miller.

BOX 12.1 continued

Computation

1. $\sum Y_1 = 159 + \cdots + 210 = 2347$

2. $\sum Y_1^2 = 159^2 + \cdots + 210^2 = 583{,}403$

3. $\sum Y_2 = 14.40 + \cdots + 9.52 = 144.57$

4. $\sum Y_2^2 = (14.40)^2 + \cdots + (9.52)^2 = 2204.1853$

5. $\sum Y_1 Y_2 = 14.40(159) + \cdots + 9.52(210) = 34{,}837.10$

6. Sum of squares of $Y_1 = \sum y_1^2 = \sum Y_1^2 - \frac{(\sum Y_1)^2}{n}$

$$= \text{quantity } \mathbf{2} - \frac{(\text{quantity } \mathbf{1})^2}{n} = 583{,}403 - \frac{(2347)^2}{12}$$

$$= 124{,}368.9167$$

7. Sum of squares of $Y_2 = \sum y_2^2 = \sum Y_2^2 - \frac{(\sum Y_2)^2}{n}$

$$= \text{quantity } \mathbf{4} - \frac{(\text{quantity } \mathbf{3})^2}{n} = 2204.1853 - \frac{(144.57)^2}{12}$$

$$= 462.4782$$

8. Sum of products $= \sum y_1 y_2 = \sum Y_1 Y_2 - \frac{(\sum Y_1)(\sum Y_2)}{n}$

$$= \text{quantity } \mathbf{5} - \frac{\text{quantity } \mathbf{1} \times \text{quantity } \mathbf{3}}{n}$$

$$= 34{,}837.10 - \frac{(2347)(144.57)}{12} = 6561.6175$$

9. Product-moment correlation coefficient [by Expression (12.3)] =

$$r_{12} = \frac{\sum y_1 y_2}{\sqrt{\sum y_1^2 \sum y_2^2}} = \frac{\text{quantity } \mathbf{8}}{\sqrt{\text{quantity } \mathbf{6} \times \text{quantity } \mathbf{7}}}$$

$$= \frac{6561.6175}{\sqrt{(124{,}368.9167)(462.4782)}} = \frac{6561.6175}{\sqrt{57{,}517{,}912.7314}}$$

$$= \frac{6561.6175}{7584.0565} = 0.8652 \approx 0.87$$

know whether there is a correlation between the weight of the gill and that of the body, the latter representing a measure of overall size. The existence of a positive correlation might lead you to conclude that a bigger-bodied crab with its resulting greater amount of metabolism would require larger gills in order to provide the necessary oxygen. The computations are illustrated in Box 12.1. The correlation coefficient of 0.87 agrees with the clear slope and narrow elliptical outline of the scattergram for these data in Figure 12.4.

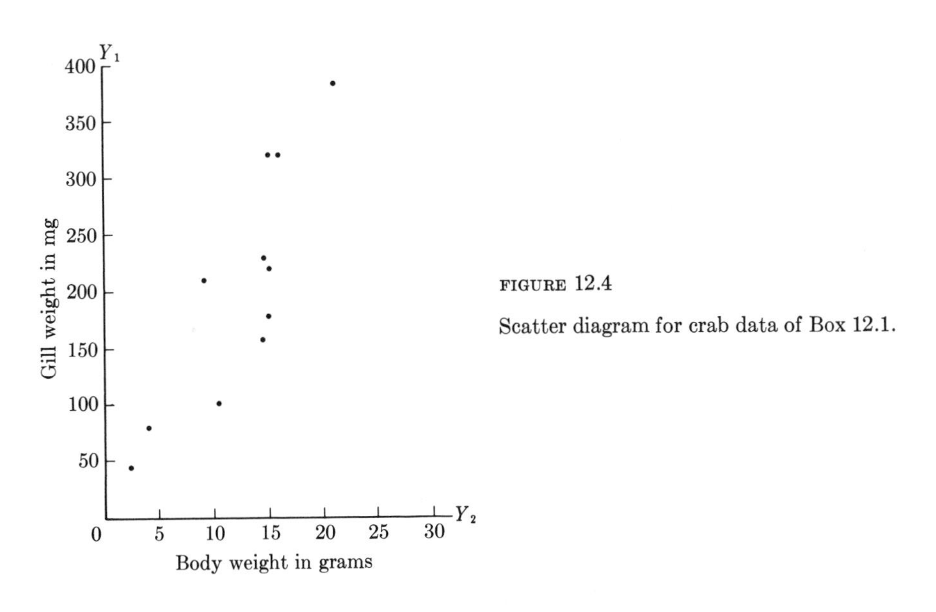

FIGURE 12.4

Scatter diagram for crab data of Box 12.1.

12.3 Significance tests in correlation

The most common significance test is whether a sample correlation coefficient could have come from a population with a parametric correlation coefficient of zero. The null hypothesis is, therefore, $H_0: \rho = 0$. This implies that the two variables are uncorrelated. If the sample comes from a bivariate normal distribution and $\rho = 0$, the standard error of the correlation coefficient is $s_r = \sqrt{(1 - r^2)/(n - 2)}$. The hypothesis is tested as a t-test with $n - 2$ degrees of freedom, $t_s = (r - 0)/\sqrt{(1 - r^2)/(n - 2)} = r\sqrt{(n - 2)/(1 - r^2)}$. We should emphasize that this standard error applies only when $\rho = 0$, so that it cannot be applied to testing a hypothesis that ρ is a specific value other than zero. The t-test for the significance of r is mathematically equivalent to the t-test for the significance of b, in either case measuring the strength of the association between the two variables being tested. This is somewhat analogous to the situation in Model I and Model II single classification anova where the same F-test establishes the significance regardless of the model.

Significance tests following this formula have been carried out systematically and are tabulated in Table **VIII**, which permits the direct inspection of a sample correlation coefficient for significance without further computation. Box 12.3 illustrates tests of the hypothesis $H_0: \rho = 0$, using Table **VIII** as well as the t-test discussed at first.

When $\rho \neq 0$, the distribution of sample values of r is markedly asymmetrical, and, although a standard error has been found for r in such cases, it should not be applied unless the sample is very large ($n > 500$), a most infrequent case of little interest. To overcome this difficulty, we transform

r to a function z, developed by Fisher. The formula for z is

$$z = \frac{1}{2} \ln \left(\frac{1 + r}{1 - r} \right) \tag{12.10}$$

You may recognize this as $z = \tanh^{-1} r$, the formula for the inverse hyperbolic tangent of r. This function has been tabulated in Table **X**, where values of z corresponding to absolute values of r are given. Inspection of Expression (12.10) will show that when $r = 0$, z will also equal zero, since $\frac{1}{2} \ln 1$ equals zero. However, as r approaches 1, $(1 + r)/(1 - r)$ approaches infinity; consequently, z approaches infinity. Therefore, substantial differences between r and z occur at the higher values for r. Thus, when r is 0.115, $z = 0.1155$. For $r = -0.531$, we obtain $z = -0.5915$; $r = 0.972$ yields $z = 2.1273$. Note by how much z exceeds r in this last pair of values. By finding a given value of z in Table **X**, we can also obtain the corresponding value of r. Inverse interpolation may be necessary. Thus, $z = 0.70$ corresponds to $r = 0.604$, and a value of $z = -2.76$ corresponds to $r = -0.992$.

The advantage of the z-transformation is that while correlation coefficients are distributed in skewed fashion for values of $\rho \neq 0$, the values of z are approximately normally distributed for any value of its parameter, which we call ζ (zeta), following the usual convention. The expected variance of z is

$$\sigma_z^2 = \frac{1}{n - 3} \tag{12.11}$$

This is an approximation adequate for sample sizes $n \geq 50$ and a tolerable approximation even when $n \geq 25$. An interesting aspect of the variance of z evident from Expression (12.11) is that it is independent of the magnitude of r, but is simply a function of sample size n.

As shown in Box 12.2, for sample sizes greater than 50 we can also use the z-transformation to test the significance of a sample r employing the hypothesis $H_0: \rho = 0$. In the second section of Box 12.2 we show the test of a null hypothesis that $\rho \neq 0$. We may have a hypothesis that the true correlation between two variables is a given value ρ different from zero. Such hypotheses about the expected correlation between two variables are frequent in genetic work, and we may wish to test observed data against such a hypothesis. Although there is no a priori reason to assume that the true correlation between right and left sides of the bee wing vein lengths is 0.5, we show the test of such a hypothesis to illustrate the method. Corresponding to $\rho = 0.5$, there is ζ, the parametric value of z. It is the z-transformation of ρ. We note that the probability that the sample r of 0.837 could have been sampled from a population with $\rho = 0.5$ is vanishingly small.

Next, in Box 12.2 we see how to set confidence limits to a sample correlation coefficient r. This is done by means of the z-transformation; it will result

BOX 12.2

Tests of significance and confidence limits for correlation coefficients.

Test of the null hypothesis H_0: $\rho = 0$ *versus* H_1: $\rho \neq 0$

The simplest procedure is to consult Table **VIII**, where the critical values of r are tabulated for $df = n - 2$ from 1 to 1000. If the absolute value of the observed r is greater than the tabulated value in the column for two variables, we reject the null hypothesis.

Examples.—In Box 12.1 we found the correlation between body weight and gill weight to be 0.8652, based on a sample of $n = 12$. For 10 degrees of freedom the critical values are 0.576 at the 5% level and 0.708 at the 1% level of significance. Since the observed correlation is greater than both of these, we can reject the null hypothesis, H_0: $\rho = 0$, at $P < 0.01$.

Table **VIII** is based upon the following test, which may be carried out when the table is not available or when an exact test is needed at significance levels or at degrees of freedom other than those furnished in the table. The null hypothesis is tested by means of the t-distribution (with $n - 2$ df) by using the standard error of r. When $\rho = 0$,

$$s_r = \sqrt{(1 - r^2)/(n - 2)}$$

Therefore,

$$t_s = \frac{(r - 0)}{\sqrt{(1 - r^2)/(n - 2)}} = r\sqrt{(n - 2)/(1 - r^2)}$$

For the data of Box 12.1, this would be

$$t_s = 0.8652\sqrt{(12 - 2)/(1 - 0.8652^2)} = 0.8652\sqrt{10/0.25143}$$

$$= 0.8652\sqrt{39.7725} = 0.8652\,(6.3065) = 5.4564 > t_{.001[10]}$$

For a one-tailed test the 0.10 and 0.02 values of t should be used for 5% and 1% significance tests, respectively. Such tests would apply if the alternative hypothesis were H_1: $\rho > 0$ or H_1: $\rho < 0$, rather than H_1: $\rho \neq 0$.

When n is greater than 50, we can also make use of the z-transformation described in the text. Since $\sigma_z = 1/\sqrt{n - 3}$, we test

$$t_s = \frac{z - 0}{1/\sqrt{n - 3}} = z\sqrt{n - 3}$$

Since z is normally distributed and we are using a parametric standard deviation, we compare t_s with $t_{\alpha[\infty]}$ or employ Table **II**, areas of the normal curve. If we had a sample correlation of $r = 0.837$ between length of right and left wing veins of bees based on $n = 500$, we would find $z = 1.2111$ in Table **X**.

$$t_s = 1.2111\sqrt{497} = 26.997$$

This value, when looked up in the table of areas of a normal curve (Table **II**), yields a very small probability ($< 10^{-6}$).

BOX 12.2 continued

Test of the null hypothesis H_0: $\rho = \rho_1$, where $\rho_1 \neq 0$

To test this hypothesis we cannot use Table **VIII** or the t-test given above, but must make use of the z-transformation.

Suppose we wish to test the null hypothesis H_0: $\rho = +0.50$ versus H_1: $\rho \neq +0.50$ for the case just considered. We would use the following expression:

$$t_s = \frac{z - \zeta}{1/\sqrt{n-3}} = (z - \zeta)\sqrt{n-3}$$

where z and ζ are the z-transformations of r and ρ, respectively. Again we compare t_s with $t_{\alpha[\infty]}$ or look it up in Table **II**. From Table **VIII** we find

For $r = 0.837$ $\quad z = 1.2111$

For $\rho = 0.500$ $\quad \zeta = 0.5493$

Therefore

$$t_s = (1.2111 - 0.5493)(\sqrt{497}) = 14.7538$$

The probability of obtaining such a value of t_s by random sampling is $P < 10^{-6}$ (see Table **II**). It is most unlikely that the parametric correlation between right and left wing veins is 0.5.

Confidence limits

If $n > 50$, we can set confidence limits to r using the z-transformation. We first convert the sample r to z, set confidence limits to this z, and then transform these limits back to the r-scale. We shall find 95% confidence limits for the above wing vein length data.

For $r = 0.837$, $z = 1.2111$; $\alpha = 0.05$

$$L_1 = z - t_{\alpha[\infty]}\sigma_z = z - \frac{t_{.05[\infty]}}{\sqrt{n-3}} = 1.2111 - \frac{1.960}{22.2953}$$

$$= 1.2111 - 0.0879 = 1.1232$$

$$L_2 = z + \frac{t_{.05[\infty]}}{\sqrt{n-3}} = 1.2111 + 0.0879 = 1.2990$$

We retransform these z-values to the r-scale by finding the corresponding arguments for the z-function in Table **X**.

$L_1 \approx 0.808$ $\quad$ and $\quad$ $L_2 \approx 0.862$

are the 95% confidence limits around $r = 0.837$.

Test of the difference between two correlation coefficients

For two correlation coefficients we may test H_0: $\rho_1 = \rho_2$ versus H_1: $\rho_1 \neq \rho_2$ as follows:

$$t_s = \frac{z_1 - z_2}{\sqrt{\frac{1}{n_1 - 3} + \frac{1}{n_2 - 3}}}$$

Since $z_1 - z_2$ is normally distributed and we are using a parametric standard deviation, we compare t_s with $t_{\alpha[\infty]}$ or employ Table **II**, areas of the normal curve.

BOX 12.2 continued

For example, the correlation between body weight and wing length in *Drosophila pseudoobscura* was found by Sokoloff (1966) to be 0.552 in a sample of $n_1 = 39$ at the Grand Canyon and 0.665 in a sample of $n_2 = 20$ at Flagstaff, Arizona.

Grand Canyon: $z_1 = 0.6213$ Flagstaff: $z_2 = 0.8017$

$$t_s = \frac{0.6213 - 0.8017}{\sqrt{\frac{1}{36} + \frac{1}{17}}} = \frac{-0.1804}{\sqrt{0.086601}} = \frac{-0.1804}{0.29428} = -0.6130$$

By linear interpolation in Table **II**, we find the probability of a t_s being between ± 0.6130 to be about $2(0.22941) = 0.45882$, so we clearly have no evidence on which to reject the null hypothesis.

in asymmetrical confidence limits when these are retransformed to r-scale, as when setting confidence limits with variables subjected to square root or logarithmic transformations.

A test for the significance of the difference between two sample correlation coefficients is the final example illustrated in Box 12.2. A standard error for the difference is computed and tested against a table of areas of the normal curve. In the example the correlation between body weight and wing length in two *Drosophila* populations was tested, and the difference in correlation coefficients between the two populations was found not significant. The formula given is an acceptable approximation when the smaller of the two samples is greater than 25. It is frequently used with even smaller sample size, as shown in our example in Box 12.2.

12.4 Applications of correlation

The purpose of correlation analysis is to measure the intensity of association observed between any pair of variables and to test whether it is greater than could be expected by chance alone. Once established, such an association is likely to lead to reasoning about causal relationships between the variables. Students of statistics are told at an early stage not to confuse significant correlation with causation. We are also warned about so-called nonsense correlations, a well-known case being the positive correlation between the number of Baptist ministers and the per capita liquor consumption in cities with populations of over 10,000 in the United States. Individual cases of correlation must be carefully analyzed before inferences are drawn from them. It is useful to distinguish correlations in which one variable is the entire or, more likely, the partial cause of another from others in which the two correlated variables have a common cause and from more complicated situations involving both direct influence and common causes. The establishment of a significant correlation does not tell us which of many possible structural models is appropriate. Further analysis is needed to discriminate between the various models.

The traditional distinction of real versus nonsense or illusory correlation is of little use. In supposedly legitimate correlations, causal connections are known or at least believed to be clearly understood. In so-called illusory correlations, no reasonable connection between the variables can be found; or if one is demonstrated, it is of no real interest or may be shown to be an artifact of the sampling procedure. Thus, the correlation between Baptist ministers and liquor consumption is simply a consequence of city size. The larger the city, the more Baptist ministers it will contain on the average and the greater will be the liquor consumption. The correlation is of little interest to anyone studying either the distribution of Baptist ministers or the consumption of alcohol. Some correlations have time as the common factor, and processes that change with time are frequently likely to be correlated, not because of any functional biological reasons but simply because the change with time in the two variables under consideration happens to be in the same direction. Thus, size of an insect population building up through the summer may be correlated with the height of some weeds, but this may simply be a function of the passage of time. There may be no ecological relation between the plant and the insects.

Perhaps the only correlations properly called nonsense or illusory are those assumed by popular belief or scientific intuition, which, when tested by proper statistical methodology using adequate sample sizes, are found to be not significant. Thus, if we can show that there is no significant correlation between amount of saturated fats eaten and the degree of atherosclerosis, we can consider this to be an illusory correlation. Remember also that when testing significance of correlations at conventional levels of significance, you must allow for type I error, which will lead to a certain percentage of correlations being judged significant when in fact the parametric value of $\rho = 0$.

Correlation coefficients have a history of extensive use and application dating back to the English biometric school at the turn of the century. Recent years have seen somewhat less application of this technique as increasing segments of biological research have become experimental. In experiments in which one factor is varied and the response of another variable to the deliberate variation of the first is examined, the method of regression is more appropriate, as has already been discussed. However, large areas of biology and of other sciences remain where the experimental method is not suitable because variables cannot be brought under control of the investigator. There are many areas of ecology, systematics, evolution, and other fields in which experimental methods are difficult to apply. As yet the weather cannot be controlled nor can historical evolutionary factors be altered. Nevertheless, we need an understanding of the scientific mechanisms underlying these phenomena as much as of those in biochemistry or experimental embryology. In such cases, correlation analysis serves as a first descriptive technique estimating the degrees of association among the variables involved.

12.5 Kendall's coefficient of rank correlation

Occasionally data are known not to be bivariate normally distributed, yet we wish to test for the significance of association between the two variables. One method of analyzing such data is by ranking the variates and calculating a coefficient of rank correlation. This approach belongs to the general family of nonparametric methods we encountered in Chapter 10, where we learned methods for analyses of ranked variates paralleling anova. In other cases especially suited to ranking methods, we cannot measure the variable on an absolute scale, but only on an ordinal scale. This is typical of data in which we estimate relative performance, as in assigning positions in a class. We can say that A is the best student, B the second best student, C and D are both equal to each other and next best, and so on. Two instructors may independently rank a group of students and we can then test whether these two sets of rankings are correlated, as they should be if the judgments of the instructors are based on objective evidence. Of greater biological interest are the following examples. We might wish to correlate order of emergence in a sample of insects with a ranking in size, or order of germination in a sample of plants with rank order of flowering. A geneticist might predict the rank order of performance of a series of n genotypes he synthesizes and would wish to show the correlation of his prediction with the rank orders of the realized performance of these genotypes. A taxonomist might wish to array n organisms from those most like form X to those least like it. Will a similar array prepared by a second taxonomist be significantly correlated with the first—that is, are the taxonomic judgments of the two observers correlated?

We present in Box 12.3 *Kendall's coefficient of rank correlation*, generally symbolized by τ (tau), although it is a sample statistic, not a parameter. The formula for Kendall's coefficient of rank correlation is $\tau = N/n(n-1)$, where n is the conventional sample size and N is a count of ranks, which can be obtained in a variety of ways. A second variable Y_2, perfectly correlated with the first variable Y_1, should be in the same order as the Y_1-variates. However, if the correlation is less than perfect, the order of the variates Y_2 will not entirely correspond to that of Y_1. The quantity N measures how well the second variable corresponds to the order of the first. It has a maximal value of $n(n-1)$ and a minimal value of $-n(n-1)$. The following small example will make this clear. Suppose we have a sample of five individuals that have been arrayed by rank of variable Y_1:

BOX 12.3

Kendall's coefficient of rank correlation, τ.

Computation of a rank correlation coefficient between the total length (Y_1) of 15 aphid stem mothers and the mean thorax length (Y_2) of their parthenogenetic offspring (based on measurement of four alates, or winged forms): $n = 15$ pairs of observations.

(1)	(2)	(3)	(4)	(5)	(1)	(2)	(3)	(4)	(5)
Stem mother	Y_1	R_1	Y_2	R_2	*Stem mother*	Y_1	R_1	Y_2	R_2
1	8.7	8	5.95	9	8	6.5	2	4.18	1
2	8.5	6	5.65	4	9	6.6	3	6.15	13
3	9.4	9	6.00	10	10	10.6	12	5.93	8
4	10.0	10	5.70	6.5	11	10.2	11	5.70	6.5
5	6.3	1	4.70	2	12	7.2	4	5.68	5
6	7.8	5	5.53	3	13	8.6	7	6.13	12
7	11.9	15	6.40	15	14	11.1	13	6.30	14
					15	11.6	14	6.03	11

Source: Data from a more extensive study by R. R. Sokal.

Computational steps

1. Rank variables Y_1 and Y_2 separately and then replace the original variates with the ranks (assign tied ranks if necessary). These ranks are listed in columns (*3*) and (*5*) above. There was one tie in variable Y_2, so the 4th and 11th variates were assigned an average rank of 6.5.
2. Write down the n ranks of one of the two variables in order, paired with the rank values assigned for the other variable (as shown below). If only one variable has ties, order the pairs by the variable without ties (as in the present example). If both variables have ties, it does not matter which of the variables is ordered.
3. Obtain a sum of the counts C_i, as follows. Examine the first value in the column of ranks paired with the ordered column. In our case, this is rank 2. Count all ranks subsequent to it which are higher than the rank being considered. Thus, in this case, count all ranks greater than 2. There are fourteen ranks following the 2 and all of them except rank 1 are greater than 2. Therefore, we count a score of $C_1 = 13$. Now we look at the next rank (rank 1) and find that all thirteen subsequent ranks are greater than it; therefore, C_2 is also equal to 13. However, C_3 is only equal to 2 since only ranks 14 and 15 are higher than rank 13. Continue in this manner, taking each rank of the variable in turn and count the number of higher ranks subsequent to it. This can usually be done in one's head, but we show it explicitly below so that the method will be entirely clear. In the case of ties, count a $\frac{1}{2}$. Thus, for C_{10}, there are $4\frac{1}{2}$ ranks greater than the first rank of 6.5.

BOX 12.3 continued

R_1	R_2	*Subsequent ranks greater than pivotal rank* R_2	*Counts* C_i
1	2	13, 5, 3, 4, 12, 9, 10, 6.5, 6.5, 8, 14, 11, 15	13
2	1	13, 5, 3, 4, 12, 9, 10, 6.5, 6.5, 8, 14, 11, 15	13
3	13	14, 15	2
4	5	12, 9, 10, 6.5, 6.5, 8, 14, 11, 15	9
5	3	4, 12, 9, 10, 6.5, 6.5, 8, 14, 11, 15	10
6	4	12, 9, 10, 6.5, 6.5, 8, 14, 11, 15	9
7	12	14, 15	2
8	9	10, 14, 11, 15	4
9	10	14, 11, 15	3
10	6.5	(6.5), 8, 14, 11, 15	$4\frac{1}{2}$
11	6.5	8, 14, 11, 15	4
12	8	14, 11, 15	3
13	14	15	1
14	11	15	1
15	15		0
			$78\frac{1}{2} = \sum^{n} C_i$

We then need the following quantity:

$$N = 4 \sum^{n} C_i - n(n-1) = 4(78\tfrac{1}{2}) - 15(14) = 314 - 210 = 104$$

4. The Kendall coefficient of rank correlation, τ, can be found as follows:

$$\tau = \frac{N}{\sqrt{\left[n(n-1) - \sum^{m} T_1\right]\left[n(n-1) - \sum^{m} T_2\right]}}$$

where $\sum^{m} T_1$ and $\sum^{m} T_2$ are the sums of correction terms for ties in the ranks of variable Y_1 and Y_2, respectively, defined as follows. A T-value equal to $t(t-1)$ is computed for each group of t tied variates and summed over m such groups. In our example $\sum^{m} T_1 = 0$, since there were no ties in the ranks R_1, $\sum^{m} T_2 = 2$ because there is $m = 1$ group of $t = 2$ tied ranks R_2. $T = t(t-1) = 2(2-1) = 2$. Had there been more groups of ties, we would have summed the T's.

$$\tau = \frac{104}{\sqrt{[15(14)][15(14) - 2]}} = \frac{104}{\sqrt{210(208)}} = \frac{104}{\sqrt{43{,}680}}$$

$$= \frac{104}{208.9976} = 0.4976$$

If there are no ties, the equation can be simplified to

$$\tau = \frac{N}{n(n-1)}$$

BOX 12.3 continued

5. To test significance for sample sizes >10, we can make use of a normal approximation to test the null hypothesis that the true value of $\tau = 0$:

$$t_s = \frac{\tau}{\sqrt{2(2n+5)/9n(n-1)}} = \frac{0.4976}{\sqrt{2[2(15)+5]/9(15)(14)}}$$

$$= \frac{0.4976}{\sqrt{70/1890}} = \frac{0.4976}{0.19253} = 2.59 \qquad \text{compared with } t_{\alpha[\infty]}$$

When this value is looked up in Table **II** (areas of the normal curve), we find the probability of such a t_s arising by chance to be 0.0096 (both tails).

When n is ≤ 10, the approximation given above is not adequate and the special table given below must be used. The table gives the 5 and 1% (two-tailed) critical values for N (the numerator of τ) for $n = 5$ to 10. These are exact only if there are no ties. If there are ties, a special table must be consulted (see Burr, 1960).

Critical value of N such that $P(N_s > N_\alpha) \leq \alpha$ where N_s and N_α refer to sample estimates and critical values of N, respectively. This table is based on Table Q in Siegel (1956).

	N_α	
n	$\alpha = 0.05$	$\alpha = 0.01$
5	20	—
6	26	30
7	30	38
8	36	44
9	40	52
10	46	58

$$\begin{array}{ll} Y_1 & 1\ 2\ 3\ 4\ 5 \\ Y_2 & 1\ 3\ 2\ 5\ 4 \end{array}$$

Note that the ranking by variable Y_2 is not totally concordant with that by Y_1. The technique employed in Box 12.3 is to count the number of higher ranks following any given rank, sum this quantity for all ranks, multiply the sum $\sum^n C_i$ by four, and subtract from it a correction factor $n(n-1)$ to obtain a statistic N. For variable Y_1 we find $\sum^n C_i = 4 + 3 + 2 + 1 + 0 = 10$, then compute $N = 4\sum^n C_i - n(n-1) = 40 - 5(4) = 20$, to obtain the maximum possible score $N = n(n-1) = 20$. Obviously, Y_1 being ordered is always perfectly concordant with itself. However, for Y_2 we only obtain $\sum^n C_i = 4 + 2 + 2 + 0 + 0 = 8$, and $N = 4(8) - 5(4) = 12$. Since the maximum score of N is $n(n-1) = 20$ and the observed score 12, an obvious coefficient suggests itself as $N/n(n-1) = [4\sum^{n} C_i - n(n-1)]/n(n-1) = 12/20 = 0.6$. Ties present minor complications that are dealt with in Box 12.3. The correlation in that box is between total body size of aphid stem mothers and mean thorax length of their offspring. In this case, there was no special need to turn to rank correlation except that there is some evidence that these data are bimodal and not normally distributed. The significance of τ for sample sizes greater than 10 can easily be tested by a standard error shown in Box 12.3. Where sample size is less than 10, look up critical values of N at the end of Box 12.3.

Exercises 12

12.1 Graph the following data in the form of a bivariate scatter diagram. Compute the correlation coefficient and set 95% confidence intervals to ρ. The data were collected for a study of geographic variation in the aphid *Pemphigus populi-transversus*. The values in the table represent locality means based on equal sample sizes for 23 localities in eastern North America. The variables, extracted from Sokal and Thomas (1965), are expressed in millimeters. Y_1 = tibia length, Y_2 = tarsus length. The correlation coefficient will estimate correlation of these two variables over localities.

Locality code number	Y_1	Y_2
1	0.631	0.140
2	.644	.139
3	.612	.140
4	.632	.141
5	.675	.155
6	.653	.148
7	.655	.146
8	.615	.136
9	.712	.159
10	.626	.140
11	.597	.133
12	.625	.144
13	.657	.147
14	.586	.134
15	.574	.134
16	.551	.127
17	.556	.130
18	.665	.147
19	.585	.138
20	.629	.150
21	.671	.148
22	.703	.151
23	.662	.142

ANS. $r = 0.910$.

12.2 The following data were extracted from a larger study by Brower (1959) on speciation in a group of swallowtail butterflies. Morphological measurements are in millimeters coded $\times$ 8.

Species	Specimen number	Y_1 Length of 8th tergile	Y_2 Length of superuncus
Papilio multicaudatus	1	24.0	14.0
	2	21.0	15.0
	3	20.0	17.5
	4	21.5	16.5
	5	21.5	16.0
	6	25.5	16.0
	7	25.5	17.5
	8	28.5	16.5
	9	23.5	15.0
	10	22.0	15.5
	11	22.5	17.5
	12	20.5	19.0
	13	21.0	13.5
	14	19.5	19.0
	15	26.0	18.0
	16	23.0	17.0
	17	21.0	18.0
	18	21.0	17.0
	19	20.5	16.0
	20	22.5	15.5
Papilio rutulus	21	20.0	11.5
	22	21.5	11.0
	23	18.5	10.0
	24	20.0	11.0
	25	19.0	11.0
	26	02.5	11.0
	27	19.5	11.0
	28	19.0	10.5
	29	21.5	11.0
	30	20.0	11.5
	31	21.5	10.0
	32	20.5	12.0
	33	20.0	10.5
	34	21.5	12.5
	35	17.5	12.0
	36	21.0	12.5
	37	21.0	11.5
	38	21.0	12.0
	39	19.5	10.5
	40	19.0	11.0
	41	18.0	11.5
	42	21.5	10.5
	43	23.0	11.0
	44	22.5	11.5
	45	19.0	13.0
	46	22.5	14.0
	47	21.0	12.5

Compute the correlation coefficient separately for each species and test significance of each. Test whether the two correlation coefficients differ significantly.

12.3 Test for the presence of association between tibia length and tarsus length in the data of Exercise 12.1 using Kendall's coefficient of rank correlation.

12.4 The following table of data is from an unpublished morphometric study of the cottonwood, *Populus deltoides*, by T. J. Crovello. Twenty-six leaves from one tree were measured when fresh and after drying. The variables shown are fresh leaf width (Y_1) and dry leaf width (Y_2), both in millimeters. Calculate r and test its significance.

Y_1	Y_2
90	88
88	87
55	52
100	95
86	83
90	88
82	77
78	75
115	109
100	95
110	105
84	78
76	71
100	97
110	105
95	90
99	98
104	100
92	92
80	82
110	106
105	97
101	98
95	91
80	76
103	97

13

Analysis of Frequencies

Almost all our work so far has dealt with estimation of parameters and tests of hypotheses in continuous variables. The present chapter treats an important class of cases, tests of hypotheses about frequencies. Biological variables may be distributed into two or more classes, depending on some criterion such as arbitrary class limits in a continuous variable or a set of mutually exclusive attributes. An example of the former would be a frequency distribution of birth weights (a continuous variable arbitrarily divided into a number of contiguous classes); one of the latter would be a qualitative frequency distribution such as the frequency of individuals of ten different species obtained from a soil sample. For any such distribution we may hypothesize that it was sampled from a population in which the frequencies of the various classes represent certain parametric proportions of the total frequency. We need a test of goodness of fit for our observed frequency distribution to the expected frequency distribution representing our hypothesis. You may recall that we first realized the need for such a test in Chapters

4 and 5, where we calculated expected binomial, Poisson, and normal frequency distributions but were unable to decide whether an observed sample distribution departed significantly from the theoretical one.

In Section 13.1 we introduce the idea of goodness of fit, discuss the types of significance tests that are appropriate, the basic rationale behind such tests, and develop general computational formulas for these tests.

Section 13.2 illustrates the actual computations for goodness of fit when the data are arranged by a single criterion of classification, as in a one-way quantitative or qualitative frequency distribution. This applies to cases expected to follow one of the well-known frequency distributions such as the binomial, Poisson, or normal distribution, as well as to expected distributions following some other law suggested by the scientific subject matter under investigation, such as, for example, tests of goodness of fit of observed genetic ratios against expected Mendelian frequencies.

In Section 13.3 we proceed to significance tests of frequencies in two-way classifications—called tests of independence. We shall discuss the common tests of 2×2 tables in which each of two criteria of classification divides the frequencies into two classes, yielding a four-cell table, as well as $R \times C$ tables with more rows and columns.

Throughout this chapter we carry out goodness of fit tests by the G-statistic. We briefly mention chi-square tests, but as is explained at various places throughout the text, G has general theoretical advantages over X^2, as well as being computationally simpler for tests of independence.

13.1 Tests for goodness of fit: introduction

The basic idea of a goodness of fit test is easily understood, given the extensive experience you now have with statistical hypothesis testing. Let us assume that a geneticist has carried out a crossing experiment between two F_1 hybrids and obtains an F_2 progeny of 90 offspring, 80 of which appear to be wild type and 10 are mutants. The geneticist assumes dominance and expects a 3:1 ratio of the phenotypes. When we calculate the actual ratios, however, we observe that the data are in a ratio $80/10 = 8{:}1$. Expected values for p and q are $\hat{p} = 0.75$ and $\hat{q} = 0.25$ for the wild type and mutant, respectively. Note that we use the caret (generally called "hat" in statistics) to indicate hypothetical or expected values of the binomial proportions. However, the observed proportions of these two classes are $p = 0.89$ and $q = 0.11$, respectively. Yet another way of noting the contrast between observation and expectation is to state it in frequencies: the observed frequencies are 80 and 10 for the two phenotypes. Expected frequencies should be $\hat{f}_1 = \hat{p}n = 0.75(90) = 67.5$ and $\hat{f}_2 = \hat{q}n = 0.25(90) = 22.5$, respectively, where n refers to the sample size of offspring from the cross. Note that when we sum the expected frequencies they yield $67.5 + 22.5 = n = 90$, as they should.

The obvious question that comes to mind is whether the deviation from the 3:1 hypothesis observed in our sample is of such a magnitude as to be improbable. In other words, do the observed data differ enough from the expected values to cause us to reject the null hypothesis? For the case just considered, you already know two methods for coming to a decision about the null hypothesis. Clearly, this is a binomial distribution in which p is the probability of being a wild type and q is the probability of being a mutant. It is possible to work out the probability of obtaining an outcome of 80 wild type and 10 mutants as well as all "worse" cases for $\hat{p} = 0.75$ and $\hat{q} = 0.25$, and a sample of $n = 90$ offspring. We use the conventional binomial expression here $(\hat{p} + \hat{q})^n$ except that p and q are hypothesized, and we replace the symbol k by n, which we adopted in Chapter 4 as the appropriate symbol for the sum of all the frequencies in a frequency distribution. In this example, we have only one sample, so what would ordinarily be labeled k in the binomial is, at the same time, n. Such a problem was illustrated in Table 4.3 and Section 4.2, and we can compute the cumulative probability of the tail of the binomial distribution. When this is done, we obtain a probability of 0.00085 for all outcomes as deviant or more deviant from the hypothesis. Note that this is a one-tailed test, the alternative hypothesis being that there are, in fact, more wild type offspring than the Mendelian hypothesis would postulate. Assuming $\hat{p} = 0.75$ and $\hat{q} = 0.25$, the observed sample is, consequently, a very unusual outcome, and we conclude that there is a significant deviation from expectation.

A less time-consuming approach based on the same principle is to look up confidence limits for the binomial proportions as was done for the sign test in Section 10.3. Interpolation in Table **IX** shows that for a sample of $n = 90$, an observed percentage of 89% would yield approximate 99% confidence limits of 78 and 96 for the true percentage of wild type individuals. Clearly, the hypothesized value of $\hat{p} = 0.75$ is beyond the 99% confidence bounds.

Now, let us develop a third approach by a goodness of fit test. Table 13.1 illustrates how we might proceed. In the first column are given the observed frequencies f representing the outcome of the experiment. Column (2) shows the expected frequencies $\hat{f}$ based on the particular hypothesis being tested. In this case, the hypothesis is a 3:1 ratio and we have already calculated the expected frequencies under these conditions as $\hat{f}_1 = \hat{p}n = 0.75(90) = 67.5$ and $\hat{f}_2 = \hat{q}n = 0.25(90) = 22.5$.

How can we develop a statistic for testing to what degree the observed frequencies in column (1) differ from the expected frequencies in column (2)? The following test statistic is easily understood and its structure makes intuitive sense. We first measure $f - \hat{f}$, the deviation of observed from expected frequencies. Note that the sum of these deviations equals zero, for reasons very similar to those causing the sum of deviations from a mean to add to zero. Hence in an example with two classes, the deviations are always

TABLE 13.1

Developing the chi-square test for goodness of fit. Observed and expected frequencies from the outcome of a genetic cross, assuming a 3:1 ratio of phenotypes among the offspring.

	(1)	*(2)*	*(3)*	*(4)*	*(5)*
Phenotypes	*Observed frequencies* f	*Expected frequencies* $\hat{f}$	*Deviations from expectation* $f - \hat{f}$	*Deviations squared* $(f - \hat{f})^2$	$\frac{(f - \hat{f})^2}{\hat{f}}$
Wild type	80	$\hat{p}n = 67.5$	12.5	156.25	2.315
Mutant	10	$\hat{q}n = 22.5$	−12.5	156.25	6.944
Sum	90	90.0	0		$X^2 = 9.259$

equal and opposite in sign. Following our previous approach of making all deviations positive by squaring them, we square $(f - \hat{f})$ in column (4) to yield a measure of the magnitude of the deviation from expectation. This quantity must be expressed as a proportion of the expected frequency. After all, if the expected frequency were 13.0, a deviation of 12.5 would be an extremely large one, comprising almost 100% of $\hat{f}$, but such a deviation would only represent 10% of an expected frequency of 125.0. Thus, we obtain column (5) as the quotient of the quantity in column (4) divided by that in column (2). Note that the magnitude of the quotient is greater for the second line, in which the $\hat{f}$ is smaller. The next step in developing our test statistic is to sum these quotients, which is done at the foot of column (5), yielding a value of 9.259.

What shall we call this new statistic? We have some nomenclatural problems here. Many of you will have recognized this as the so-called *chi-square test,* regularly taught to beginning genetics classes. The name of the test is too well established to make a change seem practical, but, in fact, this quantity just computed, the sum of column (5), could not possibly be a χ^2. The latter is a continuous and theoretical frequency distribution, while our quantity 9.259 is a sample statistic based on discrete frequencies. The latter point is easily seen if you visualize other possible outcomes. For instance, we could have had as few as zero mutants matched by 90 wild type individuals (assuming that the total number of offspring $n = 90$ remains constant), or we could have had 1, 2, 3, or more mutants, in each case balanced by the correct number of wild type offspring to yield a total of 90. Observed frequencies change in unit increments, and since the expected frequencies remain constant, it is clear that deviations, their squares, and the quotients are not continuous variables but can assume only certain values.

The reason why this test has been called the chi-square test and why many persons call the statistic obtained as the sum of column (5) a chi-square, is that the sampling distribution of this sum approximates that of a chi-square distribution with one degree of freedom. We can appreciate the reason for the single degree of freedom when we consider the frequencies in the two classes of the table and their sum, $80 + 10 = 90$. In such an example, the total frequency is fixed. Therefore, if we were to vary the frequency of any one class, the other class would have to compensate for changes in the first class to retain a correct total. If the first class had a frequency of 75, the second class would have to contain 15 items to make the total 90. Thus, only one class is able to vary freely, the other class being constrained by the constant sum. Here, the meaning of *one degree of freedom* becomes quite clear. One of the classes is free to vary; the other is not. However, since the sample statistic is not a chi-square, we have followed the increasingly prevalent convention of labeling the sample statistic X^2 rather than χ^2. The value of $X^2 = 9.259$ from Table 13.1, when compared with the critical value of χ^2 (Table **IV**), is highly significant ($P < 0.005$). The chi-square test is always one-tailed. Since the deviations are squared, negative and positive deviations both result in positive values of X^2. Clearly, we reject the 3:1 hypothesis and conclude that the proportion of wild type is greater than 0.75. The geneticist must, consequently, look for a mechanism explaining this departure from expectation.

The test for goodness of fit can be applied to a distribution with more than two classes. The example in part 1 of Box 13.1 is a complicated genetic cross in which an 18:6:6:2:12:4:12:4 ratio is expected. The table is set up as before with expected frequencies calculated as $\hat{f}_i = \hat{p}_i n$, where the values of $\hat{p}_i$ are the expected probabilities for the $a = 8$ classes. Clearly, $\sum^{a} \hat{p}_i = 1$, since the probabilities of all possible outcomes sum to one. The values of $\hat{p}_i$ for this example are $\hat{p}_1 = \frac{18}{64}$, $\hat{p}_2 = \frac{6}{64}$, and so forth. Again we calculate deviations from expectation, square them, and divide them by expected frequencies. The operation can be described by the formula

$$X^2 = \sum^{a} \frac{(f_i - \hat{f}_i)^2}{\hat{f}_i} \tag{13.1}$$

However, its use is tedious and requires the computation of deviations, their squaring, and their division by expected frequencies over all of the classes. A generally applicable computational formula for X^2 can be easily derived from Expression (13.1), as shown in Appendix A1.10. It is

$$X^2 = \sum^{a} \frac{f_i^2}{\hat{f}_i} - n \tag{13.2}$$

which can be easily obtained as the sum of the quotients of the squares of

BOX 13.1

Tests for goodness of fit. Single classification, expected frequencies based on hypothesis extrinsic to the sampled data.

1. Frequencies divided into $a \geq 2$ classes.

In a genetic experiment involving a cross between two varieties of the bean *Phaseolus vulgaris,* Smith (1939) obtained the following results:

Phenotypes ($a = 8$)	*Observed frequencies* f	*Expected frequencies* $\hat{f}$
Purple/buff	63	67.8
Purple/testaceous	31	22.6
Red/buff	28	22.6
Red/testaceous	12	7.5
Purple	39	45.2
Oxblood red	16	15.1
Buff	40	45.2
Testaceous	12	15.1
Total	241	241.1

The expected frequencies $\hat{f}_i$ were computed on the basis of the expected ratio of 18:6:6:2:12:4:12:4. Compute $\hat{f}_i$ as $\hat{p}_i n$. Thus, $\hat{f}_1 = \hat{p}_1 n = (\frac{18}{64}) \times 241 = 67.8$.

Employ Expressions (13.4a) when $\hat{f}$'s have already been calculated or (13.5) when $\hat{p}$'s are given. Since we have presented the $\hat{f}$'s in the bean example given above we shall use Expression (13.4a):

$$G = 2[\sum^a f_i \ln f_i - 2.30259 \sum^a f_i \log \hat{f}_i]$$

Using Table **XI** we find

$$\sum^a f_i \ln f_i = 63 \ln 63 + \cdots + 12 \ln 12 = 261.017 + \cdots + 29.819$$
$$= 855.206$$

Using a table of logarithms, we can compute

$$\sum^a f_i \log \hat{f}_i = 63 \log 67.8 + \cdots + 12 \log 15.1$$
$$= 63(1.83123) + \cdots + 12(1.17898) = 369.5282$$

G can then be computed as

$$G = 2[\sum^a f_i \ln f_i - 2.30259 \sum^a f_i \log \hat{f}_i]$$
$$= 2[855.206 - 2.30259(369.5282)] = 2[4.334] = 8.668$$

Since our observed $G = 8.668 < \chi^2_{.05[7]} = 14.067$, we may consider the data consistent with the null hypothesis

BOX 13.1 continued

2. Special case of frequencies divided into $a = 2$ classes

In an F_2 cross in drosophila the following 176 progeny were obtained, of which 130 were wild type flies and 46 ebony mutants. Assuming that the mutant is an autosomal recessive, one would expect a ratio of 3 wild type flies to each mutant fly.

An additional complication is the fact that a continuity correction should be applied when n is less than about 200. The computations are carried out as described in part **1** above, but using adjusted f's instead of the original f's if this correction is to be applied (see text). We now apply the G-test to the above drosophila data.

Flies	f	*Adjusted* f	*Hypothesis*	$\hat{f}$
Wild type	130	130.5	$\hat{p} = 0.75$	$\hat{p}n = 132.0$
Ebony mutant	46	45.5	$\hat{q} = 0.25$	$\hat{q}n = 44.0$
	176	176.0		176.0

Using Table **XII**, and adjusted frequencies f_i,

$$\sum^{a} f_i \ln f_i = 130.5 \ln 130.5 + 45.5 \ln 45.5 = 635.714 + 173.706$$
$$= 809.420$$

Using a table of logarithms,

$$\sum^{a} f_i \log \hat{f}_i = 130.5 \log 132.0 + 45.5 \log 44.0$$
$$= 130.5(2.12057) + 45.5(1.64345) = 351.5114$$
$$G_{\text{adj}} = 2[809.420 - 2.30259(351.5114)] = 0.0668$$

which is obviously not significant when compared with $\chi^2_{.05[1]}$.

the observed frequencies divided by their expected frequencies. From this sum of quotients is subtracted n, the sum of all the frequencies.

What can we conclude about our test for goodness of fit? If we assume for the moment that X^2 in this case is also distributed approximately as χ^2, we need to know how many degrees of freedom there are in this example to enable us to compare it with the appropriate χ^2-distribution. In this case with eight classes, any seven of them can vary freely; but the eighth class must constitute the difference between the total sum and the sum of the first seven. Thus, in an eight-class case we have seven degrees of freedom and, in general, when we have a classes, we have $a - 1$ degrees of freedom. In the example in part **1** Box 13.1, application of Expression (13.2) leads to

$X^2 = 9.4909$, which is less than $\chi^2_{.05[7]} = 14.067$. Thus the observed frequencies are compatible with the postulated ratios.

In recent years a new test of goodness of fit has come into wide use. It is the *G-test* based on the likelihood ratio statistic. It has a number of distinct advantages over the older chi-square test and we have therefore featured it exclusively in this chapter in place of the older method.

Let us reconsider the example of Table 13.1. Using Expression (4.1) for the expected relative frequencies in a binomial distribution, we can compute two quantities that are of interest to us here:

$$C(90, 80)\left(\frac{80}{90}\right)^{80}\left(\frac{10}{90}\right)^{10} = 0.1327$$

$$C(90, 80)\left(\frac{3}{4}\right)^{80}\left(\frac{1}{4}\right)^{10} = 0.0005514$$

The first quantity is the probability of observing the sampled results (80 wild type and 10 mutants) on the hypothesis that $\hat{p} = p$—that is, that the population parameter equals the observed sample proportion. The second is the probability of observing the sampled results assuming that $\hat{p} = \frac{3}{4}$, according to the Mendelian null hypothesis. Note that these expressions yield the probabilities for the observed outcomes only, *not for observed and all worse outcomes*. Thus, $P = 0.0005514$ is less than the earlier computed $P = 0.00085$, which is the probability of 10 *and fewer* mutants, assuming $\hat{p} = \frac{3}{4}$, $\hat{q} = \frac{1}{4}$.

The first probability (0.1327) is greater than the second (0.0005514), since the parameter is based on the observed data. If the observed proportion p is in fact equal to the proportion $\hat{p}$ postulated under the null hypothesis, then the two computed probabilities will be equal and their ratio, L, will equal 1.0. The greater the difference between p and $\hat{p}$ (the expected proportion under the null hypothesis), the higher the ratio will be (the probability based on p is divided by the probability based on $\hat{p}$ or defined by the null hypothesis). This indicates that the ratio of these two probabilities or *likelihoods* can be used as a statistic to measure the degree of agreement between sampled and expected frequencies. A test based on such a ratio is called a *likelihood ratio test*. In our case, $L = 0.1327/0.0005514 = 240.66$. The theoretical distribution of this ratio is, in general, complex and poorly known. However, it has been shown that the distribution of

$$G = 2 \ln L = 2 (\ln 10) \log L \tag{13.3}$$

can be approximated by the χ^2-distribution when sample sizes are large (for a definition of "large" in this case, see below). The appropriate degrees of freedom for a given test are the same as for the chi-square tests discussed above. In our case,

$$G = 2 \ln L = 2(\ln 10) \log L = 2(2.302585)(2.38140) = 2(5.48356) = 10.967$$

If we compare this observed value with a χ^2-distribution with one degree of freedom, we find that the result is significant ($P < 0.005$), as was found before for the chi-square test. In general, G will be numerically rather similar to X^2. Notation for the *log likelihood ratio test* is as little standardized as that for the chi-square test. The symbol $2I$ is sometimes used for G.

For developing a computational formula, Expression (13.3) for the G-statistic can be rewritten in various ways, depending upon the particular application. In Appendix A1.11 it is shown that for a general goodness of fit test it can be written as follows:

$$G = 2\sum^{a} f_i \ln\left(\frac{f_i}{\hat{f}_i}\right) \tag{13.4}$$

If an electronic calculator with a log key is available this is by far the simplest formula. In other cases it can be expressed as follows to simplify computation

$$\begin{aligned} G &= 2\left[\sum^{a} f_i \ln f_i - \sum^{a} f_i \ln \hat{f}_i\right] \\ &= 2\left[\sum^{a} f_i \ln f_i - (2.30259)\sum^{a} f_i \log \hat{f}_i\right] \end{aligned} \tag{13.4a}$$

which is convenient to use with Table **XI**. Table **XI** gives $f \ln f$ for integral values of f between 0 and 10,000. Thus, the quantities $f \ln f$ or $n \ln n$ can be looked up directly and accumulated on an adding machine. Since there is a chance of miscopying information from the tables, it is most convenient to use a printing calculator so that one need only check the printed record against the table in order to verify the results. If errors are found it is simple to correct the final results without having to repeat all of the computations. Another frequently used computational formula for G is

$$G = 2\left[\sum^{a} f_i \ln f_i - \sum^{a} f_i \ln \hat{p}_i - n \ln n\right] \tag{13.5}$$

also derived in Appendix A1.11.

13.2 Single classification goodness of fit tests

In Box 13.1 we illustrate G-tests of goodness of fit in those cases in which expected frequencies are based on a hypothesis extrinsic to the data—there are a classes and the expected proportions in each class are assumed on the basis of outside knowledge and are not functions of parameters estimated from the sample. We start with the general case for any number of classes, where the number of classes is symbolized by a, to emphasize the analogy with analysis of variance. The data are the results of a complicated genetic cross expected to result in an 18:6:6:2:12:4:12:4 ratio. A total of 241 progeny was obtained and allocated to the eight phenotypic classes. The expected frequencies can be simply computed by multiplying the total sample size n times the expected probabilities of occurrence. We note that the overall fit is quite good. This is an example of a test for goodness of fit with an extrinsic hypothesis because the genetic ratio tested is based on considerations prior and external to the sample observations tested in the example. The computation is quite simple, as indicated in the box.

We learned in the previous section that the value of G is to be compared with a critical value of χ^2 for $a - 1$ degrees of freedom. When we compare our result with the χ^2-distribution, we find that the value of G obtained from our sample is not significant. We do not have sufficient evidence to reject the null hypothesis and are led to conclude that the sample is consistent with the specified genetic ratio. However, especially in examples such as the one just analyzed, remember that we have not specified an alternative hypothesis; there are a variety of alternative hypotheses that also could not be excluded if we were to carry out a significance test for them. We have not really proven that the data are distributed as specified, and there are a variety of other genetic ratio hypotheses that could also be plausible.

No general directions are found in the literature regarding how small a sample may be and still be suitable for G- or X^2-tests of goodness of fit. However, caution should be exercised with sample sizes of n less than 50 in interpreting the results.

The example in part **2** of Box 13.1 is a monohybrid cross with an expected ratio of 3 wild type to 1 mutant. In tests of goodness of fit involving only two classes, the value of G as computed from Expressions (13.4) or (13.5) will exhibit a bias that can be modified by applying a *correction* for *continuity*, making the value of G approximate the χ^2-distribution more closely. This correction consists of adding or subtracting 0.5 from the observed frequencies in such a way as to minimize the value of G. One simply adjusts the f_i changing them to reduce the difference between them and the corresponding expected frequencies by one half. One then employs Expressions (13.4a) or (13.5) as before to obtain G_{adj}. Values of $(f + \frac{1}{2}) \ln (f + \frac{1}{2})$ necessary for the computations of G_{adj} can be found in Table **XII**. The correction for continuity is applied whenever $n < 200$. When $n < 25$ even this correction is insufficient to adjust the bias. An exact computation of the binomial probabilities in the manner of Table 4.3 is then indicated. The low value of G_{adj} found in Box 13.1 shows that the observed data fit closely to the expected ratios.

In some goodness of fit tests, we subtract more than one degree of freedom from the number of classes, a. These are instances where the parameters for the null hypothesis have been extracted from the sample data themselves, in contrast with the null hypotheses encountered so far (in Table 13.1 and Box 13.1). In these cases, the hypothesis to be tested was generated on the basis of the investigator's general knowledge of the specific problem and of Mendelian genetics. For this reason, the expected frequencies are based on an *extrinsic hypothesis*, a hypothesis external to the data. By contrast, consider the expected Poisson frequencies of yeast cells in a hemacytometer (Box 4.1). You will recall that to compute these frequencies, you needed values for μ, which you estimated from the sample mean $\bar{Y}$. Therefore the parameter of the computed Poisson distribution came from the sampled observations themselves. The expected Poisson frequencies represent an *intrinsic hypothesis*. In such a case, to obtain the correct number of degrees of freedom for the test of goodness of fit (chi-square or G) we would subtract from a, the number of classes into which the data had been grouped, not only one degree of freedom for n, the sum of the frequencies, but also one further degree of freedom for the estimate of the mean. Thus, in such a case, a sample statistic G would be compared with chi-square for $a - 2$ degrees of freedom.

When we apply the G-test to the observed and expected frequencies of Box 4.1, using whichever formula in Box 13.1 is most convenient from a computational point of view, we obtain $G = 7.529$. Two aspects make this computation different from that in part 1 of Box 13.1. As a general rule, we avoid expected frequencies less than 5. Therefore, the classes of $\hat{f}_i$ at the upper tail of the distribution are too small. We lump them by adding their frequencies to those in contiguous classes as shown in Box 4.1. Clearly, the observed frequencies must be lumped to match. The number of classes a is the number *after* lumping has taken place. In our case, $a = 6$.

The other new feature has already been discussed. It is the number of degrees of freedom considered for the significance test. We always subtract one degree of freedom for the fixed sum (in this case $n = 400$). However, we subtract an additional degree of freedom for every parameter of the expected frequency distribution estimated from the sampled distribution. In this case, we estimated μ from the sample and, therefore, a second degree of freedom is subtracted from a, making the final number of degrees of freedom $a - 2 = 6 - 2 = 4$. Comparing the sample value of $G = 7.529$ with the critical value of χ^2 at four degrees of freedom, we find it not significant. We therefore accept the null hypothesis and conclude that the yeast cells are randomly distributed.

The G-test for testing the goodness of fit of a set of data to an expected frequency distribution can be applied not only to the Poisson, but to the normal, binomial, and other distributions as well. For a normal distribution we customarily estimate two parameters μ and σ from the sampled data. Hence the appropriate degrees of freedom are $a - 3$. In the binomial, only one parameter, $\hat{p}$, must be estimated; the appropriate degrees of freedom are $a - 2$.

13.3 Tests of independence: two-way tables

The notion of statistical or probabilistic independence was first introduced in Section 4.1, where it was shown that if two events were independent, the probability of their occurring together could be computed as the product of their separate probabilities. Thus, if among the progeny of a certain genetic cross the probability of a kernel of corn being red is $\frac{1}{2}$ and the probability of the kernel being dented is $\frac{1}{3}$, the probability of obtaining a dented and red kernel would be $\frac{1}{2} \times \frac{1}{3} = \frac{1}{6}$, if the joint occurrences of these two characteristics are statistically independent.

The appropriate statistical test for this genetic problem would be to test the frequencies for goodness of fit to the expected ratios of 2 (red, not dented):2 (not red, not dented):1 (red, dented):1 (not red, dented). This would be a simultaneous test of two null hypotheses: that the expected proportions are $\frac{1}{2}$ and $\frac{1}{3}$ for red and dented, respectively, and that these two properties are independent. The first null hypothesis tests the Mendelian model in general. The second tests whether these characters assort independently—that is, whether they are determined by genes located in different linkage groups. If the second hypothesis must be rejected, this is taken as evidence that the characters are linked—that is, located on the same chromosome.

There are numerous instances in biology in which the second hypothesis concerning the independence of two properties is of great interest and the first hypothesis regarding the true proportion of one or both properties is of little interest. In fact, often no hypothesis regarding the parametric values $\hat{p}_i$ can be formulated by the investigator. We shall cite several examples of such situations, which lead to the test of independence to be learned in this section. We employ this test whenever we wish to test whether two different properties, each occurring in two states, are dependent on each other. For instance, specimens of a certain moth may occur in two color phases—light and dark. Fifty specimens of each phase may be exposed in the open, subject to predation by birds. The number of surviving moths is counted after a fixed interval of time. The proportion predated may differ in the two color phases. The two properties in this example are color and survival. We can divide our sample into four classes: light-colored survivors, light-colored prey, dark survivors, and dark prey. If the probability of being preyed upon is independent of the color of the moth, the expected frequencies of these four classes can be simply computed as independent products of the proportion of each color (in our experiment $\frac{1}{2}$) and the overall proportion preyed upon in the entire sample. Should the statistical test of independence explained below show that the two properties are not independent, we are led to conclude that one of the color phases is more susceptible to predation than the other. This is an important biological phenomenon; the exact proportions of the two properties are of little interest here. The proportion of the color phases is arbitrary, and the proportion survival is of interest only insofar as it differs for the two phases.

A second example might relate to a sampling experiment carried out by a plant ecologist. He obtains a random sample of 100 individuals of a fairly rare species of tree distributed over an area of 400 square miles. For each tree he notes whether it is rooted in a serpentine soil, or not, and whether the leaves are pubescent, or smooth. Thus the sample of $n = 100$ trees can be divided into four groups: serpentine-pubescent, serpentine-smooth, nonserpentine-pubescent, and nonserpentine-smooth. If the probability of a tree being pubescent or not is independent of its location, our null hypothesis of the independence of these properties will be upheld. If, on the other hand, the proportion of pubescence differs for the two types of soils, our statistical test will most probably result in rejection of the null hypothesis of independence. Again, the expected frequencies will simply be products of the independent proportions of the two properties—serpentine versus nonserpentine, and pubescent versus smooth. In this instance the proportions may themselves be of interest to the investigator.

The example we shall work out in detail is from immunology. A sample of 111 mice was divided into two groups, 57 that received a standard dose of pathogenic bacteria followed by an antiserum and a control group of 54 that received the bacteria, but no antiserum. After sufficient time had elapsed for an incubation period and for the disease to run its course, 38 dead mice and 73 survivors were counted. Of those that died, 13 had received bacteria *and* antiserum while 25 had received bacteria only. A question of interest is whether the antiserum had in any way protected the mice so that there were proportionally more survivors in that group. Here again the proportions of these properties are of no more interest than in the first example (predation on moths).

Such data are conveniently displayed in the form of a *two-way table* as shown below. Two-way and multiway tables (more than two criteria) are often known as *contingency tables*. This type of two-way table, in which each of the two criteria is divided into two classes, is known as a 2×2 *table*.

	Dead	*Alive*	Σ
Bacteria and antiserum	13	44	57
Bacteria only	25	29	54
Σ	38	73	111

Thus 13 mice received bacteria and antiserum but died, as seen in the table. The marginal totals give the number of mice exhibiting any one property: 57 mice received bacteria and antiserum; 73 mice survived the experiment. Altogether 111 mice were involved in the experiment and constitute the total sample.

In discussing such a table it is convenient to label the cells of the table and the row and column sums as follows:

a	b	$a + b$
c	d	$c + d$
$a + c$	$b + d$	n

From a two-way table one can systematically compute the expected frequencies (based on the null hypothesis of independence) and compare them with the observed frequencies. For example, the expected frequency for cell d (bacteria, alive) would be

$$\hat{f}_{\text{bact,alv}} = n\hat{p}_{\text{bact,alv}} = n\hat{p}_{\text{bact}} \times \hat{p}_{\text{alv}} = n\left(\frac{c+d}{n}\right)\left(\frac{b+d}{n}\right) = (c+d)(b+d)/n$$

which in our case would be $(54)(73)/111 = 35.5$, a higher value than the observed frequency of 29. We can proceed similarly to *compute the expected frequencies for each cell in the table by multiplying a row total times a column total, and dividing the product by the grand total.* The expected frequencies can be conveniently displayed in the form of a two-way table:

	Dead	*Alive*	Σ
Bacteria and antiserum	19.5	37.5	57.0
Bacteria only	18.5	35.5	54.0
Σ	38.0	73.0	111.0

You will note that the row and column sums of this table are identical to those in the table of observed frequencies, which should not surprise you since the expected frequencies were computed on the basis of these row and column totals. It should therefore be clear that a test of independence will not test whether any property occurs at a given proportion but can only test whether or not the two properties are manifested independently.

The statistical test appropriate to a given 2×2 table depends on the underlying model that it represents. There has been considerable confusion on this subject in the statistical literature. For our purposes here it is not necessary to distinguish among the three models of contingency tables. The G-test illustrated below will give at least approximately correct results with moderately to large sized samples regardless of the underlying model. One could also carry out a chi-square test on the deviations of the observed from the expected frequencies using Expression (13.2). This would yield $X^2 = 6.767$, using expected frequencies rounded to one decimal place. Let us state without explanation that the observed X^2 should be compared with χ^2 for one degree of freedom. We shall examine the reasons for this at the end of this section. The probability of finding a fit as bad, or worse, to these data is $0.005 < P < 0.01$. We conclude, therefore, that mortality in these mice is not independent of the presence of antiserum. We note that the percentage mortality among those animals given bacteria *and* antiserum is $(13)(100)/57 = 22.8\%$, considerably lower than the mortality of $(25)(100)/54 = 46.3\%$ among the mice to whom only bacteria had been administered. Clearly, the antiserum has been effective in reducing mortality.

In Box 13.2 we illustrate the G-test applied to the sampling experiment in plant ecology, dealing with trees rooted in two different soils and possessing two types of leaves. With small sample sizes ($n < 200$) we again apply a *correction for continuity* (*Yates' Correction*), the application of which is shown in the box. The result of the analysis shows clearly that we cannot reject the

BOX 13.2

2 × 2 test of independence.

A plant ecologist samples 100 trees of a rare species from a 400-square-mile area. He records for each tree whether it is rooted in serpentine soils, or not, and whether its leaves are pubescent or smooth.

Soil	*Pubescent*	*Smooth*	*Totals*
Serpentine	12	22	34
Not Serpentine	16	50	66
Totals	28	72	$100 = n$

The conventional algebraic representation of this table is as follows:

			Σ
	a	b	$a + b$
	c	d	$c + d$
Σ	$a + c$	$b + d$	$a + b + c + d = n$

If $ad - bc$ is positive, as it is in our example, since $(12 \times 50) - (16 \times 22) = 248$, subtract $\frac{1}{2}$ from a and d and add $\frac{1}{2}$ to b and c. If $ad - bc$ is negative, add $\frac{1}{2}$ to a and d and subtract $\frac{1}{2}$ from b and c. This is Yates' correction and may be ignored when $n > 200$. The new 2×2 table looks as follows.

Soil	*Pubescent*	*Smooth*	*Totals*
Serpentine	$11\frac{1}{2}$	$22\frac{1}{2}$	34
Not serpentine	$16\frac{1}{2}$	$49\frac{1}{2}$	66
Totals	28	72	100

Compute the following quantities using Tables **XI** and **XII**.

1. $\sum f \ln f$ for the cell frequencies

$$= 11\tfrac{1}{2} \ln 11\tfrac{1}{2} + 22\tfrac{1}{2} \ln 22\tfrac{1}{2} + 16\tfrac{1}{2} \ln 16\tfrac{1}{2} + 49\tfrac{1}{2} \ln 49\tfrac{1}{2}$$

$$= 28.087 + 70.054 + 46.255 + 193.148 = 337.544$$

2. $\sum f \ln f$ for the row and column totals

$$= 34 \ln 34 + 66 \ln 66 + 28 \ln 28 + 72 \ln 72$$

$$= 119.896 + 276.517 + 93.302 + 307.920 = 797.635$$

3. Look up $n \ln n = 100 \ln 100 = 460.517$

BOX 13.2 continued

4. $G_{\text{adj}} = 2[\text{quantity } \mathbf{1} - \text{quantity } \mathbf{2} + \text{quantity } \mathbf{3}]$

$= 2[337.544 - 797.635 + 460.517] = 2[0.426] = 0.852$

Compare G_{adj} with critical value of χ^2 for one degree of freedom. Since our observed G_{adj} is much less than $\chi^2_{.05[1]} = 3.841$, we accept the null hypothesis that the leaf type is independent of the type of soil in which the tree is rooted.

null hypothesis of independence between soil type and leaf type. The presence of pubescent leaves is independent of whether the tree is rooted in serpentine soils or not.

Tests of independence need not be restricted to 2×2 tables. In the two-way cases considered in this section, we are only concerned with two properties, but each of these properties may be divided into any number of classes. Thus organisms may occur in four color classes and be sampled from three localities, yielding a 4×3 test of independence. Such a test would examine whether the color proportions exhibited by the marginal totals are independent of the localities at which the individuals have been sampled. Such tests are often called *$R \times C$ tests of independence*, R and C standing for the number of rows and columns in the frequency table. Another case, examined in detail in Box 13.3, concerns bright red color patterns found in samples of a species of tiger beetle on four occasions during the spring and summer. It was of interest to know whether the percentage of bright red individuals changed significantly over the time of observation. We test whether the proportion of beetles colored bright red (55.7% for the entire study) is independent of the time of collection.

As shown in Box 13.3, the following is a simple general rule for computation of the G-test of independence:

$$G = 2[(\textstyle\sum f \ln f \text{ for the cell frequencies}) - (\textstyle\sum f \ln f \text{ for the row and column totals}) + n \ln n]$$

The transformations can be looked up in Table **XI**. In the formulas in Box 13.3 we employ a double subscript to refer to entries in a two-way table, as in the structurally similar case of two-way anova. The quantity f_{ij} in Box 13.3 refers to the observed frequency in row i and column j of the table.

The results in Box 13.3 show clearly that the frequency of bright red color patterns in these tiger beetles is dependent on the season. We note a decrease of bright red beetles in late spring and early summer, followed by an increase again in late summer.

The *degrees of freedom for tests of independence* are always the same and can be computed using the rules given earlier (Section 13.2). There are k cells in the table but we must subtract one degree of freedom for each independent

BOX 13.3

R × C test of independence using the *G*-test.

Frequencies of color patterns of a species of tiger beetle (*Cicindela fulgida*) found at various seasons

Season	*Color pattern* ($a = 2$)		*Totals*	*% Bright red*
($b = 4$)	*Bright red*	*Not bright red*		
Early spring	29	11	40	72.5
Late spring	273	191	464	58.8
Early summer	8	31	39	20.5
Late summer	64	64	128	50.0
Totals	374	297	$671 = n$	55.7

Source: Unpublished data from H. L. Willis.

Compute the following sums, using Table **XI** for $f \ln f$.

1. Sum of transforms of the frequencies in the body of the contingency table

$$= \sum^b \sum^a f_{ij} \ln f_{ij} = 29 \ln 29 + 11 \ln 11 + \cdots + 64 \ln 64$$

$$= 97.652 + 26.377 + \cdots + 266.169 = 3314.027$$

2. Sum of transforms of the row totals $= \sum^b \left(\sum^a f_{ij}\right) \ln \left(\sum^a f_{ij}\right)$

$$= 40 \ln 40 + \cdots + 128 \ln 128 = 147.555 + \cdots + 621.060$$

$$= 3760.400$$

3. Sum of the transforms of the column totals $= \sum^a \left(\sum^b f_{ij}\right) \ln \left(\sum^b f_{ij}\right)$

$$= 374 \ln 374 + 297 \ln 297 = 2215.672 + 1691.038 = 3906.710$$

4. Transform of the grand total $= n \ln n = 671 \ln 671 = 4367.384$

5. $G = 2$[quantity **1** − quantity **2** − quantity **3** + quantity **4**]

$$= 2[3314.027 - 3760.400 - 3906.710 + 4367.384 = 2[14.301] = 28.602$$

This value is to be compared with a χ^2-distribution with $(a - 1)(b - 1)$ degrees of freedom, where a is the number of columns and b the number of rows in the table. In our case, $df = (2 - 1)(4 - 1) = 3$.

Since $\chi^2_{.005[3]} = 12.838$, our G-value is significant at $P < 0.005$, and we must reject our null hypothesis that frequency of color pattern is independent of season.

parameter we have estimated from the data. We must, of course, subtract one degree of freedom for the observed total sample size, n. We have also estimated $a - 1$ row probabilities and $b - 1$ column probabilities, where a and b are the number of rows and columns in the table, respectively. Thus, there are $k - (a - 1) - (b - 1) - 1 = k - a - b + 1$ degrees of freedom for the test. But since $k = a \times b$, this expression becomes $(a \times b) - a - b + 1 = (a - 1) \times (b - 1)$, the conventional expression for the degrees of freedom in a two-way test of independence. Thus, the degrees of freedom in the example of Box 13.3, a 4×2 case, was $(4 - 1) \times (2 - 1) = 3$. In all 2×2 cases there is clearly only $(2 - 1) \times (2 - 1) = 1$ degree of freedom.

Another name for test of independence is *test of association*. If two properties are not independent of each other they are *associated*. Thus, in the example testing relative frequency of two leaf types on two different soils we can speak of an association between leaf types and soils. In the immunology experiment there is a negative association between presence of antiserum and mortality. *Association* is thus similar to correlation, but it is a more general term, applying to attributes as well as continuous variables. In the 2×2 tests of independence of this section, one way of looking for suspected lack of independence was to examine the percentage occurrence of one of the properties in the two classes based on the other property. Thus we compared the percentage of smooth leaves on the two types of soils, or we studied the percentage mortality with or without antiserum. This way of looking at a test of independence suggests another interpretation of these tests as tests for the significance of differences between two percentages.

Exercises 13

13.1 In an experiment to determine the mode of inheritance of a *green* mutant, 146 wild type and 30 mutant offspring were obtained when F_1 generation houseflies were crossed. Test whether the data agree with the hypothesis that the ratio of wild type to mutants is 3:1. ANS. $G = 6.4624$.

13.2 Locality A has been exhaustively collected for snakes of species S. An examination of the 167 adult males that have been collected reveals that 35 of these have pale-colored bands around their necks. From locality B, 90 miles away, we obtain a sample of 27 adult males of the same species, 6 of which show the bands. What is the chance that both samples are from the same statistical population with respect to frequency of bands?

13.3 Of 445 specimens of the butterfly *Erebia epipsodea* from mountainous areas, 2.5% have light color patches on their wings. Of 65 specimens from the prairie 70.8% have such patches (unpublished data by P. R. Ehrlich). Is this difference significant? ANS. $G_{adj} = 170.998$.

13.4 In a study of polymorphism of chromosomal inversions in the grasshopper *Moraba scurra*, Lewontin and White (1960) gave the following results for the composition of a population at Royalla "B" in 1958.

			Chromosome CD	
		St/St	St/Bl	Bl/Bl
Chromosome EF	Td/Td	22	96	75
	St/Td	8	56	64
	St/St	0	6	6

Are the frequencies of the three different combinations of chromosome EF independent of those of the frequencies of the three combinations of chromosome CD? ANS. $G = 7.396$.

13.5 Test whether the percentage of nymphs of the aphid *Myzus persicae* that developed into winged forms depends on the type of diet provided. Stem mothers had been placed on the diets one day before the birth of the nymphs (data by Mittler and Dadd, 1966).

Type of diet	*% winged forms*	*n*
Synthetic diet	100	216
Cotyledon "sandwich"	92	230
Free cotyledon	36	75

13.6 Test agreement of observed frequencies to those expected on the basis of a binomial distribution for the data given in Tables 4.1 and 4.2.

13.7 Test agreement of observed frequencies to those expected on the basis of a Poisson distribution for the data given in Table 4.5 and Table 4.6.

A1

Mathematical Appendix

A1.1 Demonstration that the sum of the deviations from the mean is equal to zero.

We have to learn two common rules of statistical algebra. We can open a pair of parentheses with a Σ sign in front of them by treating the Σ as though it were a common factor.

We have

$$\sum_{i=1}^{n} (A_i + B_i) = (A_1 + B_1) + (A_2 + B_2) + \cdots + (A_n + B_n)$$
$$= (A_1 + A_2 + \cdots + A_n) + (B_1 + B_2 + \cdots + B_n)$$

Therefore

$$\sum_{i=1}^{n} (A_i + B_i) = \sum_{i=1}^{n} A_i + \sum_{i=1}^{n} B_i$$

Also, when $\sum_{i=1}^{n} C$ is developed during an algebraic operation, where C is a constant, this can be computed as follows:

$$\sum_{i=1}^{n} C = C + C + \cdots + C \qquad (n \text{ terms})$$
$$= nC$$

Since in a given problem a mean is a constant value, $\sum^{n} \overline{Y} = n\overline{Y}$. If desired, you may check these rules, using simple numbers. In the subsequent demonstration and others to follow, whenever all summations are over n items we have simplified the notation by dropping subscripts for variables and superscripts above summation signs.

We wish to prove that $\sum y = 0$. By definition,

$$\begin{aligned} \sum y &= \sum(Y - \bar{Y}) \\ &= \sum Y - n\bar{Y} \\ &= \sum Y - \frac{n\sum Y}{n} \qquad \left(\text{since } \bar{Y} = \frac{\sum Y}{n}\right) \\ &= \sum Y - \sum Y \end{aligned}$$

Therefore $\sum y = 0$.

A1.2 Demonstration of the effects of additive, multiplicative, and combination coding on means, variances, and standard deviations.

For this proof we have to learn one more convention of statistical algebra. We have seen in Appendix A1.1 that $\sum C = nC$. However, when the $\sum$ precedes both a constant and a variable, as in $\sum CY$, the constant can be placed before the $\sum$, since

$$\begin{aligned} \sum_{i=1}^{n} CY_i &= CY_1 + CY_2 + \cdots + CY_n \\ &= C(Y_1 + Y_2 + \cdots + Y_n) \\ &= C\left(\sum_{i=1}^{n} Y_i\right) \end{aligned}$$

Therefore

$$\sum_{i=1}^{n} CY_i = C\sum_{i=1}^{n} Y_i$$

Thus $\sum CY = C\sum Y$, $\sum C^2y^2 = C^2\sum y^2$, and $\sum 2\bar{Y}Y = 2\bar{Y}\sum Y$, because both 2 and $\bar{Y}$ are constants.

Means of coded data

Additive coding.—The variable is coded $Y_c = Y + C$, where C is a constant, the additive code. Therefore

$$\sum Y_c = \sum(Y + C) = \sum Y + nC$$

and

$$\bar{Y}_c = \frac{\sum Y_c}{n} = \frac{\sum Y}{n} + C = \bar{Y} + C$$

To decode $\bar{Y}_c$, subtract C from it and you will obtain $\bar{Y}$; that is, $\bar{Y} = \bar{Y}_c - C$.

Multiplicative coding.—The variable is coded $Y_c = DY$, where D is a constant, the multiplicative code. Therefore

$$\sum Y_c = D\sum Y$$

and

$$\overline{Y}_c = \frac{\sum Y_c}{n} = D\frac{\sum Y}{n} = D\overline{Y}$$

To decode $\overline{Y}_c$ divide it by D and you will obtain $\overline{Y}$; that is, $\overline{Y} = \overline{Y}_c/D$.

Combination coding.—The variable is coded $Y_c = D(Y + C)$, where C and D are constants, the additive and multiplicative codes, respectively. Therefore

$$\sum Y_c = D\sum(Y + C) = D\sum Y + nDC$$

and

$$\overline{Y}_c = \frac{\sum Y_c}{n} = D\frac{\sum Y}{n} + DC = D\overline{Y} + DC$$

To decode $\overline{Y}_c$, divide it by D, then subtract C and you will obtain $\overline{Y}$; that is,

$$\overline{Y} = \frac{\overline{Y}_c}{D} - C.$$

Variances and standard deviations of coded data

Additive coding.—The variable is coded $Y_c = Y + C$, where C is a constant, the additive code. By definition, $y = Y - \overline{Y}$, and

$$\begin{aligned} y_c &= Y_c - \overline{Y}_c \\ &= [(Y + C) - (\overline{Y} + C)] \qquad \text{(as shown for means above)} \\ &= [Y + C - \overline{Y} - C] \\ &= [Y - \overline{Y}] \\ &= y \end{aligned}$$

Therefore $\sum y_c^2 = \sum y^2$, and

$$\frac{\sum y_c^2}{n - 1} = \frac{\sum y^2}{n - 1}$$

Thus additive coding has no effect on sums of squares, variances, or standard deviations.

Multiplicative coding.—The variable is coded $Y_c = DY$, where D is a constant, the multiplicative code. By definition, $y = Y - \overline{Y}$, and

$$\begin{aligned} y_c &= Y_c - \overline{Y}_c \\ &= DY - D\overline{Y} \qquad \text{(as shown for means above)} \\ &= D(Y - \overline{Y}) \\ &= Dy \end{aligned}$$

Therefore $y_c^2 = D^2y^2$, $\sum y_c^2 = D^2\sum y^2$, and

$$\frac{\sum y_c^2}{n - 1} = D^2\frac{\sum y^2}{n - 1}$$

$$s_c^2 = D^2 s^2$$

Thus, when data have been subjected to multiplicative coding, a sum of squares or variance can be decoded by dividing it by the *square* of the multiplicative code; a standard deviation can be decoded by dividing it by the code itself, that is, $s^2 = s_c^2/D^2$ and $s = s_c/D$.

Combination coding.—The variable is coded $Y_c = D(Y + C)$, where C and D are constants, the additive and multiplicative codes, respectively. By definition $y = Y - \bar{Y}$, and

$$\begin{aligned} y_c &= Y_c - \bar{Y}_c \\ &= [D(Y + C) - (D\bar{Y} + DC)] \qquad \text{(as shown for means above)} \\ &= [DY + DC - D\bar{Y} - DC] \\ &= D[Y - \bar{Y}] \end{aligned}$$

Therefore $y_c = Dy$, as before.

Thus in combination coding only the multiplicative code needs to be considered when decoding sums of squares, variances, or standard deviations.

A1.3 Demonstration that Expression (3.7), the computational formula for the sum of squares, equals Expression (3.6), the expression originally developed for this statistic.

We wish to prove that $\sum(Y - \bar{Y})^2 = \sum Y^2 - ((\sum Y)^2/n)$. We have

$$\begin{aligned} \sum(Y - \bar{Y})^2 &= \sum(Y^2 - 2Y\bar{Y} + \bar{Y}^2) \\ &= \sum Y^2 - 2\bar{Y}\sum Y + n\bar{Y}^2 \\ &= \sum Y^2 - \frac{2(\sum Y)^2}{n} + \frac{n(\sum Y)^2}{n^2} \qquad \left(\text{since } \bar{Y} = \frac{\sum Y}{n}\right) \\ &= \sum Y^2 - \frac{2(\sum Y)^2}{n} + \frac{(\sum Y)^2}{n} \end{aligned}$$

Hence

$$\sum(Y - \bar{Y})^2 = \sum Y^2 - \frac{(\sum Y)^2}{n}$$

A1.4 Simplified formulas for standard error of the difference between two means.

The standard error squared from Expression (8.2) is

$$\left[\frac{(n_1 - 1)s_1^2 + (n_2 - 1)s_2^2}{n_1 + n_2 - 2}\right]\left(\frac{n_1 + n_2}{n_1 n_2}\right).$$

When $n_1 = n_2 = n$, this simplifies to

$$\left[\frac{(n - 1)s_1^2 + (n - 1)s_2^2}{2n - 2}\right]\left(\frac{2n}{n^2}\right) = \left[\frac{(n - 1)(s_1^2 + s_2^2)(2)}{2(n - 1)(n)}\right] = \frac{1}{n}(s_1^2 + s_2^2)$$

which is the standard error squared of Expression (8.3).

When $n_1 \neq n_2$, but each is large so that $(n_1 - 1) \approx n_1$ and $(n_2 - 1) \approx n_2$, the standard error squared of Expression (8.2) simplifies to

$$\left[\frac{n_1 s_1^2 + n_2 s_2^2}{n_1 + n_2}\right]\left(\frac{n_1 + n_2}{n_1 n_2}\right) = \left[\frac{n_1 s_1^2}{n_1 n_2} + \frac{n_2 s_2^2}{n_1 n_2}\right] = \frac{s_1^2}{n_2} + \frac{s_2^2}{n_1}$$

which is the standard error squared of Expression (8.4).

A1.5 Demonstration that t_s^2 obtained from a test of significance of the difference between two means (as in Box 8.2) is identical to the F_s-value obtained in a single classification anova of two equal-sized groups (in the same box).

$$t_s \text{ (from Box 8.2)} = \frac{\bar{Y}_1 - \bar{Y}_2}{\sqrt{\dfrac{1}{n(n-1)}\left(\sum^n y_1^2 + \sum^n y_2^2\right)}}$$

$$t_s^2 = \frac{(\bar{Y}_1 - \bar{Y}_2)^2}{\dfrac{1}{n(n-1)}\left(\sum^n y_1^2 + \sum^n y_2^2\right)} = \frac{n(n-1)(\bar{Y}_1 - \bar{Y}_2)^2}{\sum^n y_1^2 + \sum^n y_2^2}$$

In the 2-sample anova,

$$MS_{\text{means}} = \frac{1}{2-1}\sum^2 (\bar{Y}_i - \bar{\bar{Y}})^2$$

$$= (\bar{Y}_1 - \bar{\bar{Y}})^2 + (\bar{Y}_2 - \bar{\bar{Y}})^2$$

$$= \left(\bar{Y}_1 - \frac{\bar{Y}_1 + \bar{Y}_2}{2}\right)^2 + \left(\bar{Y}_2 - \frac{\bar{Y}_1 + \bar{Y}_2}{2}\right)^2 \qquad \text{(since } \bar{\bar{Y}} = (\bar{Y}_1 + \bar{Y}_2)/2\text{)}$$

$$= \left(\frac{\bar{Y}_1 - \bar{Y}_2}{2}\right)^2 + \left(\frac{\bar{Y}_2 - \bar{Y}_1}{2}\right)^2$$

$$= \tfrac{1}{2}(\bar{Y}_1 - \bar{Y}_2)^2$$

since the squares of the numerators are identical. Then

$$MS_{\text{groups}} = n \times MS_{\text{means}} = n[\tfrac{1}{2}(\bar{Y}_1 - \bar{Y}_2)^2]$$

$$= \frac{n}{2}(\bar{Y}_1 - \bar{Y}_2)^2$$

$$MS_{\text{within}} = \frac{\sum^n y_1^2 + \sum^n y_2^2}{2(n-1)}$$

$$F_s = \frac{MS_{\text{groups}}}{MS_{\text{within}}}$$

$$= \frac{\dfrac{n}{2}(\bar{Y}_1 - \bar{Y}_2)^2}{\left(\sum^n y_1^2 + \sum^n y_2^2\right) \Big/ [2(n-1)]}$$

$$= \frac{n(n-1)(\bar{Y}_1 - \bar{Y}_2)^2}{\sum^n y_1^2 + \sum^n y_2^2}$$

$$= t_s^2$$

A1.6 Demonstration that Expression (11.5), the computational formula for the sum of products, equals $\sum(X - \bar{X})(Y - \bar{Y})$, the expression originally developed for this quantity.

All summations are over n items. We have

$$\begin{aligned}
\sum xy &= \sum(X - \bar{X})(Y - \bar{Y}) \\
&= \sum XY - \bar{X}\sum Y - \bar{Y}\sum X + n\bar{X}\bar{Y} \quad (\text{since } \sum \bar{X}\bar{Y} = n\bar{X}\bar{Y}) \\
&= \sum XY - \bar{X}n\bar{Y} - \bar{Y}n\bar{X} + n\bar{X}\bar{Y} \quad (\text{since } \sum Y/n = \bar{Y}, \\
&\qquad \sum Y = n\bar{Y}; \text{similarly, } \sum X = n\bar{X}) \\
&= \sum XY - n\bar{X}\bar{Y} \\
&= \sum XY - n\bar{X}\sum Y/n \\
&= \sum XY - \bar{X}\sum Y
\end{aligned}$$

Similarly

$$\sum xy = \sum XY - \bar{Y}\sum X$$

and

$$\sum xy = \sum XY - \frac{(\sum X)(\sum Y)}{n} \tag{11.5}$$

A1.7 Derivation of computational formula for $\sum d_{Y \cdot X}^2 = \sum y^2 - ((\sum xy)^2/\sum x^2)$.

By definition, $d_{Y \cdot X} = Y - \hat{Y}$. Since $\bar{Y} = \bar{\hat{Y}}$, we can subtract $\bar{Y}$ from both Y and $\hat{Y}$ to obtain

$$d_{Y \cdot X} = y - \hat{y} = y - bx \qquad (\text{since } \hat{y} = bx)$$

Therefore

$$\begin{aligned}
\sum d_{Y \cdot X}^2 &= \sum(y - bx)^2 = \sum y^2 - 2b\sum xy + b^2\sum x^2 \\
&= \sum y^2 - 2\frac{\sum xy}{\sum x^2}\sum xy + \frac{(\sum xy)^2}{(\sum x^2)^2}\sum x^2 = \sum y^2 - 2\frac{(\sum xy)^2}{\sum x^2} + \frac{(\sum xy)^2}{\sum x^2}
\end{aligned}$$

or

$$\sum d_{Y \cdot X}^2 = \sum y^2 - \frac{(\sum xy)^2}{\sum x^2} \tag{11.7}$$

A1.8 Demonstration that the sum of squares of the dependent variable in regression can be partitioned exactly into explained and unexplained sums of squares, the cross products canceling out.

By definition (Section 11.5)

$$\begin{aligned}
y &= \hat{y} + d_{Y \cdot X} \\
\sum y^2 &= \sum(\hat{y} + d_{Y \cdot X})^2 = \sum \hat{y}^2 + \sum d_{Y \cdot X}^2 + 2\sum \hat{y} d_{Y \cdot X}
\end{aligned}$$

If we can show that $\sum \hat{y} d_{Y \cdot X} = 0$, then we have demonstrated the required identity. We have

$$\sum \hat{y} d_{Y \cdot X} = \sum bx(y - bx) \qquad \text{[since } \hat{y} = bx \text{ from Expression (11.3) and } d_{Y \cdot X} = y - bx \text{ from Appendix A1.7]}$$

$$= b\sum xy - b^2 \sum x^2$$

$$= b\sum xy - b\frac{\sum xy}{\sum x^2}\sum x^2 \qquad (\text{since } b = \sum xy/\sum x^2)$$

$$= b\sum xy - b\sum xy$$

$$= 0$$

Therefore $\sum y^2 = \sum \hat{y}^2 + \sum d^2_{Y \cdot X}$
or, written out in terms of variates,

$$\sum(Y - \bar{Y})^2 = \sum(\hat{Y} - \bar{Y})^2 + \sum(Y - \hat{Y})^2$$

A1.9 Proof that the variance of the sum of two variables is

$$\sigma^2_{(Y_1+Y_2)} = \sigma_1^2 + \sigma_2^2 + 2\rho_{12}\sigma_1\sigma_2$$

where σ_1 and σ_2 are standard deviations of Y_1 and Y_2, respectively, and ρ_{12} is the parametric correlation coefficient between Y_1 and Y_2.

If $Z = Y_1 + Y_2$, then

$$\sigma_Z^2 = \frac{1}{n}\sum(Z - \bar{Z})^2 = \frac{1}{n}\sum\left[(Y_1 + Y_2) - \frac{1}{n}\sum(Y_1 + Y_2)\right]^2$$

$$= \frac{1}{n}\sum\left[(Y_1 + Y_2) - \frac{1}{n}\sum Y_1 - \frac{1}{n}\sum Y_2\right]^2 = \frac{1}{n}\sum\left[(Y_1 + Y_2) - \bar{Y}_1 - \bar{Y}_2\right]^2$$

$$= \frac{1}{n}\sum\left[(Y_1 - \bar{Y}_1) + (Y_2 - \bar{Y}_2)\right]^2 = \frac{1}{n}\sum\left[y_1 + y_2\right]^2$$

$$= \frac{1}{n}\sum\left[y_1^2 + y_2^2 + 2y_1y_2\right] = \frac{1}{n}\sum y_1^2 + \frac{1}{n}\sum y_2^2 + \frac{2}{n}\sum y_1y_2$$

$$= \sigma_1^2 + \sigma_2^2 + 2\sigma_{12}$$

But, since $\rho_{12} = \sigma_{12}/\sigma_1\sigma_2$, we have

$$\sigma_{12} = \rho_{12}\sigma_1\sigma_2$$

Therefore

$$\sigma_Z^2 = \sigma^2_{(Y_1+Y_2)} = \sigma_1^2 + \sigma_2^2 + 2\rho_{12}\sigma_1\sigma_2$$

Similarly,

$$\sigma_D^2 = \sigma^2_{(Y_1-Y_2)} = \sigma_1^2 + \sigma_2^2 - 2\rho_{12}\sigma_1\sigma_2$$

The analogous expressions apply to sample statistics. Thus

$$s^2_{(Y_1+Y_2)} = s_1^2 + s_2^2 + 2r_{12}s_1s_2 \qquad (12.8)$$

$$s^2_{(Y_1-Y_2)} = s_1^2 + s_2^2 - 2r_{12}s_1s_2 \qquad (12.9)$$

A1.10 Derivation of computational formula for X^2 [Expression (13.2)] from Expression (13.1).

Expanding Expression (13.1),

$$X^2 = \sum^a \frac{(f_i - \hat{f}_i)^2}{\hat{f}_i}$$

$$= \sum^a \frac{f_i^2}{\hat{f}_i} + \sum^a \frac{\hat{f}_i^2}{\hat{f}_i} - 2\sum^a \frac{f_i\hat{f}_i}{\hat{f}_i}$$

$$= \sum^a \frac{f_i^2}{\hat{f}_i} + \sum^a \hat{f}_i - 2\sum^a f_i$$

But, since $\sum^a f_i = \sum^a \hat{f}_i = n$

$$X^2 = \sum^a \frac{f_i^2}{\hat{f}_i} - n \tag{13.2}$$

A1.11 Proof that the general expression for the G-test can be simplified to Expressions (13.4) and (13.5).

In general, G is twice the natural logarithm of the ratio of the probability of the sample with all parameters estimated from the data and the probability of the sample assuming the null hypothesis is true. Assuming a multinomial distribution, this ratio is

$$L = \frac{\dfrac{n!}{f_1!f_2!\cdots f_a!}\, p_1^{f_1} p_2^{f_2} \cdots p_a^{f_a}}{\dfrac{n!}{f_1!f_2!\cdots f_a!}\, \hat{p}_1^{f_1} \hat{p}_2^{f_2} \cdots \hat{p}_a^{f_a}}$$

$$= \prod_{i=1}^{a} \left(\frac{p_i}{\hat{p}_i}\right)^{f_i}$$

$$G = 2 \ln L$$

$$= 2\sum^a f_i \ln\left(\frac{p_i}{\hat{p}_i}\right)$$

Since $f_i = np_i$ and $\hat{f}_i = n\hat{p}_i$,

$$G = 2\sum^a f_i \ln\left(\frac{f_i}{\hat{f}_i}\right) \tag{13.4}$$

If we now replace $\hat{f}_i$ by $n\hat{p}_i$,

$$G = 2\sum^a f_i \ln\left(\frac{f_i}{n\hat{p}_i}\right) = 2\left[\sum^a f_i \ln f_i - \sum^a f_i \ln \hat{p}_i - \sum^a f_i \ln n\right]$$

$$= 2\left[\sum^a f_i \ln f_i - \sum^a f_i \ln \hat{p}_i - n \ln n\right] \tag{13.5}$$

A2

Statistical Tables

TABLE **I.** Twenty-five hundred random digits.

	1	2	3	4	5	6	7	8	9	10
1	48461	14952	72619	73689	52059	37086	60050	86192	67049	64739
2	76534	38149	49692	31366	52093	15422	20498	33901	10319	43397
3	70437	25861	38504	14752	23757	59660	67844	78815	23758	86814
4	59584	03370	42806	11393	71722	93804	09095	07856	55589	46020
5	04285	58554	16085	51555	27501	73883	33427	33343	45507	50063
6	77340	10412	69189	85171	29082	44785	83638	02583	96483	76553
7	59183	62687	91778	80354	23512	97219	65921	02035	59847	91403
8	91800	04281	39979	03927	82564	28777	59049	97532	54540	79472
9	12066	24817	81099	48940	69554	55925	48379	12866	51232	21580
10	69907	91751	53512	23748	65906	91385	84983	27915	48491	91068
11	80467	04873	54053	25955	48518	13815	37707	68687	15570	08890
12	78057	67835	28302	45048	56761	97725	58438	91528	24645	18544
13	05648	39387	78191	88415	60269	94880	58812	42931	71898	61534
14	22304	39246	01350	99451	61862	78688	30339	60222	74052	25740
15	61346	50269	67005	40442	33100	16742	61640	21046	31909	72641
16	66793	37696	27965	30459	91011	51426	31006	77468	61029	57108
17	86411	48809	36698	42453	83061	43769	39948	87031	30767	13953
18	62098	12825	81744	28882	27369	88183	65846	92545	09065	22655
19	68775	06261	54265	16203	23340	84750	16317	88686	86842	00879
20	52679	19595	13687	74872	89181	01939	18447	10787	76246	80072
21	84096	87152	20719	25215	04349	54434	72344	93008	83282	31670
22	63964	55937	21417	49944	38356	98404	14850	17994	17161	98981
23	31191	75131	72386	11689	95727	05414	88727	45583	22568	77700
24	30545	68523	29850	67833	05622	89975	79042	27142	99257	32349
25	52573	91001	52315	26430	54175	30122	31796	98842	37600	26025
26	16586	81842	01076	99414	31574	94719	34656	80018	86988	79234
27	81841	88481	61191	25013	30272	23388	22463	65774	10029	58376
28	43563	66829	72838	08074	57080	15446	11034	98143	74989	26885
29	19945	84193	57581	77252	85604	45412	43556	27518	90572	00563
30	79374	23796	16919	99691	80276	32818	62953	78831	54395	30705
31	48503	26615	43980	09810	38289	66679	73799	48418	12647	40044
32	32049	65541	37937	41105	70106	89706	40829	40789	59547	00783
33	18547	71562	95493	34112	76895	46766	96395	31718	48302	45893
34	03180	96742	61486	43305	34183	99605	67803	13491	09243	29557
35	94822	24738	67749	83748	59799	25210	31093	62925	72061	69991
36	34330	60599	85828	19152	68499	27977	35611	96240	62747	89529
37	43770	81537	59527	95674	76692	86420	69930	10020	72881	12532
38	56908	77192	50623	41215	14311	42834	80651	93750	59957	31211
39	32787	07189	80539	75927	75475	73965	11796	72140	48944	74156
40	52441	78392	11733	57703	29133	71164	55355	31006	25526	55790
41	22377	54723	18227	28449	04570	18882	00023	67101	06895	08915
42	18376	73460	88841	39602	34049	20589	05701	08249	74213	25220
43	53201	28610	87957	21497	64729	64983	71551	99016	87903	63875
44	34919	78901	59710	27396	02593	05665	11964	44134	00273	76358
45	33617	92159	21971	16901	57383	34262	41744	60891	57624	06962
46	70010	40964	98780	72418	52571	18415	64362	90636	38034	04909
47	19282	68447	35665	31530	59832	49181	21914	65742	89815	39231
48	91429	73328	13266	54898	68795	40948	80808	63887	89939	47938
49	97637	78393	33021	05867	86520	45363	43066	00988	64040	09803
50	95150	07625	05255	83254	93943	52325	93230	62668	79529	65964

BLE II. Areas of the normal curve.

Standard deviation units	0	1	2	3	4	5	6	7	8	9	Standard deviation units
0.0	.0000	.0040	.0080	.0120	.0160	.0199	.0239	.0279	.0319	.0359	0.0
0.1	.0398	.0438	.0478	.0517	.0557	.0596	.0636	.0675	.0714	.0753	0.1
0.2	.0793	.0832	.0871	.0910	.0948	.0987	.1026	.1064	.1103	.1141	0.2
0.3	.1179	.1217	.1255	.1293	.1331	.1368	.1406	.1443	.1480	.1517	0.3
0.4	.1554	.1591	.1628	.1664	.1700	.1736	.1772	.1808	.1844	.1879	0.4
0.5	.1915	.1950	.1985	.2019	.2054	.2088	.2123	.2157	.2190	.2224	0.5
0.6	.2257	.2291	.2324	.2357	.2389	.2422	.2454	.2486	.2517	.2549	0.6
0.7	.2580	.2611	.2642	.2673	.2704	.2734	.2764	.2794	.2823	.2852	0.7
0.8	.2881	.2910	.2939	.2967	.2995	.3023	.3051	.3078	.3106	.3133	0.8
0.9	.3159	.3186	.3212	.3238	.3264	.3289	.3315	.3340	.3365	.3389	0.9
1.0	.3413	.3438	.3461	.3485	.3508	.3531	.3554	.3577	.3599	.3621	1.0
1.1	.3643	.3665	.3686	.3708	.3729	.3749	.3770	.3790	.3810	.3830	1.1
1.2	.3849	.3869	.3888	.3907	.3925	.3944	.3962	.3980	.3997	.4015	1.2
1.3	.4032	.4049	.4066	.4082	.4099	.4115	.4131	.4147	.4162	.4177	1.3
1.4	.4192	.4207	.4222	.4236	.4251	.4265	.4279	.4292	.4306	.4319	1.4
1.5	.4332	.4345	.4357	.4370	.4382	.4394	.4406	.4418	.4429	.4441	1.5
1.6	.4452	.4463	.4474	.4484	.4495	.4505	.4515	.4525	.4535	.4545	1.6
1.7	.4554	.4564	.4573	.4582	.4591	.4599	.4608	.4616	.4625	.4633	1.7
1.8	.4641	.4649	.4656	.4664	.4671	.4678	.4686	.4693	.4699	.4706	1.8
1.9	.4713	.4719	.4726	.4732	.4738	.4744	.4750	.4756	.4761	.4767	1.9
2.0	.4772	.4778	.4783	.4788	.4793	.4798	.4803	.4808	.4812	.4817	2.0
2.1	.4821	.4826	.4830	.4834	.4838	.4842	.4846	.4850	.4854	.4857	2.1
2.2	.4861	.4864	.4868	.4871	.4875	.4878	.4881	.4884	.4887	.4890	2.2
2.3	.4893	.4896	.4898	.4901	.4904	.4906	.4909	.4911	.4913	.4916	2.3
2.4	.4918	.4920	.4922	.4925	.4927	.4929	.4931	.4932	.4934	.4936	2.4
2.5	.4938	.4940	.4941	.4943	.4945	.4946	.4948	.4949	.4951	.4952	2.5
2.6	.4953	.4955	.4956	.4957	.4959	.4960	.4961	.4962	.4963	.4964	2.6
2.7	.4965	.4966	.4967	.4968	.4969	.4970	.4971	.4972	.4973	.4974	2.7
2.8	.4974	.4975	.4976	.4977	.4977	.4978	.4979	.4979	.4980	.4981	2.8
2.9	.4981	.4982	.4982	.4983	.4984	.4984	.4985	.4985	.4986	.4986	2.9
3.0	.4987	.4987	.4987	.4988	.4988	.4989	.4989	.4989	.4990	.4990	3.0
3.1	.4990	.4991	.4991	.4991	.4992	.4992	.4992	.4992	.4993	.4993	3.1
3.2	.4993	.4993	.4994	.4994	.4994	.4994	.4994	.4995	.4995	.4995	3.2
3.3	.4995	.4995	.4995	.4996	.4996	.4996	.4996	.4996	.4996	.4997	3.3
3.4	.4997	.4997	.4997	.4997	.4997	.4997	.4997	.4997	.4997	.4998	3.4
3.5	.499767										
3.6	.499841										
3.7	.499892										
3.8	.499928										
3.9	.499952										
4.0	.499968										
4.1	.499979										
4.2	.499987										
4.3	.499991										
4.4	.499995										
4.5	.499997										
4.6	.499998										
4.7	.499999										
4.8	.499999										
4.9	.500000										

Tabled area

$\frac{1}{2}$-α

α

Argument $= \frac{Y - \mu_Y}{\sigma}$

Note: The quantity given is the area under the standard normal density function between the mean and the critical point. The area is generally labeled $\frac{1}{2} - \alpha$ (as shown in the figure). By inverse interpolation one can find the number of standard deviations corresponding to a given area.

TABLE **III. Critical values of Student's t-distribution.**

$\nu \backslash \alpha$	0.9	0.5	0.4	0.2	0.1
1	.158	1.000	1.376	3.078	6.314
2	.142	.816	1.061	1.886	2.920
3	.137	.765	.978	1.638	2.353
4	.134	.741	.941	1.533	2.132
5	.132	.727	.920	1.476	2.015
6	.131	.718	.906	1.440	1.943
7	.130	.711	.896	1.415	1.895
8	.130	.706	.889	1.397	1.860
9	.129	.703	.883	1.383	1.833
10	.129	.700	.879	1.372	1.812
11	.129	.697	.876	1.363	1.796
12	.128	.695	.873	1.356	1.782
13	.128	.694	.870	1.350	1.771
14	.128	.692	.868	1.345	1.761
15	.128	.691	.866	1.341	1.753
16	.128	.690	.865	1.337	1.746
17	.128	.689	.863	1.333	1.740
18	.127	.688	.862	1.330	1.734
19	.127	.688	.861	1.328	1.729
20	.127	.687	.860	1.325	1.725
21	.127	.686	.859	1.323	1.721
22	.127	.686	.858	1.321	1.717
23	.127	.685	.858	1.319	1.714
24	.127	.685	.857	1.318	1.711
25	.127	.684	.856	1.316	1.708
26	.127	.684	.856	1.315	1.706
27	.127	.684	.855	1.314	1.703
28	.127	.683	.855	1.313	1.701
29	.127	.683	.854	1.311	1.699
30	.127	.683	.854	1.310	1.697
40	.126	.681	.851	1.303	1.684
60	.126	.679	.848	1.296	1.671
120	.126	.677	.845	1.289	1.658
∞	.126	.674	.842	1.282	1.645

Note: If a one-tailed test is desired, the probabilities at the head of the table must be halved. For degrees of freedom $\nu > 30$, interpolate between the values of the argument ν. The table is designed for harmonic interpolation. Thus, to obtain $t_{.05[43]}$, interpolate between $t_{.05[40]} = 2.021$ and $t_{.05[60]} = 2.000$, which are furnished in the table. Transform the arguments into $120/\nu = 120/43 = 2.791$ and interpolate between $120/60 = 2.000$ and $120/40 = 3.000$ by ordinary linear interpolation:

$$t_{.05[43]} = (0.791 \times 2.021) + [(1 - 0.791) \times 2.000]$$
$$= 2.017$$

When $\nu > 120$, interpolate between $120/\infty = 0$ and $120/120 = 1$. Values in this table have been taken from a more extensive one (table III) in R. A. Fisher and F. Yates, *Statistical Tables for Biological, Agricultural and Medical Research*, 5th ed. (Oliver & Boyd, Edinburgh, 1958) with permission of the authors and their publishers.

0.05	0.02	0.01	0.001	α / ν
12.706	31.821	63.657	636.619	1
4.303	6.965	9.925	31.598	2
3.182	4.541	5.841	12.924	3
2.776	3.747	4.604	8.610	4
2.571	3.365	4.032	6.869	5
2.447	3.143	3.707	5.959	6
2.365	2.998	3.499	5.408	7
2.306	2.896	3.355	5.041	8
2.262	2.821	3.250	4.781	9
2.228	2.764	3.169	4.587	10
2.201	2.718	3.106	4.437	11
2.179	2.681	3.055	4.318	12
2.160	2.650	3.012	4.221	13
2.145	2.624	2.977	4.140	14
2.131	2.602	2.947	4.073	15
2.120	2.583	2.921	4.015	16
2.110	2.567	2.898	3.965	17
2.101	2.552	2.878	3.922	18
2.093	2.539	2.861	3.883	19
2.086	2.528	2.845	3.850	20
2.080	2.518	2.831	3.819	21
2.074	2.508	2.819	3.792	22
2.069	2.500	2.807	3.767	23
2.064	2.492	2.797	3.745	24
2.060	2.485	2.787	3.725	25
2.056	2.479	2.779	3.707	26
2.052	2.473	2.771	3.690	27
2.048	2.467	2.763	3.674	28
2.045	2.462	2.756	3.659	29
2.042	2.457	2.750	3.646	30
2.021	2.423	2.704	3.551	40
2.000	2.390	2.660	3.460	60
1.980	2.358	2.617	3.373	120
1.960	2.326	2.576	3.291	∞

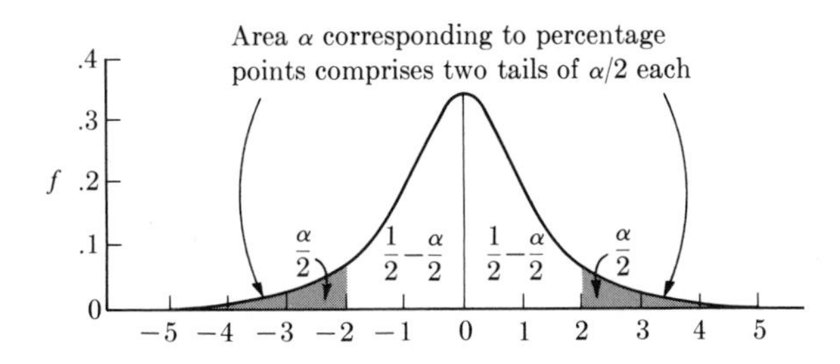

TABLE **IV.** Critical values of the chi-square distribution.

$\nu \backslash \alpha$	0.995	0.975	0.9	0.5	0.1
1	.000	.000	0.016	0.455	2.706
2	0.010	0.051	0.211	1.386	4.605
3	0.072	0.216	0.584	2.366	6.251
4	0.207	0.484	1.064	3.357	7.779
5	0.412	0.831	1.610	4.351	9.236
6	0.676	1.237	2.204	5.348	10.645
7	0.989	1.690	2.833	6.346	12.017
8	1.344	2.180	3.490	7.344	13.362
9	1.735	2.700	4.168	8.343	14.684
10	2.156	3.247	4.865	9.342	15.987
11	2.603	3.816	5.578	10.341	17.275
12	3.074	4.404	6.304	11.340	18.549
13	3.565	5.009	7.042	12.340	19.812
14	4.075	5.629	7.790	13.339	21.064
15	4.601	6.262	8.547	14.339	22.307
16	5.142	6.908	9.312	15.338	23.542
17	5.697	7.564	10.085	16.338	24.769
18	6.265	8.231	10.865	17.338	25.989
19	6.844	8.907	11.651	18.338	27.204
20	7.434	9.591	12.443	19.337	28.412
21	8.034	10.283	13.240	20.337	29.615
22	8.643	10.982	14.042	21.337	30.813
23	9.260	11.688	14.848	22.337	32.007
24	9.886	12.401	15.659	23.337	33.196
25	10.520	13.120	16.473	24.337	34.382
26	11.160	13.844	17.292	25.336	35.563
27	11.808	14.573	18.114	26.336	36.741
28	12.461	15.308	18.939	27.336	37.916
29	13.121	16.047	19.768	28.336	39.088
30	13.787	16.791	20.599	29.336	40.256
31	14.458	17.539	21.434	30.336	41.422
32	15.134	18.291	22.271	31.336	42.585
33	15.815	19.047	23.110	32.336	43.745
34	16.501	19.806	23.952	33.336	44.903
35	17.192	20.569	24.797	34.336	46.059
36	17.887	21.336	25.643	35.336	47.212
37	18.586	22.106	26.492	36.335	48.363
38	19.289	22.878	27.343	37.335	49.513
39	19.996	23.654	28.196	38.335	50.660
40	20.707	24.433	29.051	39.335	51.805
41	21.421	25.215	29.907	40.335	52.949
42	22.138	25.999	30.765	41.335	54.090
43	22.859	26.785	31.625	42.335	55.230
44	23.584	27.575	32.487	43.335	56.369
45	24.311	28.366	33.350	44.335	57.505
46	25.042	29.160	34.215	45.335	58.641
47	25.775	29.956	35.081	46.335	59.774
48	26.511	30.755	35.949	47.335	60.907
49	27.249	31.555	36.818	48.335	62.038
50	27.991	32.357	37.689	49.335	63.167

Note: For values of $\nu > 100$, compute approximate critical values of χ^2 by formula as follows: $\chi^2_{\alpha[\nu]} \approx \frac{1}{2}(t_{2\alpha[\infty]} + \sqrt{2\nu - 1})^2$, where $t_{2\alpha[\infty]}$ can be looked up in Table **III**. Thus $\chi^2_{.05[120]}$ is computed as $\frac{1}{2}(t_{.10[\infty]} + \sqrt{240 - 1})^2 = \frac{1}{2}(1.645 + \sqrt{239})^2 = \frac{1}{2}(17.10462)^2 = 146.284$. For $\alpha > 0.5$ employ $t_{1-2\alpha[\infty]}$ in the above formula. When $\alpha = 0.5$, $t_{2\alpha} = 0$. Values of chi-square from 1 to 30 degrees of freedom have been taken from a more extensive table by C. M. Thompson (*Biometrika* **32**:188-189, 1941) with permission of the publisher.

0.05	0.025	0.01	0.005	ν
3.841	5.024	6.635	7.879	1
5.991	7.378	9.210	10.597	2
7.815	9.348	11.345	12.838	3
9.488	11.143	13.277	14.860	4
11.070	12.832	15.086	16.750	5
12.592	14.449	16.812	18.548	6
14.067	16.013	18.475	20.278	7
15.507	17.535	20.090	21.955	8
16.919	19.023	21.666	23.589	9
18.307	20.483	23.209	25.188	10
19.675	21.920	24.725	26.757	11
21.026	23.337	26.217	28.300	12
22.362	24.736	27.688	29.819	13
23.685	26.119	29.141	31.319	14
24.996	27.488	30.578	32.801	15
26.296	28.845	32.000	34.267	16
27.587	30.191	33.409	35.718	17
28.869	31.526	34.805	37.156	18
30.144	32.852	36.191	38.582	19
31.410	34.170	37.566	39.997	20
32.670	35.479	38.932	41.401	21
33.924	36.781	40.289	42.796	22
35.172	38.076	41.638	44.181	23
36.415	39.364	42.980	45.558	24
37.652	40.646	44.314	46.928	25
38.885	41.923	45.642	48.290	26
40.113	43.194	46.963	49.645	27
41.337	44.461	48.278	50.993	28
42.557	45.722	49.588	52.336	29
43.773	46.979	50.892	53.672	30
44.985	48.232	52.191	55.003	31
46.194	49.480	53.486	56.329	32
47.400	50.725	54.776	57.649	33
48.602	51.966	56.061	58.964	34
49.802	53.203	57.342	60.275	35
50.998	54.437	58.619	61.582	36
52.192	55.668	59.892	62.884	37
53.384	56.896	61.162	64.182	38
54.572	58.120	62.428	65.476	39
55.758	59.342	63.691	66.766	40
56.942	60.561	64.950	68.053	41
58.124	61.777	66.206	69.336	42
59.304	62.990	67.459	70.616	43
60.481	64.202	68.710	71.893	44
61.656	65.410	69.957	73.166	45
62.830	66.617	71.201	74.437	46
64.001	67.821	72.443	75.704	47
65.171	69.023	73.683	76.969	48
66.339	70.222	74.919	78.231	49
67.505	71.420	76.154	79.490	50

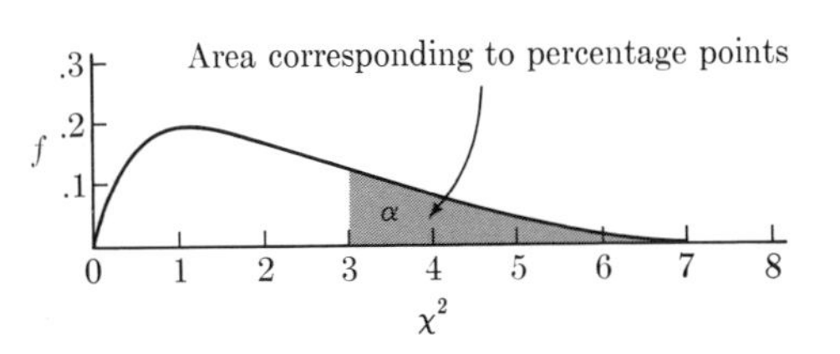

TABLE **IV** continued.

$\nu \backslash \alpha$	0.995	0.975	0.9	0.5	0.1
51	28.735	33.162	38.560	50.335	64.295
52	29.481	33.968	39.433	51.335	65.422
53	30.230	34.776	40.308	52.335	66.548
54	30.981	35.586	41.183	53.335	67.673
55	31.735	36.398	42.060	54.335	68.796
56	32.490	37.212	42.937	55.335	69.918
57	33.248	38.027	43.816	56.335	71.040
58	34.008	38.844	44.696	57.335	72.160
59	34.770	39.662	45.577	58.335	73.279
60	35.534	40.482	46.459	59.335	74.397
61	36.300	41.303	47.342	60.335	75.514
62	37.068	42.126	48.226	61.335	76.630
63	37.838	42.950	49.111	62.335	77.745
64	38.610	43.776	49.996	63.335	78.860
65	39.383	44.603	50.883	64.335	79.973
66	40.158	45.431	51.770	65.335	81.085
67	40.935	46.261	52.659	66.335	82.197
68	41.713	47.092	53.548	67.334	83.308
69	42.494	47.924	54.438	68.334	84.418
70	43.275	48.758	55.329	69.334	85.527
71	44.058	49.592	56.221	70.334	86.635
72	44.843	50.428	57.113	71.334	87.743
73	45.629	51.265	58.006	72.334	88.850
74	46.417	52.103	58.900	73.334	89.956
75	47.206	52.942	59.795	74.334	91.061
76	47.997	53.782	60.690	75.334	92.166
77	48.788	54.623	61.586	76.334	93.270
78	49.582	55.466	62.483	77.334	94.373
79	50.376	56.309	63.380	78.334	95.476
80	51.172	57.153	64.278	79.334	96.578
81	51.969	57.998	65.176	80.334	97.680
82	52.767	58.845	66.076	81.334	98.780
83	53.567	59.692	66.976	82.334	99.880
84	54.368	60.540	67.876	83.334	100.98
85	55.170	61.389	68.777	84.334	102.08
86	55.973	62.239	69.679	85.334	103.18
87	56.777	63.089	70.581	86.334	104.28
88	57.582	63.941	71.484	87.334	105.37
89	58.389	64.793	72.387	88.334	106.47
90	59.196	65.647	73.291	89.334	107.56
91	60.005	66.501	74.196	90.334	108.66
92	60.815	67.356	75.101	91.334	109.76
93	61.625	68.211	76.006	92.334	110.85
94	62.437	69.068	76.912	93.334	111.94
95	63.250	69.925	77.818	94.334	113.04
96	64.063	70.783	78.725	95.334	114.13
97	64.878	71.642	79.633	96.334	115.22
98	65.694	72.501	80.541	97.334	116.32
99	66.510	73.361	81.449	98.334	117.41
100	67.328	74.222	82.358	99.334	118.50

0.05	0.025	0.01	0.005	ν
68.669	72.616	77.386	80.747	51
69.832	73.810	78.616	82.001	52
70.993	75.002	79.843	83.253	53
72.153	76.192	81.069	84.502	54
73.311	77.380	82.292	85.749	55
74.468	78.567	83.513	86.994	56
75.624	79.752	84.733	88.237	57
76.778	80.936	85.950	89.477	58
77.931	82.117	87.166	90.715	59
79.082	83.298	88.379	91.952	60
80.232	84.476	89.591	93.186	61
81.381	85.654	90.802	94.419	62
82.529	86.830	92.010	95.649	63
83.675	88.004	93.217	96.878	64
84.821	89.177	94.422	98.105	65
85.965	90.349	95.626	99.331	66
87.108	91.519	96.828	100.55	67
88.250	92.689	98.028	101.78	68
89.391	93.856	99.228	103.00	69
90.531	95.023	100.43	104.21	70
91.670	96.189	101.62	105.43	71
92.808	97.353	102.82	106.65	72
93.945	98.516	104.01	107.86	73
95.081	99.678	105.20	109.07	74
96.217	100.84	106.39	110.29	75
97.351	102.00	107.58	111.50	76
98.484	103.16	108.77	112.70	77
99.617	104.32	109.96	113.91	78
100.75	105.47	111.14	115.12	79
101.88	106.63	112.33	116.32	80
103.01	107.78	113.51	117.52	81
104.14	108.94	114.69	118.73	82
105.27	110.09	115.88	119.93	83
106.39	111.24	117.06	121.13	84
107.52	112.39	118.24	122.32	85
108.65	113.54	119.41	123.52	86
109.77	114.69	120.59	124.72	87
110.90	115.84	121.77	125.91	88
112.02	116.99	122.94	127.11	89
113.15	118.14	124.12	128.30	90
114.27	119.28	125.29	129.49	91
115.39	120.43	126.46	130.68	92
116.51	121.57	127.63	131.87	93
117.63	122.72	128.80	133.06	94
118.75	123.86	129.97	134.25	95
119.87	125.00	131.14	135.43	96
120.99	126.14	132.31	136.62	97
122.11	127.28	133.48	137.80	98
123.23	128.42	134.64	138.99	99
124.34	129.56	135.81	140.17	100

TABLE **V. Critical values of the F-distribution.**

ν_2 (degrees of freedom of denominator mean square)	α	ν_1 (degrees of freedom of numerator mean square) 1	2	3	4	5	6
1	.05	161	199	216	225	230	234
	.025	648	800	864	900	922	937
	.01	4050	5000	5400	5620	5760	5860
2	.05	18.5	19.0	19.2	19.2	19.3	19.3
	.025	38.5	39.0	39.2	39.2	39.3	39.3
	.01	98.5	99.0	99.2	99.2	99.3	99.3
3	.05	10.1	9.55	9.28	9.12	9.01	8.94
	.025	17.4	16.0	15.4	15.1	14.9	14.7
	.01	34.1	30.8	29.5	28.7	28.2	27.9
4	.05	7.71	6.94	6.59	6.39	6.26	6.16
	.025	12.2	10.6	9.98	9.60	9.36	9.20
	.01	21.2	18.0	16.7	16.0	15.5	15.2
5	.05	6.61	5.79	5.41	5.19	5.05	4.95
	.025	10.0	8.43	7.76	7.39	7.15	6.98
	.01	16.3	13.3	12.1	11.4	11.0	10.7
6	.05	5.99	5.14	4.76	4.53	4.39	4.28
	.025	8.81	7.26	6.60	6.23	5.99	5.82
	.01	13.7	10.9	9.78	9.15	8.75	8.47
7	.05	5.59	4.74	4.35	4.12	3.97	3.87
	.025	8.07	6.54	5.89	5.52	5.29	5.12
	.01	12.2	9.55	8.45	7.85	7.46	7.19
8	.05	5.32	4.46	4.07	3.84	3.69	3.58
	.025	7.57	6.06	5.42	5.05	4.82	4.65
	.01	11.3	8.65	7.59	7.01	6.63	6.37
9	.05	5.12	4.26	3.86	3.63	3.48	3.37
	.025	7.21	5.71	5.08	4.72	4.48	4.32
	.01	10.6	8.02	6.99	6.42	6.06	5.80
10	.05	4.96	4.10	3.71	3.48	3.33	3.22
	.025	6.94	5.46	4.83	4.47	4.24	4.07
	.01	10.0	7.56	6.55	5.99	5.64	5.39

Note: Interpolation for number of degrees of freedom not furnished in the arguments is by means of harmonic interpolation (see footnote for Table **III**). If both ν_1 and ν_2 require interpolation, one needs to interpolate for each of these arguments in turn. Thus to obtain $F_{.05[55,80]}$ one first interpolates between $F_{.05[50,60]}$ and $F_{.05[60,60]}$ and between $F_{.05[50,120]}$ and $F_{.05[60,120]}$, to estimate $F_{.05[55,60]}$ and $F_{.05[55,120]}$, respectively. One then interpolates between these two values to obtain the desired quantity. Entries for α = 0.05, 0.025, 0.01, and 0.005 and for ν_1 and ν_2 = 1 to 10, 12, 15, 20, 24, 30, 40, 60, 120, and ∞ were copied from a table by M. Merrington and C. M. Thompson (*Biometrika* **33**:73–88, 1943) with permission of the publisher.

ν_2 (degrees of freedom of denominator mean square)	ν_1 (degrees of freedom of numerator mean square) 7	8	9	10	11	α
1	237	239	241	241	243	.05
	948	957	963	969	973	.025
	5930	5980	6020	6060	6080	.01
2	19.4	19.4	19.4	19.4	19.4	.05
	39.4	39.4	39.4	39.4	39.4	.025
	99.4	99.4	99.4	99.4	99.4	.01
3	8.89	8.85	8.81	8.79	8.76	.05
	14.6	14.5	14.5	14.4	14.3	.025
	27.7	27.5	27.3	27.2	27.1	.01
4	6.09	6.04	6.00	5.96	5.93	.05
	9.07	8.98	8.90	8.84	8.79	.025
	15.0	14.8	14.7	14.5	14.4	.01
5	4.88	4.82	4.77	4.74	4.71	.05
	6.85	6.76	6.68	6.62	6.57	.025
	10.5	10.3	10.2	10.1	9.99	.01
6	4.21	4.15	4.10	4.06	4.03	.05
	5.70	5.60	5.52	5.46	5.41	.025
	8.26	8.10	7.98	7.87	7.79	.01
7	3.77	3.73	3.68	3.64	3.60	.05
	4.99	4.89	4.82	4.76	4.71	.025
	6.99	6.84	6.72	6.62	6.54	.01
8	3.50	3.44	3.39	3.35	3.31	.05
	4.53	4.43	4.36	4.30	4.25	.025
	6.18	6.03	5.91	5.81	5.73	.01
9	3.29	3.23	3.18	3.14	3.10	.05
	4.20	4.10	4.03	3.96	3.91	.025
	5.61	5.47	5.35	5.26	5.18	.01
10	3.14	3.07	3.02	2.98	2.94	.05
	3.95	3.85	3.78	3.72	3.67	.025
	5.20	5.06	4.94	4.85	4.77	.01

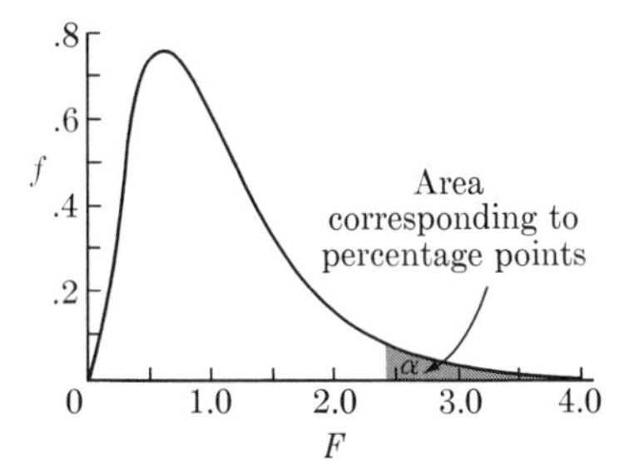

TABLE V continued.

ν_2 (degrees of freedom of denominator mean square)	α	ν_1 (degrees of freedom of numerator mean square) 12	15	20	24	30
1	.05	244	246	248	249	250
	.025	977	985	993	997	1000
	.01	6110	6160	6210	6230	6260
2	.05	19.4	19.4	19.4	19.5	19.5
	.025	39.4	39.4	39.4	39.5	39.5
	.01	99.4	99.4	99.4	99.5	99.5
3	.05	8.74	8.70	8.66	8.64	8.62
	.025	14.3	14.3	14.2	14.1	14.1
	.01	27.1	26.9	26.7	26.6	26.5
4	.05	5.91	5.86	5.80	5.77	5.75
	.025	8.75	8.66	8.56	8.51	8.46
	.01	14.4	14.2	14.0	13.9	13.8
5	.05	4.68	4.62	4.56	4.53	4.50
	.025	6.52	6.43	6.33	6.28	6.23
	.01	9.89	9.72	9.55	9.47	9.38
6	.05	4.00	3.94	3.87	3.84	3.81
	.025	5.37	5.27	5.17	5.12	5.07
	.01	7.72	7.56	7.40	7.31	7.23
7	.05	3.57	3.51	3.44	3.41	3.38
	.025	4.67	4.57	4.47	4.42	4.36
	.01	6.47	6.31	6.16	6.07	5.99
8	.05	3.28	3.22	3.15	3.12	3.08
	.025	4.20	4.10	4.00	3.95	3.89
	.01	5.67	5.52	5.36	5.28	5.20
9	.05	3.07	3.01	2.94	2.90	2.86
	.025	3.87	3.77	3.67	3.61	3.56
	.01	5.11	4.96	4.81	4.73	4.65
10	.05	2.91	2.85	2.77	2.74	2.70
	.025	3.62	3.52	3.42	3.37	3.31
	.01	4.71	4.56	4.41	4.33	4.25

ν_1 (degrees of freedom of numerator mean square) 40	60	120	∞	α	ν_2 (degrees of freedom of denominator mean square)
251	252	253	254	.05	1
1010	1010	1010	1020	.025	
6290	6310	6340	6370	.01	
19.5	19.5	19.5	19.5	.05	2
39.5	39.5	39.5	39.5	.025	
99.5	99.5	99.5	99.5	.01	
8.59	8.57	8.55	8.53	.05	3
14.0	14.0	13.9	13.9	.025	
26.4	26.3	26.2	26.1	.01	
5.72	5.69	5.66	5.63	.05	4
8.41	8.36	8.31	8.26	.025	
13.7	13.7	13.6	13.5	.01	
4.46	4.43	4.40	4.36	.05	5
6.18	6.12	6.07	6.02	.025	
9.29	9.20	9.11	9.02	.01	
3.77	3.74	3.70	3.67	.05	6
5.01	4.96	4.90	4.85	.025	
7.14	7.06	6.97	6.88	.01	
3.34	3.30	3.27	3.23	.05	7
4.31	4.25	4.20	4.14	.025	
5.91	5.82	5.74	5.65	.01	
3.04	3.01	2.97	2.93	.05	8
3.84	3.78	3.73	3.67	.025	
5.12	5.03	4.95	4.86	.01	
2.83	2.79	2.75	2.71	.05	9
3.51	3.45	3.39	3.33	.025	
4.57	4.48	4.40	4.31	.01	
2.66	2.62	2.58	2.54	.05	10
3.26	3.20	3.14	3.08	.025	
4.17	4.08	4.00	3.91	.01	

TABLE V continued.

ν_2 (degrees of freedom of denominator mean square)	α	ν_1 (degrees of freedom of numerator mean square) 1	2	3	4	5	6
11	.05	4.84	3.98	3.59	3.36	3.20	3.09
	.025	6.72	5.26	4.63	4.28	4.04	3.88
	.01	9.65	7.21	6.22	5.67	5.32	5.07
12	.05	4.75	3.89	3.49	3.26	3.11	3.00
	.025	6.55	5.10	4.47	4.12	3.89	3.73
	.01	9.33	6.93	5.95	5.41	5.06	4.82
15	.05	4.54	3.68	3.29	3.06	2.90	2.79
	.025	6.20	4.77	4.15	3.80	3.58	3.41
	.01	8.68	6.36	5.42	4.89	4.56	4.32
20	.05	4.35	3.49	3.10	2.87	2.71	2.60
	.025	5.87	4.46	3.86	3.51	3.29	3.13
	.01	8.10	5.85	4.94	4.43	4.10	3.87
24	.05	4.26	3.40	3.01	2.78	2.62	2.51
	.025	5.72	4.32	3.72	3.38	3.15	2.99
	.01	7.82	5.61	4.72	4.22	3.90	3.67
30	.05	4.17	3.32	2.92	2.69	2.53	2.42
	.025	5.57	4.18	3.59	3.25	3.03	2.87
	.01	7.56	5.39	4.51	4.02	3.70	3.47
40	.05	4.08	3.23	2.84	2.61	2.45	2.34
	.025	5.42	4.05	3.46	3.13	2.90	2.74
	.01	7.31	5.18	4.31	3.83	3.51	3.29
60	.05	4.00	3.15	2.76	2.53	2.37	2.25
	.025	5.29	3.93	3.34	3.01	2.79	2.63
	.01	7.08	4.98	4.13	3.65	3.34	3.12
120	.05	3.92	3.07	2.68	2.45	2.29	2.17
	.025	5.15	3.80	3.23	2.89	2.67	2.52
	.01	6.85	4.79	3.95	3.48	3.17	2.96
∞	.05	3.84	3.00	2.60	2.37	2.21	2.10
	.025	5.02	3.69	3.11	2.79	2.57	2.41
	.01	6.63	4.61	3.78	3.32	3.02	2.80

ν_2 (degrees of freedom of denominator mean square)	α	ν_1 (degrees of freedom of numerator mean square) 7	8	9	10	11
11	.05	3.01	2.95	2.90	2.85	2.82
	.025	3.76	3.66	3.59	3.53	3.48
	.01	4.89	4.74	4.63	4.54	4.46
12	.05	2.91	2.85	2.80	2.75	2.72
	.025	3.61	3.51	3.44	3.37	3.32
	.01	4.64	4.50	4.39	4.30	4.22
15	.05	2.71	2.64	2.59	2.54	2.51
	.025	3.29	3.20	3.12	3.06	3.01
	.01	4.14	4.00	3.89	3.80	3.73
20	.05	2.51	2.45	2.39	2.35	2.31
	.025	3.01	2.91	2.84	2.77	2.72
	.01	3.70	3.56	3.46	3.37	3.29
24	.05	2.42	2.36	2.30	2.25	2.22
	.025	2.87	2.78	2.70	2.64	2.59
	.01	3.50	3.36	3.26	3.17	3.09
30	.05	2.33	2.27	2.21	2.16	2.13
	.025	2.75	2.65	2.57	2.51	2.46
	.01	3.30	3.17	3.07	2.98	2.90
40	.05	2.25	2.18	2.12	2.08	2.04
	.025	2.62	2.53	2.45	2.39	2.33
	.01	3.12	2.99	2.89	2.80	2.73
60	.05	2.17	2.10	2.04	1.99	1.95
	.025	2.51	2.41	2.33	2.27	2.22
	.01	2.95	2.82	2.72	2.63	2.56
120	.05	2.09	2.02	1.96	1.91	1.87
	.025	2.39	2.30	2.22	2.16	2.10
	.01	2.79	2.66	2.56	2.47	2.40
∞	.05	2.01	1.94	1.88	1.83	1.79
	.025	2.29	2.19	2.11	2.05	1.99
	.01	2.64	2.51	2.41	2.32	2.25

TABLE V continued.

ν_2 (degrees of freedom of denominator mean square)	α	ν_1 (degrees of freedom of numerator mean square) 12	15	20	24	30
11	.05	2.79	2.72	2.65	2.61	2.57
	.025	3.43	3.33	3.23	3.17	3.12
	.01	4.40	4.25	4.10	4.02	3.94
12	.05	2.69	2.62	2.54	2.51	2.47
	.025	3.28	3.18	3.07	3.02	2.96
	.01	4.16	4.01	3.86	3.78	3.70
15	.05	2.48	2.40	2.33	2.39	2.25
	.025	2.96	2.86	2.76	2.70	2.64
	.01	3.67	3.52	3.37	3.29	3.21
20	.05	2.28	2.20	2.12	2.08	2.04
	.025	2.68	2.57	2.46	2.41	2.35
	.01	3.23	3.09	2.94	2.86	2.78
24	.05	2.18	2.11	2.03	1.98	1.94
	.025	2.54	2.44	2.33	2.27	2.21
	.01	3.03	2.89	2.74	2.66	2.58
30	.05	2.09	2.01	1.93	1.89	1.84
	.025	2.41	2.31	2.20	2.14	2.07
	.01	2.84	2.70	2.55	2.47	2.39
40	.05	2.04	1.92	1.84	1.79	1.74
	.025	2.29	2.18	2.07	2.01	1.94
	.01	2.66	2.52	2.37	2.29	2.20
60	.05	1.92	1.84	1.75	1.70	1.65
	.025	2.17	2.06	1.94	1.88	1.82
	.01	2.50	2.35	2.20	2.12	2.03
120	.05	1.83	1.75	1.66	1.61	1.55
	.025	2.05	1.95	1.82	1.76	1.69
	.01	2.34	2.19	2.03	1.95	1.86
∞	.05	1.75	1.67	1.57	1.52	1.46
	.025	1.94	1.83	1.71	1.64	1.57
	.01	2.18	2.04	1.88	1.79	1.70

ν_1 (degrees of freedom of numerator mean square)					ν_2 (degrees of freedom of denominator mean square)
40	60	120	∞	α	
2.53	2.49	2.45	2.40	.05	11
3.06	3.00	2.94	2.88	.025	
3.86	3.78	3.69	3.60	.01	
2.43	2.38	2.34	2.30	.05	12
2.91	2.85	2.79	2.72	.025	
3.62	3.54	3.45	3.36	.01	
2.20	2.16	2.11	2.07	.05	15
2.59	2.52	2.46	2.40	.025	
3.13	3.05	2.96	2.87	.01	
1.99	1.95	1.90	1.84	.05	20
2.29	2.22	2.16	2.09	.025	
2.69	2.61	2.52	2.42	.01	
1.89	1.84	1.79	1.73	.05	24
2.15	2.08	2.01	1.94	.025	
2.49	2.40	2.31	2.21	.01	
1.79	1.74	1.68	1.62	.05	30
2.01	1.94	1.87	1.79	.025	
2.30	2.21	2.11	2.01	.01	
1.69	1.64	1.58	1.51	.05	40
1.88	1.80	1.72	1.64	.025	
2.11	2.02	1.92	1.80	.01	
1.59	1.53	1.47	1.39	.05	60
1.74	1.67	1.58	1.48	.025	
1.94	1.84	1.73	1.60	.01	
1.50	1.43	1.35	1.25	.05	120
1.61	1.53	1.43	1.31	.025	
1.76	1.66	1.53	1.38	.01	
1.39	1.32	1.22	1.00	.05	∞
1.48	1.39	1.27	1.00	.025	
1.59	1.47	1.32	1.00	.01	

TABLE **VI. Critical values of F_{max}.**

$\nu \backslash a$	2	3	4	5	6
2	39.0 199.	87.5 448.	142. 729.	202. 1036	266. 1362
3	15.4 47.5	27.8 85.	39.2 120.	50.7 151.	62.0 184.
4	9.60 23.2	15.5 37.	20.6 49.	25.2 59.	29.5 69.
5	7.15 14.9	10.8 22.	13.7 28.	16.3 33.	18.7 38.
6	5.82 11.1	8.38 15.5	10.4 19.1	12.1 22.	13.7 25.
7	4.99 8.89	6.94 12.1	8.44 14.5	9.70 16.5	10.8 18.4
8	4.43 7.50	6.00 9.9	7.18 11.7	8.12 13.2	9.03 14.5
9	4.03 6.54	5.34 8.5	6.31 9.9	7.11 11.1	7.80 12.1
10	3.72 5.85	4.85 7.4	5.67 8.6	6.34 9.6	6.92 10.4
12	3.28 4.91	4.16 6.1	4.79 6.9	5.30 7.6	5.72 8.2
15	2.86 4.07	3.54 4.9	4.01 5.5	4.37 6.0	4.68 6.4
20	2.46 3.32	2.95 3.8	3.29 4.3	3.54 4.6	3.76 4.9
30	2.07 2.63	2.40 3.0	2.61 3.3	2.78 3.4	2.91 3.6
60	1.67 1.96	1.85 2.2	1.96 2.3	2.04 2.4	2.11 2.4
∞	1.00 1.00	1.00 1.0	1.00 1.0	1.00 1.0	1.00 1.0

Note: Corresponding to each value of a (number of samples) and ν (degrees of freedom) are two critical values of F_{max} representing the upper 5% and 1% percentage points. The corresponding probabilities $\alpha = 0.05$ and 0.01 represent *one tail* of the F_{max}-distribution. This table was copied from H. A. David (*Biometrika* **39**:422–424, 1952) with permission of the publisher and author.

7	8	9	10	11	12
333. 1705	403. 2063	475. 2432	550. 2813	626. 3204	704. 3605
72.9 21(6)	83.5 24(9)	93.9 28(1)	104. 31(0)	114. 33(7)	124. 36(1)
33.6 79.	37.5 89.	41.1 97.	44.6 106.	48.0 113.	51.4 120.
20.8 42.	22.9 46.	24.7 50.	26.5 54.	28.2 57.	29.9 60.
15.0 27.	16.3 30.	17.5 32.	18.6 34.	19.7 36.	20.7 37.
11.8 20.	12.7 22.	13.5 23.	14.3 24.	15.1 26.	15.8 27.
9.78 15.8	10.5 16.9	11.1 17.9	11.7 18.9	12.2 19.8	12.7 21.
8.41 13.1	8.95 13.9	9.45 14.7	9.91 15.3	10.3 16.0	10.7 16.6
7.42 11.1	7.87 11.8	8.28 12.4	8.66 12.9	9.01 13.4	9.34 13.9
6.09 8.7	6.42 9.1	6.72 9.5	7.00 9.9	7.25 10.2	7.48 10.6
4.95 6.7	5.19 7.1	5.40 7.3	5.59 7.5	5.77 7.8	5.93 8.0
3.94 5.1	4.10 5.3	4.24 5.5	4.37 5.6	4.49 5.8	4.59 5.9
3.02 3.7	3.12 3.8	3.21 3.9	3.29 4.0	3.36 4.1	3.39 4.2
2.17 2.5	2.22 2.5	2.26 2.6	2.30 2.6	2.33 2.7	2.36 2.7
1.00 1.0	1.00 1.0	1.00 1.0	1.00 1.0	1.00 1.0	1.00 1.0

TABLE **VII. Shortest unbiased confidence limits for the variance.**

	Confidence coefficients			Confidence coefficients			Confidence coefficients	
ν	0.95	0.99	ν	0.95	0.99	ν	0.95	0.99
2	.2099	.1505	14	.5135	.4289	26	.6057	.5261
	23.605	114.489		2.354	3.244		1.825	2.262
3	.2681	.1983	15	.5242	.4399	27	.6110	.5319
	10.127	29.689		2.276	3.091		1.802	2.223
4	.3125	.2367	16	.5341	.4502	28	.6160	.5374
	6.590	15.154		2.208	2.961		1.782	2.187
5	.3480	.2685	17	.5433	.4598	29	.6209	.5427
	5.054	10.076		2.149	2.848		1.762	2.153
6	.3774	.2956	18	.5520	.4689	30	.6255	.5478
	4.211	7.637		2.097	2.750		1.744	2.122
7	.4025	.3192	19	.5601	.4774	40	.6636	.5900
	3.679	6.238		2.050	2.664		1.608	1.896
8	.4242	.3400	20	.5677	.4855	50	.6913	.6213
	3.314	5.341		2.008	2.588		1.523	1.760
9	.4432	.3585	21	.5749	.4931	60	.7128	.6458
	3.048	4.720		1.971	2.519		1.464	1.668
10	.4602	.3752	22	.5817	.5004	70	.7300	.6657
	2.844	4.265		1.936	2.458		1.421	1.607
11	.4755	.3904	23	.5882	.5073	80	.7443	.6824
	2.683	3.919		1.905	2.402		1.387	1.549
12	.4893	.4043	24	.5943	.5139	90	.7564	.6966
	2.553	3.646		1.876	2.351		1.360	1.508
13	.5019	.4171	25	.6001	.5201	100	.7669	.7090
	2.445	3.426		1.850	2.305		1.338	1.475

Note: The factors in this table have been obtained by dividing the quantity $n - 1$ by the values found in a table prepared by D. V. Lindley, D. A. East, and P. A. Hamilton (*Biometrika* **47**:433–437, 1960).

TABLE **VIII. Critical values for correlation coefficients.**

ν	α	r	ν	α	r	ν	α	r
1	.05	.997	16	.05	.468	35	.05	.325
	.01	1.000		.01	.590		.01	.418
2	.05	.950	17	.05	.456	40	.05	.304
	.01	.990		.01	.575		.01	.393
3	.05	.878	18	.05	.444	45	.05	.288
	.01	.959		.01	.561		.01	.372
4	.05	.811	19	.05	.433	50	.05	.273
	.01	.917		.01	.549		.01	.354
5	.05	.754	20	.05	.423	60	.05	.250
	.01	.874		.01	.537		.01	.325
6	.05	.707	21	.05	.413	70	.05	.232
	.01	.834		.01	.526		.01	.302
7	.05	.666	22	.05	.404	80	.05	.217
	.01	.798		.01	.515		.01	.283
8	.05	.632	23	.05	.396	90	.05	.205
	.01	.765		.01	.505		.01	.267
9	.05	.602	24	.05	.388	100	.05	.195
	.01	.735		.01	.496		.01	.254
10	.05	.576	25	.05	.381	125	.05	.174
	.01	.708		.01	.487		.01	.228
11	.05	.553	26	.05	.374	150	.05	.159
	.01	.684		.01	.478		.01	.208
12	.05	.532	27	.05	.367	200	.05	.138
	.01	.661		.01	.470		.01	.181
13	.05	.514	28	.05	.361	300	.05	.113
	.01	.641		.01	.463		.01	.148
14	.05	.497	29	.05	.355	400	.05	.098
	.01	.623		.01	.456		.01	.128
15	.05	.482	30	.05	.349	500	.05	.088
	.01	.606		.01	.449		.01	.115
						1,000	.05	.062
							.01	.081

Note: Upper value is 5%, lower value is 1% critical value. This table is reproduced by permission from *Statistical Methods*, 5th edition, by George W. Snedecor, © 1956, by The Iowa State University Press.

TABLE IX. Confidence limits for percentages

This table furnishes confidence limits for percentages based on the binomial distribution.

The first part of the table furnishes limits for samples up to size $n = 30$. The arguments are Y, number of items in the sample which exhibit a given property, and n, sample size. Argument Y is tabled for integral values between 0 and 15, which yield percentages up to 50%. For each sample size n and number of items Y with the given property, three lines of numerical values are shown. The first line of values gives 95% confidence limits for the percentage, the second line lists the observed percentage incidence of the property, and the third line of values furnishes the 99% confidence limits for the percentage. Thus for example, for $Y = 8$ individuals showing the property out of a sample of $n = 20$, the second line indicates that this represents an incidence of the property of 40.00%, the first line yields the 95% confidence limits of this percentage as 19.10% to 63.95%, and the third line gives the 99% limits as 14.60% to 70.10%.

Interpolate in this table (up to $n = 49$) by dividing L_1^- and L_2^-, the lower and upper confidence limits at the next lower tabled sample size n^- by desired sample size n, and multiply them by the next lower tabled sample size n^-. Thus, for example, to obtain the confidence limits of the percentage corresponding to 8 individuals showing the given property in a sample of 22 individuals (which corresponds to 36.36% of the individuals showing the property), compute the lower confidence limit $L_1 = L_1^- n^-/n = (19.10)20/22 = 17.36\%$ and the upper confidence limit $L_2 = L_2^- n^-/n = (63.95)20/22 = 58.14\%$.

The second half of the table is for larger sample sizes ($n = 50$, 100, 200, 500, and 1,000). The arguments along the left margin of the table are now percentages from 0 to 50% in increments of 1%, rather than counts. The 95% and 99% confidence limits corresponding to a given percentage incidence p and sample size n are the functions given in two lines in the body of the table. For instance, the 99% confidence limits of an observed incidence of 12% in a sample of 500 are found to be 8.56–16.19%, in the second of the two lines. Interpolation in this table between the furnished sample sizes can be achieved by means of the following formula for the lower limit

$$L_1 = \frac{[L_1^- n^-(n^+ - n) + L_1^+ n^+(n - n^-)]}{n(n^+ - n^-)}$$

In the above expression, n is the size of the observed sample, n^- and n^+ are the next lower and upper tabled sample sizes, respectively, L_1^- and L_1^+ are corresponding tabled confidence limits for these sample sizes, and L_1 is the lower confidence limit to be found by interpolation. The upper confidence limit, L_2, can be obtained by a corresponding formula by substituting 2 for the subscript 1. By way of an example we shall illustrate setting 95% confidence limits to an observed percentage of 25% in a sample size of 80. The tabled 95% limits for $n = 50$ are 13.84–39.27%. For $n = 100$ the corresponding tabled limits are 16.88–34.66%. When we substitute the values for the lower limits in the above formula we obtain

$$L_1 = [(13.84)(50)(100 - 80) + (16.88)(100)(80 - 50)]/80(100 - 50) = 16.12\%$$

for the lower confidence limit. Similarly, for the upper confidence limit we compute

$$L_2 = [(39.27)(50)(100 - 80) + (34.66)(100)(80 - 50)]/80(100 - 50) = 35.81\%$$

The tabled values in parentheses are limits for percentages that could not be obtained in any real sampling problem (for example, 25% in 50 items), but are necessary for purposes of interpolation. For percentages greater than 50% look up the complementary percentage as the argument. The complements of the tabled binomial confidence limits are the desired limits.

These tables have been extracted from more extensive ones by D. Mainland, L. Herrera and M. I. Sutcliffe, *Tables for Use with Binomial Samples* (1956) with permission of the authors. The interpolation formulas cited are also due to these authors. Confidence limits of odd percentages up to 13% for $n = 50$ were computed by interpolation.

TABLE **IX**. Confidence limits for percentages.

Y	Confidence coefficients	n = 5	10	15	20	25	30	Confidence coefficients	Y
0	95	0.00 - 52.18	0.00 - 30.85	0.00 - 21.80	0.00 - 16.85	0.00 - 13.72	0.00 - 11.57	95	0
		0.00	0.00	0.00	0.00	0.00	0.00		
	99	0.00 - 65.34	0.00 - 41.13	0.00 - 29.76	0.00 - 23.27	0.00 - 19.10	0.00 - 16.19	99	
1	95	0.51 - 71.60	0.25 - 44.50	0.17 - 32.00	0.13 - 24.85	0.10 - 20.36	0.08 - 17.23	95	1
		20.00	10.00	6.67	5.00	4.00	3.33		
	99	0.10 - 81.40	0.05 - 54.40	0.03 - 40.27	0.02 - 31.70	0.02 - 26.24	0.02 - 22.33	99	
2	95	5.28 - 85.34	2.52 - 55.60	1.66 - 40.49	1.24 - 31.70	0.98 - 26.05	0.82 - 22.09	95	2
		40.00	20.00	13.33	10.00	8.00	6.67		
	99	2.28 - 91.72	1.08 - 64.80	0.71 - 48.71	0.53 - 38.70	0.42 - 32.08	0.35 - 27.35	99	
3	95		6.67 - 65.20	4.33 - 48.07	3.21 - 37.93	2.55 - 31.24	2.11 - 26.53	95	3
			30.00	20.00	15.00	12.00	10.00		
	99		3.70 - 73.50	2.39 - 56.07	1.77 - 45.05	1.40 - 37.48	1.16 - 32.03	99	
4	95		12.20 - 73.80	7.80 - 55.14	5.75 - 43.65	4.55 - 36.10	3.77 - 30.74	95	4
			40.00	26.67	20.00	16.00	13.33		
	99		7.68 - 80.91	4.88 - 62.78	3.58 - 50.65	2.83 - 42.41	2.34 - 36.39	99	
5	95		18.70 - 81.30	11.85 - 61.62	8.68 - 49.13	6.84 - 40.72	5.64 - 34.74	95	5
			50.00	33.33	25.00	20.00	16.67		
	99		12.80 - 87.20	8.03 - 68.89	5.85 - 56.05	4.60 - 47.00	3.79 - 40.44	99	
6	95			16.33 - 67.74	11.90 - 54.30	9.35 - 45.14	7.70 - 38.56	95	6
				40.00	30.00	24.00	20.00		
	99			11.67 - 74.40	8.45 - 60.95	6.62 - 51.38	5.43 - 44.26	99	
7	95			21.29 - 73.38	15.38 - 59.20	12.06 - 49.38	9.92 - 42.29	95	7
				46.67	35.00	28.00	23.33		
	99			15.87 - 79.54	11.40 - 65.70	8.90 - 55.56	7.29 - 48.01	99	

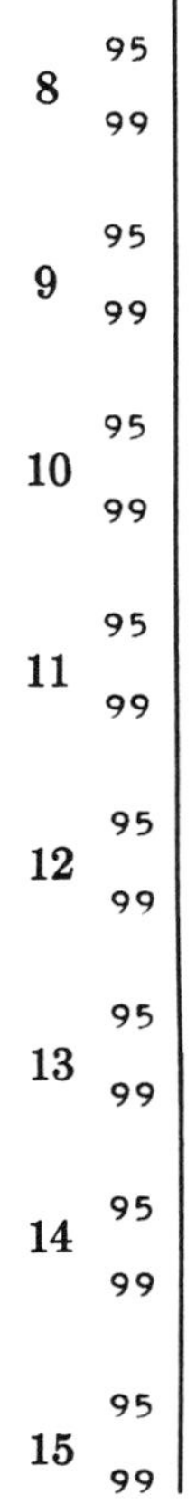

8	95	19.10 - 63.95	14.96 - 53.50	12.29 - 45.89	95	8
		40.00	32.00	26.67		
	99	14.60 - 70.10	11.36 - 59.54	9.30 - 51.58	99	
9	95	23.05 - 68.48	17.97 - 57.48	14.73 - 49.40	95	9
		45.00	36.00	30.00		
	99	18.08 - 74.30	14.01 - 63.36	11.43 - 55.00	99	
10	95	27.20 - 72.80	21.12 - 61.32	17.29 - 52.80	95	10
		50.00	40.00	33.33		
	99	21.75 - 78.25	16.80 - 67.04	13.69 - 58.35	99	
11	95		24.41 - 65.06	19.93 - 56.13	95	11
			44.00	36.67		
	99		19.75 - 70.55	16.06 - 61.57	99	
12	95		27.81 - 68.69	22.66 - 59.39	95	12
			48.00	40.00		
	99		22.84 - 73.93	18.50 - 64.69	99	
13	95			25.46 - 62.56	95	13
				43.33		
	99			21.07 - 67.72	99	
14	95			28.35 - 65.66	95	14
				46.67		
	99			23.73 - 70.66	99	
15	95			31.30 - 68.70	95	15
				50.00		
	99			26.47 - 73.53	99	

TABLE **IX** continued.

%	Confidence coefficients	n 50	100	200	500	1000
0	95	.00- 7.11	.00- 3.62	.00- 1.83	.00- 0.74	.00- 0.37
	99	.00-10.05	.00- 5.16	.00- 2.62	.00- 1.05	.00- 0.53
1	95	(.02- 8.88)	.02- 5.45	.12- 3.57	.32- 2.32	.48- 1.83
	99	(.00-12.02)	.00- 7.21	.05- 4.55	.22- 2.80	.37- 2.13
2	95	.05-10.66	.24- 7.04	.55- 5.04	1.06- 3.56	1.29- 3.01
	99	.01-13.98	.10- 8.94	.34- 6.17	.87- 4.12	1.13- 3.36
3	95	(.27-12.19)	.62- 8.53	1.11- 6.42	1.79- 4.81	2.11- 4.19
	99	(.16-15.60)	.34-10.57	.78- 7.65	1.52- 5.44	1.88- 4.59
4	95	.49-13.72	1.10- 9.93	1.74- 7.73	2.53- 6.05	2.92- 5.36
	99	.21-17.21	.68-12.08	1.31- 9.05	2.17- 6.75	2.64- 5.82
5	95	(.88-15.14)	1.64-11.29	2.43- 9.00	3.26- 7.29	3.73- 6.54
	99	(.45-18.76)	1.10-13.53	1.89-10.40	2.83- 8.07	3.39- 7.05
6	95	1.26-16.57	2.24-12.60	3.18-10.21	4.11- 8.43	4.63- 7.64
	99	.69-20.32	1.56-14.93	2.57-11.66	3.63- 9.24	4.25- 8.18
7	95	(1.74-17.91)	2.86-13.90	3.88-11.47	4.96- 9.56	5.52- 8.73
	99	(1.04-21.72)	2.08-16.28	3.17-12.99	4.43-10.42	5.12- 9.31
8	95	2.23-19.25	3.51-15.16	4.70-12.61	5.81-10.70	6.42- 9.83
	99	1.38-23.13	2.63-17.61	3.93-14.18	5.23-11.60	5.98-10.43
9	95	(2.78-20.54)	4.20-16.40	5.46-13.82	6.66-11.83	7.32-10.93
	99	(1.80-24.46)	3.21-18.92	4.61-15.44	6.04-12.77	6.84-11.56
10	95	3.32-21.82	4.90-17.62	6.22-15.02	7.51-12.97	8.21-12.03
	99	2.22-25.80	3.82-20.20	5.29-16.70	6.84-13.95	7.70-12.69
11	95	(3.93-23.06)	5.65-18.80	7.05-16.16	8.41-14.06	9.14-13.10
	99	(2.70-27.11)	4.48-21.42	6.06-17.87	7.70-15.07	8.60-13.78
12	95	4.54-24.31	6.40-19.98	7.87-17.30	9.30-15.16	10.06-14.16
	99	3.18-28.42	5.15-22.65	6.83-19.05	8.56-16.19	9.51-14.86
13	95	(5.18-27.03)	7.11-21.20	8.70-18.44	10.20-16.25	10.99-15.23
	99	(3.72-29.67)	5.77-23.92	7.60-20.23	9.42-17.31	10.41-15.95
14	95	5.82-26.75	7.87-22.37	9.53-19.58	11.09-17.34	11.92-16.30
	99	4.25-30.92	6.46-25.13	8.38-21.40	10.28-18.43	11.31-17.04
15	95	(6.50-27.94)	8.64-23.53	10.36-20.72	11.98-18.44	12.84-17.37
	99	(4.82-32.14)	7.15-26.33	9.15-22.58	11.14-19.55	12.21-18.13

TABLE **IX** continued.

%	Confidence coefficients	n				
		50	100	200	500	1000
16	95	7.17-29.12	9.45-24.66	11.22-21.82	12.90-19.50	13.79-18.42
	99	5.40-33.36	7.89-27.49	9.97-23.71	12.03-20.63	13.14-19.19
17	95	(7.88-30.28)	10.25-25.79	12.09-22.92	13.82-20.57	14.73-19.47
	99	(6.00-34.54)	8.63-28.65	10.79-24.84	12.92-21.72	14.07-20.25
18	95	8.58-31.44	11.06-26.92	12.96-24.02	14.74-21.64	15.67-20.52
	99	6.60-35.73	9.37-29.80	11.61-25.96	13.81-22.81	14.99-21.32
19	95	(9.31-32.58)	11.86-28.06	13.82-25.12	15.66-22.71	16.62-21.57
	99	(7.23-36.88)	10.10-30.96	12.43-27.09	14.71-23.90	15.92-22.38
20	95	10.04-33.72	12.66-29.19	14.69-26.22	16.58-23.78	17.56-22.62
	99	7.86-38.04	10.84-32.12	13.26-28.22	15.60-24.99	16.84-23.45
21	95	(10.79-34.84)	13.51-30.28	15.58-27.30	17.52-24.83	18.52-23.65
	99	(8.53-39.18)	11.63-33.24	14.11-29.31	16.51-26.05	17.78-24.50
22	95	11.54-35.95	14.35-31.37	16.48-28.37	18.45-25.88	19.47-24.69
	99	9.20-40.32	12.41-34.35	14.97-30.40	17.43-27.12	18.72-25.55
23	95	(12.30-37.06)	15.19-32.47	17.37-29.45	19.39-26.93	20.43-25.73
	99	(9.88-41.44)	13.60-34.82	15.83-31.50	18.34-28.18	19.67-26.59
24	95	13.07-38.17	16.03-33.56	18.27-30.52	20.33-27.99	21.39-26.77
	99	10.56-42.56	13.98-36.57	16.68-32.59	19.26-29.25	20.61-27.64
25	95	(13.84-39.27)	16.88-34.66	19.16-31.60	21.26-29.04	22.34-27.81
	99	(11.25-43.65)	14.77-37.69	17.54-33.68	20.17-30.31	21.55-28.69
26	95	14.63-40.34	17.75-35.72	20.08-32.65	22.21-30.08	23.31-28.83
	99	11.98-44.73	15.59-38.76	18.43-34.75	21.10-31.36	22.50-29.73
27	95	(15.45-41.40)	18.62-36.79	20.99-33.70	23.16-31.11	24.27-29.86
	99	(12.71-45.79)	16.42-39.84	19.31-35.81	22.04-32.41	23.46-30.76
28	95	16.23-42.48	19.50-37.85	21.91-34.76	24.11-32.15	25.24-30.89
	99	13.42-46.88	17.25-40.91	20.20-36.88	22.97-33.46	24.41-31.80
29	95	(17.06-43.54)	20.37-38.92	22.82-35.81	25.06-33.19	26.21-31.92
	99	(14.18-47.92)	18.07-41.99	21.08-37.94	23.90-34.51	25.37-32.84
30	95	17.87-44.61	21.24-39.98	23.74-36.87	26.01-34.23	27.17-32.95
	99	14.91-48.99	18.90-43.06	21.97-39.01	24.83-35.55	26.32-33.87

TABLE **IX** continued.

%	Confidence coefficients	n = 50	100	200	500	1000
31	95	(18.71-45.65)	22.14-41.02	24.67-37.90	26.97-35.25	28.15-33.97
	99	(15.68-50.02)	19.76-44.11	22.88-40.05	25.78-36.59	27.29-34.90
32	95	19.55-46.68	23.04-42.06	25.61-38.94	27.93-36.28	29.12-34.99
	99	16.46-51.05	20.61-45.15	23.79-41.09	26.73-37.62	28.25-35.92
33	95	(20.38-47.72)	23.93-43.10	26.54-39.97	28.90-37.31	30.09-36.01
	99	(17.23-52.08)	21.47-46.19	24.69-42.13	27.68-38.65	29.22-36.95
34	95	21.22-48.76	24.83-44.15	27.47-41.01	29.86-38.33	31.07-37.03
	99	18.01-53.11	22.33-47.24	25.60-43.18	28.62-39.69	30.18-37.97
35	95	(22.06-49.80)	25.73-45.19	28.41-42.04	30.82-39.36	32.04-38.05
	99	(18.78-54.14)	23.19-48.28	26.51-44.22	29.57-40.72	31.14-39.00
36	95	22.93-50.80	26.65-46.20	29.36-43.06	31.79-40.38	33.02-39.06
	99	19.60-55.13	24.08-49.30	27.44-45.24	30.53-41.74	32.12-40.02
37	95	(23.80-51.81)	27.57-47.22	30.31-44.08	32.76-41.39	34.00-40.07
	99	(20.42-56.12)	24.96-50.31	28.37-46.26	31.49-42.76	33.09-41.03
38	95	24.67-52.81	28.49-48.24	31.25-45.10	33.73-42.41	34.98-41.09
	99	21.23-57.10	25.85-51.32	29.30-47.29	32.45-43.78	34.07-42.05
39	95	(25.54-53.82)	29.41-49.26	32.20-46.12	34.70-43.43	35.97-42.10
	99	(22.05-58.09)	26.74-52.34	30.23-48.31	33.42-44.80	35.04-43.06
40	95	26.41-54.82	30.33-50.28	33.15-47.14	35.68-44.44	36.95-43.11
	99	22.87-59.08	27.63-53.35	31.16-49.33	34.38-45.82	36.02-44.08
41	95	(27.31-55.80)	31.27-51.28	34.12-48.15	36.66-45.45	37.93-44.12
	99	(23.72-60.04)	28.54-54.34	32.11-50.33	35.35-46.83	37.00-45.09
42	95	28.21-56.78	32.21-52.28	35.08-49.16	37.64-46.46	38.92-45.12
	99	24.57-60.99	29.45-55.33	33.06-51.33	36.32-47.83	37.98-46.10
43	95	(29.10-57.76)	33.15-53.27	36.05-50.16	38.62-47.46	39.91-46.13
	99	(25.42-61.95)	30.37-56.32	34.01-52.34	37.29-48.84	38.96-47.10
44	95	30.00-58.74	34.09-54.27	37.01-51.17	39.60-48.47	40.90-47.14
	99	26.27-62.90	31.28-57.31	34.95-53.34	38.27-49.85	39.95-48.11
45	95	(30.90-59.71)	35.03-55.27	37.97-52.17	40.58-49.48	41.89-48.14
	99	(27.12-63.86)	32.19-58.30	35.90-54.34	39.24-50.86	40.93-49.12
46	95	31.83-60.67	35.99-56.25	38.95-53.17	41.57-50.48	42.88-49.14
	99	28.00-64.78	33.13-59.26	36.87-55.33	40.22-51.85	41.92-50.12
47	95	(32.75-61.62)	36.95-57.23	39.93-54.16	42.56-51.48	43.87-50.14
	99	(28.89-65.69)	34.07-60.22	37.84-56.31	41.21-52.85	42.91-51.12
48	95	33.68-62.57	37.91-58.21	40.91-55.15	43.55-52.47	44.87-51.14
	99	29.78-66.61	35.01-61.19	38.80-57.30	42.19-53.85	43.90-52.12
49	95	(34.61-63.52)	38.87-59.19	41.89-56.14	44.54-53.47	45.86-52.14
	99	(30.67-67.53)	35.95-62.15	39.77-58.28	43.18-54.84	44.89-53.12
50	95	35.53-64.47	39.83-60.17	42.86-57.14	45.53-54.47	46.85-53.15
	99	31.55-68.45	36.89-63.11	40.74-59.26	44.16-55.84	45.89-54.11

TABLE **X. The z-transformation of correlation coefficient r.**

r	z	r	z
.00	0.0000	.50	0.5493
.01	0.0100	.51	0.5627
.02	0.0200	.52	0.5763
.03	0.0300	.53	0.5901
.04	0.0400	.54	0.6042
.05	0.0500	.55	0.6184
.06	0.0601	.56	0.6328
.07	0.0701	.57	0.6475
.08	0.0802	.58	0.6625
.09	0.0902	.59	0.6777
.10	0.1003	.60	0.6931
.11	0.1104	.61	0.7089
.12	0.1206	.62	0.7250
.13	0.1307	.63	0.7414
.14	0.1409	.64	0.7582
.15	0.1511	.65	0.7753
.16	0.1614	.66	0.7928
.17	0.1717	.67	0.8107
.18	0.1820	.68	0.8291
.19	0.1923	.69	0.8480
.20	0.2027	.70	0.8673
.21	0.2132	.71	0.8872
.22	0.2237	.72	0.9076
.23	0.2342	.73	0.9287
.24	0.2448	.74	0.9505
.25	0.2554	.75	0.9730
.26	0.2661	.76	0.9962
.27	0.2769	.77	1.0203
.28	0.2877	.78	1.0454
.29	0.2986	.79	1.0714
.30	0.3095	.80	1.0986
.31	0.3205	.81	1.1270
.32	0.3316	.82	1.1568
.33	0.3428	.83	1.1881
.34	0.3541	.84	1.2212
.35	0.3654	.85	1.2562
.36	0.3769	.86	1.2933
.37	0.3884	.87	1.3331
.38	0.4001	.88	1.3758
.39	0.4118	.89	1.4219
.40	0.4236	.90	1.4722
.41	0.4356	.91	1.5275
.42	0.4477	.92	1.5890
.43	0.4599	.93	1.6584
.44	0.4722	.94	1.7380
.45	0.4847	.95	1.8318
.46	0.4973	.96	1.9459
.47	0.5101	.97	2.0923
.48	0.5230	.98	2.2976
.49	0.5361	.99	2.6467

TABLE **XI.** ***f* ln *f* as a function of *f*.**

0-490

f	0	1	2	3	4
0	0.000	0.000	1.386	3.296	5.545
10	23.026	26.377	29.819	33.344	36.947
20	59.915	63.935	68.003	72.116	76.273
30	102.036	106.454	110.904	115.385	119.896
40	147.555	152.256	156.982	161.732	166.504
50	195.601	200.523	205.465	210.425	215.405
60	245.661	250.763	255.882	261.017	266.169
70	297.395	302.650	307.920	313.204	318.501
80	350.562	355.950	361.351	366.764	372.189
90	404.983	410.488	416.005	421.532	427.070
100	460.517	466.127	471.747	477.377	483.017
110	517.053	522.758	528.472	534.195	539.927
120	574.499	580.291	586.091	591.899	597.715
130	632.779	638.651	644.530	650.416	656.311
140	691.830	697.775	703.727	709.687	715.653
150	751.595	757.609	763.630	769.657	775.691
160	812.028	818.106	824.191	830.281	836.378
170	873.086	879.224	885.369	891.519	897.676
180	934.732	940.928	947.129	953.336	959.548
190	996.935	1003.184	1009.439	1015.699	1021.964
200	1059.663	1065.964	1072.270	1078.581	1084.896
210	1122.893	1129.242	1135.596	1141.955	1148.319
220	1186.598	1192.994	1199.394	1205.799	1212.209
230	1250.758	1257.198	1263.643	1270.092	1276.545
240	1315.353	1321.836	1328.323	1334.814	1341.309
250	1380.365	1386.889	1393.416	1399.948	1406.483
260	1445.777	1452.340	1458.906	1465.477	1472.051
270	1511.574	1518.174	1524.778	1531.386	1537.997
280	1577.741	1584.378	1591.018	1597.661	1604.309
290	1644.265	1650.937	1657.612	1664.291	1670.972
300	1711.135	1717.840	1724.549	1731.261	1737.976
310	1778.337	1785.076	1791.817	1798.562	1805.309
320	1845.863	1852.633	1859.406	1866.182	1872.961
330	1913.701	1920.501	1927.305	1934.111	1940.921
340	1981.842	1988.672	1995.505	2002.342	2009.181
350	2050.277	2057.136	2063.998	2070.863	2077.731
360	2118.997	2125.885	2132.775	2139.668	2146.564
370	2187.996	2194.911	2201.829	2208.749	2215.672
380	2257.265	2264.207	2271.151	2278.097	2285.047
390	2326.797	2333.765	2340.735	2347.707	2354.682
400	2396.586	2403.579	2410.574	2417.571	2424.572
410	2466.624	2473.642	2480.662	2487.684	2494.709
420	2536.907	2543.948	2550.992	2558.038	2565.087
430	2607.428	2614.493	2621.560	2628.629	2635.701
440	2678.181	2685.269	2692.359	2699.451	2706.546
450	2749.161	2756.272	2763.384	2770.499	2777.616
460	2820.364	2827.496	2834.631	2841.768	2848.906
470	2891.784	2898.938	2906.094	2913.252	2920.412
480	2963.417	2970.592	2977.769	2984.948	2992.129
490	3035.259	3042.454	3049.652	3056.851	3064.053

Note: When $f > 999$, and can be factored $f = ab$, where both a and $b < 999$, the following identity is useful for computing $f \ln f$: $ab \ln ab = a(b \ln b) + b(a \ln a)$.

5	6	7	8	9	f
8.047	10.751	13.621	16.636	19.775	0
40.621	44.361	48.165	52.027	55.944	10
80.472	84.711	88.988	93.302	97.652	20
124.437	129.007	133.604	138.228	142.879	30
171.300	176.118	180.957	185.818	190.699	40
220.403	225.420	230.454	235.506	240.575	50
271.335	276.517	281.714	286.927	292.153	60
323.812	329.136	334.473	339.823	345.186	70
377.625	383.074	388.534	394.006	399.489	80
432.618	438.177	443.747	449.327	454.917	90
488.666	494.325	499.993	505.670	511.357	100
545.667	551.416	557.174	562.941	568.716	110
603.539	609.372	615.212	621.060	626.916	120
662.212	668.121	674.037	679.961	685.892	130
721.626	727.607	733.594	739.587	745.588	140
781.731	787.778	793.831	799.890	805.956	150
842.481	848.590	854.705	860.826	866.953	160
903.838	910.005	916.179	922.357	928.542	170
965.766	971.989	978.217	984.451	990.690	180
1028.235	1034.510	1040.791	1047.077	1053.368	190
1091.217	1097.542	1103.873	1110.208	1116.548	200
1154.687	1161.060	1167.438	1173.820	1180.207	210
1218.623	1225.041	1231.464	1237.891	1244.322	220
1283.003	1289.464	1295.930	1302.400	1308.875	230
1347.808	1354.312	1360.819	1367.330	1373.846	240
1413.022	1419.565	1426.113	1432.664	1439.218	250
1478.628	1485.210	1491.795	1498.385	1504.977	260
1544.612	1551.231	1557.853	1564.479	1571.108	270
1610.959	1617.614	1624.271	1630.933	1637.597	280
1677.658	1684.346	1691.038	1697.734	1704.433	290
1744.695	1751.417	1758.142	1764.871	1771.602	300
1812.060	1818.815	1825.572	1832.332	1839.096	310
1879.743	1886.529	1893.317	1900.108	1906.903	320
1947.734	1954.549	1961.368	1968.190	1975.014	330
2016.023	2022.868	2029.716	2036.566	2043.420	340
2084.602	2091.475	2098.352	2105.231	2112.113	350
2153.463	2160.364	2167.268	2174.175	2181.084	360
2222.597	2229.526	2236.456	2243.390	2250.326	370
2291.999	2298.953	2305.910	2312.870	2319.832	380
2361.660	2368.640	2375.623	2382.608	2389.596	390
2431.574	2438.579	2445.587	2452.597	2459.609	400
2501.736	2508.765	2515.797	2522.831	2529.868	410
2572.138	2579.191	2586.247	2593.305	2600.365	420
2642.776	2649.852	2656.931	2664.012	2671.095	430
2713.643	2720.742	2727.844	2734.947	2742.053	440
2784.735	2791.857	2798.980	2806.106	2813.234	450
2856.047	2863.191	2870.336	2877.483	2884.633	460
2927.575	2934.739	2941.905	2949.074	2956.245	470
2999.312	3006.497	3013.685	3020.874	3028.065	480
3071.256	3078.462	3085.669	3092.879	3100.090	490

TABLE **XI** continued.

500-990

f	0	1	2	3	4
500	3107.304	3114.520	3121.737	3128.957	3136.178
510	3179.549	3186.785	3194.022	3201.262	3208.503
520	3251.991	3259.246	3266.502	3273.761	3281.022
530	3324.625	3331.899	3339.174	3346.452	3353.731
540	3397.447	3404.740	3412.034	3419.330	3426.628
550	3470.455	3477.766	3485.079	3492.393	3499.709
560	3543.645	3550.973	3558.304	3565.636	3572.971
570	3617.013	3624.359	3631.708	3639.058	3646.409
580	3690.556	3697.920	3705.286	3712.653	3720.022
590	3764.272	3771.653	3779.036	3786.420	3793.806
600	3838.158	3845.556	3852.955	3860.356	3867.759
610	3912.210	3919.624	3927.040	3934.458	3941.877
620	3986.426	3993.857	4001.289	4008.722	4016.158
630	4060.803	4068.250	4075.698	4083.148	4090.599
640	4135.340	4142.802	4150.266	4157.731	4165.198
650	4210.032	4217.510	4224.989	4232.470	4239.952
660	4284.878	4292.371	4299.866	4307.362	4314.859
670	4359.876	4367.384	4374.894	4382.405	4389.917
680	4435.023	4442.546	4450.070	4457.596	4465.123
690	4510.317	4517.855	4525.393	4532.934	4540.476
700	4585.756	4593.308	4600.861	4608.416	4615.972
710	4661.328	4668.904	4676.471	4684.040	4691.610
720	4737.061	4744.641	4752.222	4759.805	4767.389
730	4812.923	4820.516	4828.111	4835.708	4843.306
740	4888.921	4896.528	4904.137	4911.747	4919.359
750	4965.055	4972.676	4980.298	4987.921	4995.546
760	5041.322	5048.956	5056.591	5064.228	5071.866
770	5117.721	5125.368	5133.016	5140.666	5148.317
780	5194.249	5201.909	5209.570	5217.233	5224.897
790	5270.906	5278.579	5286.253	5293.928	5301.604
800	5347.689	5355.375	5363.061	5370.749	5378.438
810	5424.598	5432.295	5439.994	5447.694	5455.396
820	5501.630	5509.339	5517.051	5524.763	5532.477
830	5578.783	5586.505	5594.229	5601.953	5609.679
840	5656.058	5663.792	5671.527	5679.263	5687.001
850	5733.451	5741.197	5748.944	5756.692	5764.441
860	5810.962	5818.719	5826.478	5834.238	5841.999
870	5888.589	5896.358	5904.128	5911.900	5919.672
880	5966.331	5974.112	5981.893	5989.676	5997.460
890	6044.187	6051.979	6059.772	6067.566	6075.361
900	6122.155	6129.958	6137.762	6145.567	6153.374
910	6200.235	6208.049	6215.864	6223.680	6231.497
920	6278.424	6286.249	6294.075	6301.902	6309.730
930	6356.722	6364.557	6372.394	6380.232	6388.071
940	6435.127	6442.973	6450.821	6458.670	6466.519
950	6513.639	6521.496	6529.354	6537.213	6545.073
960	6592.256	6600.123	6607.992	6615.861	6623.732
970	6670.977	6678.855	6686.734	6694.614	6702.495
980	6749.802	6757.690	6765.579	6773.469	6781.360
990	6828.728	6836.626	6844.525	6852.426	6860.327

5	6	7	8	9	f
3143.402	3150.628	3157.855	3165.085	3172.316	500
3215.746	3222.991	3230.238	3237.487	3244.738	510
3288.284	3295.548	3302.815	3310.083	3317.353	520
3361.013	3368.296	3375.581	3382.868	3390.157	530
3433.928	3441.230	3448.533	3455.839	3463.146	540
3507.027	3514.347	3521.669	3528.992	3536.318	550
3580.307	3587.644	3594.984	3602.325	3609.668	560
3653.763	3661.118	3668.475	3675.834	3683.194	570
3727.393	3734.765	3742.140	3749.515	3756.893	580
3801.194	3808.583	3815.975	3823.367	3830.762	590
3875.163	3882.569	3889.977	3897.386	3904.797	600
3949.298	3956.720	3964.144	3971.570	3978.997	610
4023.595	4031.033	4038.473	4045.915	4053.359	620
4098.052	4105.506	4112.962	4120.420	4127.879	630
4172.666	4180.136	4187.608	4195.081	4202.556	640
4247.436	4254.921	4262.408	4269.897	4277.387	650
4322.358	4329.859	4337.361	4344.864	4352.370	660
4397.431	4404.947	4412.463	4419.982	4427.502	670
4472.652	4480.182	4487.714	4495.247	4502.781	680
4548.019	4555.563	4563.109	4570.657	4578.206	690
4623.529	4631.088	4638.649	4646.210	4653.774	700
4699.182	4706.755	4714.329	4721.905	4729.482	710
4774.974	4782.561	4790.150	4797.739	4805.330	720
4850.905	4858.505	4866.107	4873.711	4881.315	730
4926.971	4934.585	4942.201	4949.817	4957.435	740
5003.172	5010.799	5018.428	5026.058	5033.689	750
5079.505	5087.146	5094.787	5102.431	5110.075	760
5155.969	5163.622	5171.277	5178.933	5186.591	770
5232.562	5240.228	5247.896	5255.564	5263.235	780
5309.282	5316.961	5324.641	5332.323	5340.005	790
5386.128	5393.819	5401.512	5409.206	5416.901	800
5463.098	5470.802	5478.507	5486.213	5493.921	810
5540.191	5547.907	5555.624	5563.343	5571.063	820
5617.405	5625.134	5632.863	5640.593	5648.325	830
5694.739	5702.479	5710.220	5717.963	5725.706	840
5772.192	5779.943	5787.696	5795.450	5803.206	850
5849.761	5857.524	5865.289	5873.054	5880.821	860
5927.446	5935.221	5942.997	5950.774	5958.552	870
6005.245	6013.031	6020.818	6028.607	6036.396	880
6083.157	6090.955	6098.753	6106.553	6114.353	890
6161.181	6168.990	6176.799	6184.610	6192.422	900
6239.316	6247.135	6254.956	6262.777	6270.600	910
6317.559	6325.390	6333.221	6341.053	6348.887	920
6395.911	6403.752	6411.594	6419.437	6427.282	930
6474.370	6482.221	6490.074	6497.928	6505.783	940
6552.934	6560.797	6568.660	6576.524	6584.390	950
6631.604	6639.476	6647.350	6655.225	6663.100	960
6710.377	6718.259	6726.143	6734.028	6741.914	970
6789.252	6797.145	6805.039	6812.935	6820.831	980
6868.229	6876.132	6884.037	6891.942	6899.848	990

TABLE **XII.** $(f + \frac{1}{2})\ln(f + \frac{1}{2})$ as a function of f.

f	0	1	2	3	4
0	-0.347	0.608	2.291	4.385	6.768
10	24.689	28.087	31.572	35.136	38.775
20	61.919	65.963	70.054	74.190	78.367
30	104.241	108.675	113.140	117.637	122.163
40	149.903	154.616	159.354	164.115	168.899
50	198.060	202.991	207.943	212.913	217.902
60	248.210	253.321	258.448	263.591	268.750
70	300.021	305.283	310.560	315.850	321.155
80	353.255	358.649	364.056	369.475	374.906
90	407.734	413.245	418.767	424.299	429.843
100	463.321	468.936	474.561	480.196	485.840
110	519.904	525.614	531.332	537.060	542.796
120	577.394	583.190	588.994	594.806	600.626
130	635.714	641.589	647.472	653.363	659.260
140	694.802	700.750	706.706	712.669	718.639
150	754.601	760.619	766.643	772.673	778.710
160	815.066	821.148	827.235	833.329	839.429
170	876.154	882.296	888.444	894.597	900.756
180	937.829	944.028	950.232	956.441	962.656
190	1000.059	1006.311	1012.569	1018.831	1025.099
200	1062.813	1069.117	1075.425	1081.738	1088.056
210	1126.067	1132.419	1138.775	1145.136	1151.502
220	1189.795	1196.194	1202.596	1209.003	1215.415
230	1253.978	1260.420	1266.867	1273.318	1279.773
240	1318.594	1325.079	1331.568	1338.061	1344.558
250	1383.626	1390.152	1396.681	1403.215	1409.752
260	1449.058	1455.623	1462.191	1468.763	1475.339
270	1514.874	1521.476	1528.082	1534.691	1541.304
280	1581.059	1587.697	1594.339	1600.985	1607.634
290	1647.601	1654.274	1660.951	1667.631	1674.315
300	1714.487	1721.194	1727.905	1734.618	1741.335
310	1781.706	1788.446	1795.189	1801.935	1808.684
320	1849.247	1856.019	1862.793	1869.571	1876.352
330	1917.101	1923.903	1930.708	1937.516	1944.327
340	1985.256	1992.088	1998.923	2005.761	2012.601
350	2053.706	2060.567	2067.430	2074.297	2081.166
360	2122.441	2129.330	2136.221	2143.116	2150.013
370	2191.453	2198.369	2205.288	2212.210	2219.134
380	2260.735	2267.678	2274.624	2281.572	2288.522
390	2330.281	2337.249	2344.221	2351.194	2358.171
400	2400.082	2407.076	2414.072	2421.071	2428.073
410	2470.133	2477.151	2484.172	2491.196	2498.222
420	2540.427	2547.470	2554.515	2561.562	2568.612
430	2610.960	2618.026	2625.094	2632.165	2639.238
440	2681.725	2688.814	2695.905	2702.998	2710.094
450	2752.716	2759.828	2766.941	2774.057	2781.175
460	2823.930	2831.063	2838.199	2845.337	2852.477
470	2895.361	2902.516	2909.673	2916.832	2923.993
480	2967.004	2974.180	2981.358	2988.538	2995.720
490	3038.856	3046.053	3053.251	3060.452	3067.654

5	6	7	8	9	f
9.376	12.167	15.112	18.191	21.387	0
42.483	46.255	50.089	53.979	57.923	10
82.586	86.844	91.140	95.472	99.840	20
126.718	131.302	135.913	140.550	145.214	30
173.706	178.535	183.385	188.256	193.148	40
222.909	227.935	232.978	238.038	243.116	50
273.924	279.114	284.319	289.538	294.772	60
326.472	331.803	337.147	342.503	347.873	70
380.348	385.802	391.268	396.746	402.234	80
435.397	440.961	446.536	452.121	457.716	90
491.494	497.157	502.830	508.512	514.204	100
548.541	554.294	560.057	565.827	571.606	110
606.454	612.291	618.135	623.987	629.847	120
665.166	671.078	676.998	682.926	688.860	130
724.616	730.599	736.590	742.587	748.591	140
784.753	790.803	796.860	802.922	808.991	150
845.535	851.647	857.765	863.889	870.019	160
906.921	913.091	919.267	925.449	931.636	170
968.877	975.102	981.334	987.570	993.812	180
1031.372	1037.650	1043.933	1050.222	1056.515	190
1094.379	1100.707	1107.040	1113.377	1119.720	200
1157.873	1164.248	1170.628	1177.013	1183.402	210
1221.831	1228.252	1234.677	1241.106	1247.540	220
1286.233	1292.697	1299.165	1305.637	1312.114	230
1351.059	1357.565	1364.074	1370.588	1377.105	240
1416.293	1422.839	1429.388	1435.941	1442.497	250
1481.919	1488.502	1495.089	1501.680	1508.275	260
1547.921	1554.541	1561.165	1567.793	1574.424	270
1614.286	1620.942	1627.602	1634.265	1640.931	280
1681.002	1687.692	1694.386	1701.083	1707.783	290
1748.056	1754.779	1761.506	1768.236	1774.970	300
1815.437	1822.193	1828.952	1835.714	1842.479	310
1883.135	1889.922	1896.712	1903.505	1910.301	320
1951.141	1957.958	1964.778	1971.601	1978.427	330
2019.445	2026.291	2033.141	2039.993	2046.848	340
2088.038	2094.913	2101.791	2108.671	2115.555	350
2156.913	2163.815	2170.721	2177.629	2184.540	360
2226.061	2232.991	2239.923	2246.858	2253.795	370
2295.476	2302.431	2309.390	2316.351	2323.314	380
2365.150	2372.131	2379.115	2386.101	2393.090	390
2435.077	2442.083	2449.092	2456.103	2463.117	400
2505.250	2512.281	2519.314	2526.349	2533.387	410
2575.664	2582.719	2589.775	2596.835	2603.896	420
2646.313	2653.391	2660.471	2667.553	2674.638	430
2717.192	2724.293	2731.395	2738.500	2745.607	440
2788.296	2795.418	2802.543	2809.670	2816.799	450
2859.619	2866.763	2873.909	2881.058	2888.208	460
2931.156	2938.322	2945.489	2952.659	2959.831	470
3002.905	3010.091	3017.279	3024.469	3031.662	480
3074.859	3082.065	3089.274	3096.484	3103.697	490

TABLE **XIII. Critical values of U, the Mann-Whitney statistic.**

n_1	n_2 \ α	0.10	0.05	0.025	0.01	0.005	0.001
3	2	6					
	3	8	9				
4	2	8					
	3	11	12				
	4	13	15	16			
5	2	9	10				
	3	13	14	15			
	4	16	18	19	20		
	5	20	21	23	24	25	
6	2	11	12				
	3	15	16	17			
	4	19	21	22	23	24	
	5	23	25	27	28	29	
	6	27	29	31	33	34	
7	2	13	14				
	3	17	19	20	21		
	4	22	24	25	27	28	
	5	27	29	30	32	34	
	6	31	34	36	38	39	42
	7	36	38	41	43	45	48
8	2	14	15	16			
	3	19	21	22	24		
	4	25	27	28	30	31	
	5	30	32	34	36	38	40
	6	35	38	40	42	44	47
	7	40	43	46	49	50	54
	8	45	49	51	55	57	60
9	1	9					
	2	16	17	18			
	3	22	23	25	26	27	
	4	27	30	32	33	35	
	5	33	36	38	40	42	44
	6	39	42	44	47	49	52
	7	45	48	51	54	56	60
	8	50	54	57	61	63	67
	9	56	60	64	67	70	74
10	1	10					
	2	17	19	20			
	3	24	26	27	29	30	
	4	30	33	35	37	38	40
	5	37	39	42	44	46	49
	6	43	46	49	52	54	57
	7	49	53	56	59	61	65
	8	56	60	63	67	69	74
	9	62	66	70	74	77	82
	10	68	73	77	81	84	90

Note: Critical values are tabulated for two samples of sizes n_1 and n_2, where $n_1 \geq n_2$, up to $n_1 = n_2 = 20$. The upper bounds of the critical values are furnished so that the sample statistic U_s has to be greater than a given critical value. The probabilities at the heads of the columns are based on a one-tailed test and represent the proportion of the area of the distribution of U in one tail

TABLE **XIII** continued.

n_1	n_2 \ α	0.10	0.05	0.025	0.01	0.005	0.001
11	1	11					
	2	19	21	22			
	3	26	28	30	32	33	
	4	33	36	38	40	42	44
	5	40	43	46	48	50	53
	6	47	50	53	57	59	62
	7	54	58	61	65	67	71
	8	61	65	69	73	75	80
	9	68	72	76	81	83	89
	10	74	79	84	88	92	98
	11	81	87	91	96	100	106
12	1	12					
	2	20	22	23			
	3	28	31	32	34	35	
	4	36	39	41	42	45	48
	5	43	47	49	52	54	58
	6	51	55	58	61	63	68
	7	58	63	66	70	72	77
	8	66	70	74	79	81	87
	9	73	78	82	87	90	96
	10	81	86	91	96	99	106
	11	88	94	99	104	108	115
	12	95	102	107	113	117	124
13	1	13					
	2	22	24	25	26		
	3	30	33	35	37	38	
	4	39	42	44	47	49	51
	5	47	50	53	56	58	62
	6	55	59	62	66	68	73
	7	63	67	71	75	78	83
	8	71	76	80	84	87	93
	9	79	84	89	94	97	103
	10	87	93	97	103	106	113
	11	95	101	106	112	116	123
	12	103	109	115	121	125	133
	13	111	118	124	130	135	143
14	1	14					
	2	24	25	27	28		
	3	32	35	37	40	41	
	4	41	45	47	50	52	55
	5	50	54	57	60	63	67
	6	59	63	67	71	73	78
	7	67	72	76	81	83	89
	8	76	81	86	90	94	100
	9	85	90	95	100	104	111
	10	93	99	104	110	114	121
	11	102	108	114	120	124	132
	12	110	117	123	130	134	143
	13	119	126	132	139	144	153
	14	127	135	141	149	154	164

beyond the critical value. For a two-tailed test use the same critical values but double the probability at the heads of the columns. This table was extracted from a more extensive one (table 11.4) in D. B. Owen, *Handbook of Statistical Tables* (Addison-Wesley Publishing Co., Reading, Mass., 1962): Courtesy of U.S. Atomic Energy Commission, with permission of the publishers.

TABLE **XIII** continued.

n_1	n_2 \ α	0.10	0.05	0.025	0.01	0.005	0.001
15	1	15					
	2	25	27	29	30		
	3	35	38	40	42	43	
	4	44	48	50	53	55	59
	5	53	57	61	64	67	71
	6	63	67	71	75	78	83
	7	72	77	81	86	89	95
	8	81	87	91	96	100	106
	9	90	96	101	107	111	118
	10	99	106	111	117	121	129
	11	108	115	121	128	132	141
	12	117	125	131	138	143	152
	13	127	134	141	148	153	163
	14	136	144	151	159	164	174
	15	145	153	161	169	174	185
16	1	16					
	2	27	29	31	32		
	3	37	40	42	45	46	
	4	47	50	53	57	59	62
	5	57	61	65	68	71	75
	6	67	71	75	80	83	88
	7	76	82	86	91	94	101
	8	86	92	97	102	106	113
	9	96	102	107	113	117	125
	10	106	112	118	124	129	137
	11	115	122	129	135	140	149
	12	125	132	139	146	151	161
	13	134	143	149	157	163	173
	14	144	153	160	168	174	185
	15	154	163	170	179	185	197
	16	163	173	181	190	196	208
17	1	17					
	2	28	31	32	34		
	3	39	42	45	47	49	51
	4	50	53	57	60	62	66
	5	60	65	68	72	75	80
	6	71	76	80	84	87	93
	7	81	86	91	96	100	106
	8	91	97	102	108	112	119
	9	101	108	114	120	124	132
	10	112	119	125	132	136	145
	11	122	130	136	143	148	158
	12	132	140	147	155	160	170
	13	142	151	158	166	172	183
	14	153	161	169	178	184	195
	15	163	172	180	189	195	208
	16	173	183	191	201	207	220
	17	183	193	202	212	219	232

TABLE **XIII** continued.

n_1	n_2 \ α	0.10	0.05	0.025	0.01	0.005	0.001
18	1	18					
	2	30	32	34	36		
	3	41	45	47	50	52	54
	4	52	56	60	63	66	69
	5	63	68	72	76	79	84
	6	74	80	84	89	92	98
	7	85	91	96	102	105	112
	8	96	103	108	114	118	126
	9	107	114	120	126	131	139
	10	118	125	132	139	143	153
	11	129	137	143	151	156	166
	12	139	148	155	163	169	179
	13	150	159	167	175	181	192
	14	161	170	178	187	194	206
	15	172	182	190	200	206	219
	16	182	193	202	212	218	232
	17	193	204	213	224	231	245
	18	204	215	225	236	243	258
19	1	18	19				
	2	31	34	36	37	38	
	3	43	47	50	53	54	57
	4	55	59	63	67	69	73
	5	67	72	76	80	83	88
	6	78	84	89	94	97	103
	7	90	96	101	107	111	118
	8	101	108	114	120	124	132
	9	113	120	126	133	138	146
	10	124	132	138	146	151	161
	11	136	144	151	159	164	175
	12	147	156	163	172	177	188
	13	158	167	175	184	190	202
	14	169	179	188	197	203	216
	15	181	191	200	210	216	230
	16	192	203	212	222	230	244
	17	203	214	224	235	242	257
	18	214	226	236	248	255	271
	19	226	238	248	260	268	284
20	1	19	20				
	2	33	36	38	39	40	
	3	45	49	52	55	57	60
	4	58	62	66	70	72	77
	5	70	75	80	84	87	93
	6	82	88	93	98	102	108
	7	94	101	106	112	116	124
	8	106	113	119	126	130	139
	9	118	126	132	140	144	154
	10	130	138	145	153	158	168
	11	142	151	158	167	172	183
	12	154	163	171	180	186	198
	13	166	176	184	193	200	212
	14	178	188	197	207	213	226
	15	190	200	210	220	227	241
	16	201	213	222	233	241	255
	17	213	225	235	247	254	270
	18	225	237	248	260	268	284
	19	237	250	261	273	281	298
	20	249	262	273	286	295	312

TABLE **XIV. Critical values of the Wilcoxon rank sum.**

n \ nominal α	0.05		0.025		0.01		0.005	
	T	α	T	α	T	α	T	α
5	0	.0312						
	1	.0625						
6	2	.0469	0	.0156				
	3	.0781	1	.0312				
7	3	.0391	2	.0234	0	.0078		
	4	.0547	3	.0391	1	.0156		
8	5	.0391	3	.0195	1	.0078	0	.0039
	6	.0547	4	.0273	2	.0117	1	.0078
9	8	.0488	5	.0195	3	.0098	1	.0039
	9	.0645	6	.0273	4	.0137	2	.0059
10	10	.0420	8	.0244	5	.0098	3	.0049
	11	.0527	9	.0322	6	.0137	4	.0068
11	13	.0415	10	.0210	7	.0093	5	.0049
	14	.0508	11	.0269	8	.0122	6	.0068
12	17	.0461	13	.0212	9	.0081	7	.0046
	18	.0549	14	.0261	10	.0105	8	.0061
13	21	.0471	17	.0239	12	.0085	9	.0040
	22	.0549	18	.0287	13	.0107	10	.0052
14	25	.0453	21	.0247	15	.0083	12	.0043
	26	.0520	22	.0290	16	.0101	13	.0054
15	30	.0473	25	.0240	19	.0090	15	.0042
	31	.0535	26	.0277	20	.0108	16	.0051
16	35	.0467	29	.0222	23	.0091	19	.0046
	36	.0523	30	.0253	24	.0107	20	.0055
17	41	.0492	34	.0224	27	.0087	23	.0047
	42	.0544	35	.0253	28	.0101	24	.0055
18	47	.0494	40	.0241	32	.0091	27	.0045
	48	.0542	41	.0269	33	.0104	28	.0052
19	53	.0478	46	.0247	37	.0090	32	.0047
	54	.0521	47	.0273	38	.0102	33	.0054
20	60	.0487	52	.0242	43	.0096	37	.0047
	61	.0527	53	.0266	44	.0107	38	.0053

Note: This table furnishes critical values for the one-tailed test of significance of the rank sum T_s obtained in Wilcoxon's matched-pairs signed-ranks test. Since the exact probability level desired cannot be obtained with integral critical values of T, two such values and their attendant probabilities bracketing the desired significance level are furnished. Thus, to find the significant 1% values for $n = 19$ we note the two critical values of T, 37 and 38, in the table. The probabilities corresponding to these two values of T are 0.0090 and 0.0102. Clearly a rank sum of $T_s = 37$ would have a probability of less than 0.01 and would be considered significant by the stated

TABLE **XIV** continued.

n \ nominal α	0.05		0.025		0.01		0.005	
	T	α	T	α	T	α	T	α
21	67	.0479	58	.0230	49	.0097	42	.0045
	68	.0516	59	.0251	50	.0108	43	.0051
22	75	.0492	65	.0231	55	.0095	48	.0046
	76	.0527	66	.0250	56	.0104	49	.0052
23	83	.0490	73	.0242	62	.0098	54	.0046
	84	.0523	74	.0261	63	.0107	55	.0051
24	91	.0475	81	.0245	69	.0097	61	.0048
	92	.0505	82	.0263	70	.0106	62	.0053
25	100	.0479	89	.0241	76	.0094	68	.0048
	101	.0507	90	.0258	77	.0101	69	.0053
26	110	.0497	98	.0247	84	.0095	75	.0047
	111	.0524	99	.0263	85	.0102	76	.0051
27	119	.0477	107	.0246	92	.0093	83	.0048
	120	.0502	108	.0260	93	.0100	84	.0052
28	130	.0496	116	.0239	101	.0096	91	.0048
	131	.0521	117	.0252	102	.0102	92	.0051
29	140	.0482	126	.0240	110	.0095	100	.0049
	141	.0504	127	.0253	111	.0101	101	.0053
30	151	.0481	137	.0249	120	.0098	109	.0050
	152	.0502	138	.0261	121	.0104	110	.0053
31	163	.0491	147	.0239	130	.0099	118	.0049
	164	.0512	148	.0251	131	.0105	119	.0052
32	175	.0492	159	.0249	140	.0097	128	.0050
	176	.0512	160	.0260	141	.0103	129	.0053
33	187	.0485	170	.0242	151	.0099	138	.0049
	188	.0503	171	.0253	152	.0104	139	.0052
34	200	.0488	182	.0242	162	.0098	148	.0048
	201	.0506	183	.0252	163	.0103	149	.0051
35	213	.0484	195	.0247	173	.0096	159	.0048
	214	.0501	196	.0257	174	.0100	160	.0051

criterion. For two-tailed tests in which the alternative hypothesis is that the pairs could differ in either direction, double the probabilities stated at the head of the table. For sample sizes $n > 50$ compute

$$t_{\alpha[\infty]} = \left[T_s - \frac{n(n+1)}{4}\right] \Big/ \sqrt{\frac{n(n+1)(2n+1)}{24}}$$

for a two-tailed test.

TABLE **XIV** continued.

n \ nominal α	0.05		0.025		0.01		0.005	
	T	α	T	α	T	α	T	α
36	227	.0489	208	.0248	185	.0096	171	.0050
	228	.0505	209	.0258	186	.0100	172	.0052
37	241	.0487	221	.0245	198	.0099	182	.0048
	242	.0503	222	.0254	199	.0103	183	.0050
38	256	.0493	235	.0247	211	.0099	194	.0048
	257	.0509	236	.0256	212	.0104	195	.0050
39	271	.0493	249	.0246	224	.0099	207	.0049
	272	.0507	250	.0254	225	.0103	208	.0051
40	286	.0486	264	.0249	238	.0100	220	.0049
	287	.0500	265	.0257	239	.0104	221	.0051
41	302	.0488	279	.0248	252	.0100	233	.0048
	303	.0501	280	.0256	253	.0103	234	.0050
42	319	.0496	294	.0245	266	.0098	247	.0049
	320	.0509	295	.0252	267	.0102	248	.0051
43	336	.0498	310	.0245	281	.0098	261	.0048
	337	.0511	311	.0252	282	.0102	262	.0050
44	353	.0495	327	.0250	296	.0097	276	.0049
	354	.0507	328	.0257	297	.0101	277	.0051
45	371	.0498	343	.0244	312	.0098	291	.0049
	372	.0510	344	.0251	313	.0101	292	.0051
46	389	.0497	361	.0249	328	.0098	307	.0050
	390	.0508	362	.0256	329	.0101	308	.0052
47	407	.0490	378	.0245	345	.0099	322	.0048
	408	.0501	379	.0251	346	.0102	323	.0050
48	426	.0490	396	.0244	362	.0099	339	.0050
	427	.0500	397	.0251	363	.0102	340	.0051
49	446	.0495	415	.0247	379	.0098	355	.0049
	447	.0505	416	.0253	380	.0100	356	.0050
50	466	.0495	434	.0247	397	.0098	373	.0050
	467	.0506	435	.0253	398	.0101	374	.0051

Bibliography

Allee, W. C., and E. Bowen. 1932. Studies in animal aggregations: Mass protection against colloidal silver among goldfishes. *J. Exp. Zool.*, **61**:185–207.

Archibald, E. E. A. 1950. Plant populations. II. The estimation of the number of individuals per unit area of species in heterogeneous plant populations. *Ann. Bot. N.S.*, **14**:7–21.

Banta, A. M. 1939. Studies on the physiology, genetics, and evolution of some Cladocera. Carnegie Institution of Washington, Dept. Genetics, Paper 39. 285 pp.

Blakeslee, A. F. 1921. The globe mutant in the Jimson Weed (*Datura stramonium*). *Genetics*, **6**:241–264.

Block, B. C. 1966. The relation of temperature to the chirp-rate of male snowy tree crickets, *Oecanthus fultoni* (Orthoptera: Gryllidae). *Ann. Entomol. Soc. Amer.*, **59**:56–59.

Brower, L. P. 1959. Speciation in butterflies of the *Papilio glaucus* group. I. Morphological relationships and hybridization. *Evolution*, **13**:40–63.

Brown, B. E., and A. W. A. Brown. 1956. The effects of insecticidal poisoning on the level of cytochrome oxidase in the American cockroach. *J. Econ. Entomol.*, **49**:675–679.

Burr, E. J. 1960. The distribution of Kendall's score S for a pair of tied rankings. *Biometrika*, **47**:151–171.

Carter, G. R., and C. A. Mitchell. 1958. Methods for adapting the virus of Rinderpest to rabbits. *Science*, **128**:252–253.

Cowan, I. M., and P. A. Johnston. 1962. Blood serum protein variations at the species and subspecies level in deer of the genus *Odocoilcus*. *Syst. Zool.*, **11**:131–138.

Davis, E. A., Jr. 1955. Seasonal changes in the energy balance of the English sparrow. *Auk*, **72**:385–411.

Fröhlich, F. W. 1921. *Grundzüge einer Lehre vom Licht- und Farbensinn. Ein Beitrag zur allgemeinen Physiologie der Sinne.* Fischer, Jena. 86 pp.

Gabriel, K. R. 1964. A procedure for testing the homogeneity of all sets of means in analysis of variance. *Biometrics*, **20**:459–477.

Gartler, S. M., I. L. Firschein, and T. Dobzhansky. 1956. Chromatographic investigation of urinary amino-acids in the great apes. *Am. J. Phys. Anthropol.*, **14**:41–57.

Geissler, A. 1889. Beiträge zur Frage des Geschlechtsverhältnisses der Geborenen. *Z. K. Sächs. Stat. Bur.*, **35**:1–24.

Hunter, P. E. 1959. Selection of *Drosophila melanogaster* for length of larval period. *Z. Vererbungsl.*, **90**:7–28.

Johnson, N. K. 1966. Bill size and the question of competition in allopatric and sympatic populations of Dusky and Gray Flycatchers. *Syst. Zool.*, **15**:70–87.

Kouskolekas, C. A., and G. C. Decker. 1966. The effect of temperature on the rate of development of the potato leafhopper, *Empoasca fabae* (Homoptera: Cicadellidae). *Ann. Entomol. Soc. Amer.*, **59**:292–298.

Lewontin, R. C., and M. J. D. White. 1960. Interaction between inversion polymorphisms of two chromosome pairs in the grasshopper, *Moraba scurra*. *Evolution*, **14**:116–129.

Littlejohn, M. J. 1965. Premating isolation in the *Hyla ewingi* complex. *Evolution*, **19**:234–243.

Millis, J., and Y. P. Seng. 1954. The effect of age and parity of the mother on birth weight of the offspring. *Ann. Human Genetics*, **19**:58–73.

Mittler, T. E., and R. H. Dadd. 1966. Food and wing determination in *Myzus persicae* (Homoptera: Aphidae). *Ann. Entomol. Soc. Amer.*, **59**:1162–1166.

Mosimann, J. E. 1968. *Elementary Probability for the Biological Sciences.* Appleton-Century-Crofts, New York. 255 pp.

Nelson, V. E. 1964. The effects of starvati on and humidity on water content in *Tribolium confusum* Duval (Coleoptera). Unpublished Ph.D. thesis, University of Colorado. 111 pp.

Newman, K. J., and H. V. Meredith. 1956. Individual growth in skeletal bigonial diameter during the childhood period from 5 to 11 years of age. *Am. J. Anatomy*, **99**:157–187.

Olson, E. C., and R. L. Miller. 1958. *Morphological Integration.* University of Chicago Press, Chicago. 317 pp.

Park, W. H., A. W. Williams, and C. Krumwiede. 1924. *Pathogenic Microöganisms.* Lea & Febiger, Philadelphia and New York. 811 pp.

Phillips, J. R., and L. D. Newsom. 1966. Diapause in *Heliothis zea* and *Heliothis virescens* (Lepidoptera: Noctuidae). *Ann. Entomol. Soc. Amer.*, **59**:154–159.

Siegel, S. 1956. *Nonparametric Statistics for the Behavioral Sciences.* McGraw-Hill, New York, Toronto, and London. 312 pp.

Sinnott, E. W. and D. Hammond. 1935. Factorial balance in the determination of fruit shape in *Cucurbita*. *Amer. Nat.*, **64**:509–524.

Smith, F. L. 1939. A genetic analysis of red seed-coat color in *Phaseolus vulgaris*. *Hilgardia*, **12**:553–621.

Sokal, R. R. 1952. Variation in a local population of *Pemphigus*. *Evolution*, **6**:296–315.

Sokal, R. R. 1967. A comparison of fitness characters and their responses to density in stock and selected cultures of wild type and black *Tribolium castaneum*. *Tribolium Inf. Bull.*, **10**:142–147.

Sokal, R. R., and P. E. Hunter. 1955. A morphometric analysis of DDT-resistant and non-resistant housefly strains. *Ann. Entomol. Soc. Amer.* **48**:499–507.

Sokal, R. R., and I. Karten. 1964. Competition among genotypes in *Tribolium castaneum* at varying densities and gene frequencies (the black locus). *Genetics*, **49**:195–211.

Sokal, R. R., and F. J. Rohlf. 1969. *Biometry*. W. H. Freeman and Company, San Francisco. 776 pp.

Sokal, R. R., and P. A. Thomas. 1965. Geographic variation of *Pemphigus populi-transversus* in Eastern North America: Stem mothers and new data on alates. *Univ. Kansas Sci. Bull.*, **46**:201–252.

Sokoloff, A. 1955. Competition between sibling species of the *Pseudoobscura* subgroup of *Drosophila*. *Ecol. Monogr.*, **25**:387–409.

Sokoloff, A. 1966. Morphological variation in natural and experimental populations of *Drosophila pseudoobscura* and *Drosophila persimilis*. *Evolution*, **20**:49–71.

Student (W. S. Gossett). 1907. On the error of counting with a haemacytometer. *Biometrika*, **5**:351–360.

Swanson, C. O., W. L. Latshaw, and E. L. Tague. 1921. Relation of the calcium content of some Kansas soils to soil reaction by the electrometric titration. *J. Agr. Res.*, **20**:855–868.

Tate, R. F., and G. W. Klett. 1959. Optimal confidence intervals for the variance of a normal distribution. *J. Am. Stat. Assoc.*, **54**:674–682.

Utida, S. 1943. Studies on experimental population of the Azuki bean weevil, *Callosobruchus chinensis* (L.). VIII. Statistical analysis of the frequency distribution of the emerging weevils on beans. *Mem. Coll. Agr. Kyoto Imp. Univ.*, **54**:1–22.

Vollenweider, R. A., and M. Frei. 1953. Vertikale und zeitliche Verteilung der Leitfähigkeit in einem eutrophen Gewässer während der Sommerstagnation. *Schweiz. Z. Hydrol.* **15**:158–167.

Whittaker, R. H. 1952. A study of summerfoliage insect communities in the Great Smoky Mountains. *Ecol. Monogr.*, **22**:1–44.

Willis, E. R., and N. Lewis. 1957. The longevity of starved cockroaches. *J. Econ. Entomol.*, **50**:438–440.

Woodson, R. E., Jr. 1964. The geography of flower color in butterflyweed. *Evolution*, **18**:143–163.

Index